Advanced Aluminium Composites and Alloys

Edited by Leszek A. Dobrzański

Published in London, United Kingdom

IntechOpen

Supporting open minds since 2005

Advanced Aluminium Composites and Alloys
http://dx.doi.org/10.5772/intechopen.87723
Edited by Leszek A. Dobrzański

Contributors
Jorge Salguero, Moises Batista, Irene Del Sol, Alvaro Gomez-Parra, Uyime Donatus, Mariana Milagre, João Araujo, Caruline Machado, Isolda Costa, Michael Bodunrin, Ayotunde Olayinka, Olamilekan Oloyode, Sunday Aribo, Zygmunt Mikno, Francisco Mata, Issam Hanafi, Halil Ibrahim Kurt, Engin Ergul, Murat Oduncuoglu, Necip Fazil Yilmaz, Xiaoguang Sun, Xiaogang Li, Chaofang Dong, Maria Richert, Rafał Hubicki, Fawzy H. Samuel, Mohamed H. Abdelaziz, Jacobo Hernandez-Sandoval, Agnes Marie Samuel, Herbert Doty, Leszek A. Dobrzański

First published in London, United Kingdom, 2021 by IntechOpen
IntechOpen is the global imprint of INTECHOPEN LIMITED, registered in England and Wales, registration number: 11086078, 5 Princes Gate Court, London, SW7 2QJ, United Kingdom
Printed in Croatia

British Library Cataloguing-in-Publication Data
A catalogue record for this book is available from the British Library

Additional hard and PDF copies can be obtained from orders@intechopen.com

Advanced Aluminium Composites and Alloys
Edited by Leszek A. Dobrzański
p. cm.
Print ISBN 978-1-83880-450-3
Online ISBN 978-1-83880-451-0
eBook (PDF) ISBN 978-1-83880-456-5

For EU product safety concerns:
IN TECH d.o.o., Prolaz Marije Krucifikse Kozulić 3, 51000 Rijeka, Croatia, info@intechopen.com or visit our website at intechopen.com.

Meet the editor

Prof. Leszek Adam Dobrzański, DSc, Ph.D., MSc, Eng, Hon Prof, M Dr HC is a full professor and the director of the Science Centre ASKLEPIOS of the Medical and Dental Engineering Centre for Research, Design and Production ASKLEPIOS Ltd., Gliwice, Poland. He is president of World Academy WAMME, vice president of the Academy of Engineers in Poland, and a foreign fellow of engineering academies in Ukraine and Slovakia. He is also editor in chief of JAMME and AMSE journals. He was awarded the title of Professor in 1995 by the President of the Republic of Poland and granted abroad a title of Honorary Professor in Lviv, Ukraine. He was also awarded the title of Doctor Honoris Causa in Ruse, Bulgaria; Khmelnytskyi, Ukraine; and Miskolc, Hungary. He is also a recipient of the Commander's Crosses of Rebirth of Poland by the President of Poland, Merite de L'Innvention by King of Belgium, and many other awards. From 1971 to 2017, Dr. Dobrzański was a full professor, director, dean of faculty, and vice-rector at Silesian University of Technology, Gliwice, Poland. He was also vice-rector of the Academy in Bielsko-Biała, Poland, and vice-director of the Institute in Kraków, Poland. He was a visiting professor at more than 100 universities worldwide and an invited lecturer at more than 100 global scientific conferences. Dr. Dobrzański has supervised sixty-two Ph.D. thesis and promoted around 1000 MSc and BSc theses. He is the author or co-author of sixty books and fifty book chapters, sixty patents, and around 3000 publications. He is a scientific editor for many international scientific journals and conferences. He has been cited ca. 15,500 times in scientific journals.

Contents

Preface

Several essential elements characterize the turn of 2020 and 2021. Although the COVID-19 disease pandemic related to the transmission of the SARS-CoV-2 coronavirus has dominated all global activities, the directions of all undertakings are generally determined by the sustainable development goals set by the United Nations [1]. It is about providing people around the world with permanent access to products and consumer goods that directly affect the level and quality of life, the quality and potential of health protection, protection of the climate and natural goods, dissemination and improvement of the level of education, information exchange and other aspects. An important role in this respect is played by the scientific and engineering communities, serving to activate the development of societies. Undoubtedly, an important determinant of the present and future prosperity and high quality of life is the continuous development of engineering materials, closely related to the development of nanotechnology and surface engineering [2].

Industrial production is an important determinant of achieving the goals mentioned above, emphasizing ecological conditions, climate protection, and human health and life. The achieved stage of the scientific and technical revolution Industry 4.0 includes smart factories manufacturing smart products for which raw materials are supplied by smart suppliers and [3]. Physical production processes are monitored by production systems and make smart decisions. It is possible to do experiments with the use of "digital twins" in virtual reality by simulating the actual conditions of production, operation and maintenance of products. To achieve these goals, cyber-physical systems (CPS), Internet of Things (IoT) and cloud computing are used, with the use of large data sets while ensuring cybersecurity. Automation, robotization and digitization are the current essences of industrial activities. The concept of the developed information society Society 5.0 [4] corresponds to this.

However, it turned out that this model is one-sided and requires a correction and a significant extension. The simplified approach in the classic Industry 4.0 model gives the erroneous impression that progress is only about monitoring, controlling, coordinating and integrating information and communication technologies that make up cyber-physical systems, without the need to make real progress in the field of technological machines, manufacturing technologies and the engineering materials necessary for the manufacturing of any product. A far-reaching simplification is also reducing technological issues only to additive production, which is not competitive to many other manufacturing technologies absolutely necessary in overall manufacturing processes. Therefore, a holistic, extended and supplemented model of industry 4.0 was developed [5–9]. Only such a model of Industry 4.0 is adequate to the actual situation in the developing industry.

The general development of material culture and human civilization in general and the associated level and quality of life of societies largely depend on the development of technical materials, mainly engineering [10]. For thousands of years, materials have been implemented from which all products useful for life were made. The selection of materials for these goals was made by trial and error. Nowadays, the possibilities of using cyber-physical systems and large data sets and advanced methods of artificial intelligence and machine learning are the essence of the systematically implemented Materials 4.0 approach [5–10]. It is accompanied by the use of materials on-demand with properties designed and required by designers when still two decades ago, it was only possible to choose from the materials offered by manufacturers. Achieving the material engineering paradigm according to the six expectations rule (6xE) [2, 11] is fully ensured because the operational functions of the product are ensured by designing the expected material, processed using the expected technology, to give the expected geometric features and shape of the product, enabling the expected structure to be obtained order to get a set of expected properties, ensuring the expected utility functions of the designed product.

Beginning in 10,000 years BC, humankind successively mastered the sourcing from nature and manufacturing of gold, copper, bronze, and finally iron and the manufacturing of products expected from these metals to meet the everyday needs of contemporary people [11]. It is also how the successive epochs of civilization development are defined. This progress was slow but systematic. In ancient Egypt, thousands of years BC, an engineering composite material, because artificially manufactured, was invented where reinforcement was made of straw fibres surrounded by a clay matrix, dried in the sun [1]. An important stage in this development was the invention of steel, which probably already took place around the 3rd century BC in ancient India. Still, it is believed that Sir Henry Bessemer, in 1856, was the first to devise a method that is considered the first step in the modern development of steelmaking and starting to steel mass production.

In 1825, Hans Christian Oersted discovered aluminium, and in 1856, thanks to the efforts of Henri Étienne Sainte-Claire Deville, industrial production of aluminium began, which in 1884 reached 3 tons in the world. Over time, numerous aluminium alloys have been developed, currently classified into eight series, differentiated by the alloying additives used, affecting significant but differentiated properties improvements.

Since the 1980s, AMC Aluminum Matrix Composites have become known, mainly due to their applications in the automotive industry. Due to the proliferation of carbon composites, AMC initially did not gain popularity. Breakthrough progress in this area dates back to the last 30 years. It was due to the attractive properties of AMC, including their density and functionality, as well as stiffness, strength, thermal and electrical properties.

The increase in aluminium production, its alloys and composites with aluminium matrix were compared with steel world production. Despite the 30 times lower production of aluminium [12] than steel, aluminium alloys and composites with the aluminium matrix are significant due to their lower density than that of steel. The challenges posed by the development in the Industry 4.0 phase, especially the expectations of the automotive and aviation industries, force constant progress in the development of new materials with the participation of aluminium.

The book "*Advanced Aluminium Composites and Alloys*" is my another book published in my personal academic career and third prepared with IntechOpen. The topic is very familiar to me because I am interested in it as it is one of the main areas of my scientific interest for a long-time. This book contains a collection of studies by authors from 12 different countries. Despite the asymmetrical number of chapters on each fundamental topic, it means advanced aluminium-based composite materials and alloys of this metal, roughly half in volume was devoted to each of these topics.

The book opens with my original study on advanced composites based on aluminium alloys and their production processes. Composite materials were manufactured by gas pressure infiltration with liquid aluminium alloys of suitably formed porous skeletons sintered from a mixture of Al2O3 powder and carbon fibres then are thermally degraded, using halloysite HNTs nanotubes by mechanical milling, consolidation in press and sintering and selective SLS laser sintering of titanium powders. Another group of manufacturing technologies is the mechanical synthesis of a mixture of aluminium alloy powder and HNTs halloysite nanotubes or MWCNTs multi-wall carbon nanotubes, respectively, and subsequent consolidation with plas tic deformation. The third group concerns composite surface layers on substrates of aluminium alloys produced by laser feathering of WC/W_2C or SiC carbides.

The next chapter in this part of the book, "The Theoretical Overview of the Selected Optimization and Prediction Models Useful in the Design of Aluminum Alloys and Aluminum Matrix Composites," was written by Halil Ibrahim Kurt et al. from Turkey. This chapter presents original research results from their own work and cited from the literature on the theory of artificial neural network (ANN), adaptive neural fuzzy inference systems (ANFIS) and Taguchi method and their applications in engineering design and manufacturing of aluminium alloys and AMC composites.

All other chapters deal with various aspects of aluminium alloys. The chapter titled "Effect of Zr Addition and Aging Treatment on the Tensile Properties of Al-Si-Cu-Mg Cast Alloys" is prepared by the international team of Jacobo Hernandez-Sandoval et al. from Canada, Mexico, Egypt and the USA. The chapter concerns the tensile strength of analysed materials containing aluminium at room and elevated temperatures. Zirconium forms phases with the participation of Ti, Si and Al, and their coagulation leads to a decrease in strength.

Rafał Hubicki and Maria Richert from Poland have prepared a chapter entitled "The High-Speed 6xxx Aluminum Alloys in Shape Extrusion Industry", where they analyzed alloys used in the automotive and construction industries.

The 9-person team of Uyime Donatus et al. from Brazil, South Africa, USA, UK and Canada wrote the next chapter entitled "Corrosion Resistance of Precipitation-Hardened Al Alloys: A Comparison between New Generation Al-Cu-Li and Conventional Alloys". The corrosion resistance of conventional alloys and new alloys of precipitation hardening alloys was compared. The AA6082-T6 alloy became the most resistant to corrosion, while the AA2024-T3 alloy showed the highest density of pitting spots.

The chapter "Machining of Al-Cu and Al-Zn Alloys for Aeronautical Components" by the team of Jorge Salguero et al. from Spain focused on the analysis of the relationship between drilling, milling and turning conditions, quality characteristics and the main wear mechanism during machining as factors influencing performance improvement and micro and macro geometric deviations.

In turn, in the chapter "Analysis of Surface Roughness of EN AW 2024 and EN AW 2030 Alloys after Micromachining" developed by Francisco Mata and Issam Hanafi from Spain and Morocco, the focus was on this technology suitable for the production of very small components in the industry. Very good surface properties can be achieved when turning aluminium alloys with a diameter of not less than 0.05 mm.

Zygmunt Mikno from Poland has prepared a chapter on "Resistance Welding of Aluminum Alloys with an Electromechanical Electrode Force System", which concerns the operation and depends on the new clamping solution in the welding machine and the optimization of the welding process of aluminium rods. The research consisted of the numerical analysis of two electrode pressure systems, i.e. conventional pneumatic and electromechanical, using the SORPAS software.

Last but not least is the chapter entitled "Applications of Aluminum Alloys in Rail Transportation" and was prepared in China by Xiaoguang Sun et al. This chapter focuses on the latest applications of aluminium alloys, including for the car body, gearbox and steering rack, and analyze key manufacturing techniques such as casting, forming and welding.

This book is a continuation of several books previously published in the last decade by InTech on various theoretical aspects, production, application and research of aluminium, its alloys and composites based on aluminium alloys, edited successively by Tibor Kvackaj (2011) [13], Zaki Ahmad (2011, 2012) [14, 15], Subbarayan Sivasankaran (2017) [16], Kavian Cooke (2020) [17].

Aluminium, its alloys and composites with aluminium participation undoubtedly belong to engineering materials of strategic importance for development in many areas. For about 150 years of practical use, they have found numerous applications, often competitive but in many cases unrivalled. Annual world production in 2015 has exceeded 62,500,000 metric tons. The main recipients are the automotive, aviation and transport industries, but also using these materials can be manufactured precision microelements.

At this point, I would like to thank the Authors for preparing individual chapters and the IntechOpen publisher for many months of cooperation in preparing this book for printing.

I am deeply convinced that this book is valuable and will be of interest to numerous PT Readers. Therefore, it remains for me to wish that my previsions meet with a friendly reception. I wish the PT Readers enjoy reading this book and hope it serves them in solving real engineering problems.

Leszek A. Dobrzański, Hon., Prof., Dr HC,
Director of Science Centre,
Medical and Dental Engineering Centre for Research,
Design and Production ASKLEPIOS,
Gliwice, Poland

References

[1] United Nations General Assembly. Transforming Our World: The 2030 Agenda for Sustainable Development: A/RES/70/1. New York: United Nations; 2015.

[2] Dobrzański LA. Engineering materials and material design. Fundamentals of materials science and metallurgy. 2nd ed changed and supplemented. Gliwice-Warszawa: WNT;2006. 1596 p. (in Polish)

[3] Kagermann H, Wahlster W, Helbig J. Recommendations for Implementing the Strategic Initiative INDUSTRIE 4.0: Final Report of the Industrie 4.0 Working Group. Bonn, Germany: Federal Ministry of Education and Research; 2013.

[4] Government of Japan Cabinet Office. Society 5.0 [Internet]. 2019. Available from: http://web.archive.org/web/20190710182953/https://www8.cao.go.jp/cstp/society5_0/index.html [Accessed: 2021-02-10]

[5] Dobrzański LA, Dobrzańska-Danikiewicz AD. Why are Carbon-Based Materials Important in Civilization Progress and Especially in the Industry 4.0 Stage of the Industrial Revolution?. Materials Performance and Characterization. 2019;8:337-370. DOI: 910.1520/MPC20190145

[6] Dobrzański LA, Dobrzański LB, Dobrzańska-Danikiewicz AD, Dobrzańska J. The Concept of Sustainable Development of Modern Dentistry. Processes. 2020;8:1605. DOI: 10.3390/pr8121605

[7] Dobrzański LA, Dobrzański LB. Approach to the design and manufacturing of prosthetic dental restorations according to the rules of the Industry 4.0. Materials Performance and Characterization. 2020;9:394-476. DOI: 10.1520/MPC20200020

[8] Dobrzański LA, Dobrzański LB. Dentistry 4.0 Concept in the Design and Manufacturing of Prosthetic Dental Restorations. Processes. 2020;8:525. DOI: 10.3390/pr8050525

[9] Dobrzański LA. Role of materials design in maintenance engineering in the context of industry 4.0 idea. Journal of Achievements in Materials and Manufacturing Engineering. 2019;96: 12-49. DOI: 10.5604/01.3001.0013.7932

[10] Jose R, Ramakrishna S. Materials 4.0: Materials Big Data Enabled Materials Discovery. Applied Materials Today. 2018;10: 127-132. DOI: 10.1016/j.apmt.2017.12.015

[11] Dobrzański LA. Significance of materials science for the future development of societies. Journal of Materials Processing Technology. 2006; 175:133-148. DOI: 10.1016/j.jmatprotec.2005.04.003

[12] Aluminum Market - Industry Analysis, Market Size, Share, Trends, Application Analysis, Growth And Forecast 2019-2025 [Internet]. Available from: https://www.industryarc.com/Research/Aluminium-Market-Research-503190 [Accessed: 2021-02-10]

[13] Kvackaj T. (Ed.), Aluminium Alloys, Theory and Applications (ID: 44), IntechOpen, 2011, DOI: 10.5772/576, https://www.intechopen.com/books/44

[14] Ahmad Z. (Ed.), Recent Trends in Processing and Degradation of Aluminium Alloys (ID: 217), IntechOpen, 2011, DOI: 10.5772/741, https://www.intechopen.com/books/217

[15] Ahmad Z. (Ed.), Aluminium Alloys - New Trends in Fabrication and Applications (ID: 3053), IntechOpen, 2012, DOI: 10.5772/3354, https://www.intechopen.com/books/3053

[16] Sivasankaran S. (Ed.), Aluminium Alloys – Recent Trends in Processing, Characterization, Mechanical behavior and Applications (ID: 6071), IntechOpen, 2017, DOI: 10.5772/68032, https://www.intechopen.com/books/6071

[17] Cooke K. (Ed.), Aluminium Alloys and Composites (ID: 8862), IntechOpen, 2020, DOI: 10.5772/intechopen.81519, https://www.intechopen.com/books/8862

Section 1

Composites Materials with Aluminium Matrix

Chapter 1

Advanced Composites with Aluminum Alloys Matrix and Their Fabrication Processes

Leszek A. Dobrzański

Abstract

This chapter introduces advanced aluminum alloy matrix composites and their manufacturing processes. In the beginning, the state of the art is characterized and the general characteristics of aluminum and its practical applications are presented, starting with the history of aluminum. The current approximate distribution of bauxite resources in the world and the production of bauxite and alumina in the leading countries of the world, as well as the production of primary and secondary aluminum and the range of aluminum end products, are presented. Aluminum alloys intended for plastic deformation and castings, and composite materials in general and with a matrix of aluminum alloys in particular, have been characterized in general. Against this background, a detailed review of the results of the Author's own research included in numerous projects and own publications on advanced composite materials, their production technology, their structure, and properties were done. The range of aluminum alloy matrices of composite materials was adequately characterized, which include AlSi12, AlSi7Mg0.3, AlMg1SiCu, AlMg3, AlMg5, and AlMg9, respectively. Composite materials tested in terms of manufacturing technology include three groups. The first group includes gas pressure infiltration with liquid aluminum alloys of suitably formed porous preforms. Porous frameworks as a reinforcement for pressure-infiltrated composite materials with a matrix of aluminum alloys are produced by three methods. Al_2O_3 powder with the addition of 30–50% carbon fibers is uniaxially pressed, sintered, and heated to thermally degrade the carbon fibers and create the required pore sizes. In the second case, the ceramic porous skeleton is produced with the use of halloysite nanotubes HNTs by mechanical milling, press consolidation, and sintering. A third method is SLS selective laser sintering using titanium powders. Another group of manufacturing technologies is the mechanical synthesis of the mixture of AlMg1SiCu aluminum alloy powder and respectively, halloysite nanotubes HNTs in a volume fraction from 5 to 15% or multi-wall carbon nanotubes MWCNTs in a volume fraction from 0.5 to 5%, and subsequent consolidation involving plastic deformation. The third group of analyzed materials concerns composite surface layers on substrates of aluminum alloys produced by laser feathering of WC/W2C or SiC carbides. The structure and properties of the mentioned composite materials with aluminum alloys matrices are described in detail. The chapter summary provides final remarks on the importance of advanced aluminum alloy composite materials in industrial development. The importance of particular groups of engineering materials in the history and the development of the methodology for the selection of engineering materials, including the current stage of Materials 4.0, was emphasized. The importance of material design in engineering design is emphasized. Concepts of the development of societies were presented: Society 5.0 and Industry 4.0. The own concept of a holistic model of the extended Industry 4.0 was presented, taking into account advanced engineering materials and technological processes. Particular attention was paid to the importance of advanced composite materials with an aluminum alloy matrix in the context of the current stage of Industry 4.0 of the industrial revolution. Growth in the production of aluminum, its alloys, and composites with its matrix was compared with that of steel. Despite the 30 times less production, aluminum is important due to its lower density. The challenges posed by the development in the Industry 4.0 stage, including the expectations of the automotive and aviation industry, force constant progress in the development of new materials with the participation of aluminum, including the composite materials with an aluminum alloy matrix presented in this chapter.

Keywords: Aluminum, Aluminum alloys matrix composite, pressure infiltration, halloysite nanotubes, multi-wall carbon nanotubes, additive manufacturing, selective laser sintering, mullite, mechanical synthesis, laser feathering, Industry 4.0, Sustainable Development Goals, materials science paradigm, aluminum market

1. State-of-the-art and the general characteristics of aluminium and its practical applications

1.1 General characteristics of aluminum

Aluminum is a metal of very high technical importance [1–4]. It occurs in nature in many minerals and is the third element (after oxygen and silicon) in terms of its share in the earth's crust. It is, however, the second metal after iron in terms of technical applications. Its main ore is bauxite, from which pure Al_2O_3 oxide is produced, and then by electrolysis of the oxide dissolved in the molten cryolite (sodium fluoroaluminate), blast aluminum is obtained, which can be further refined.

Aluminum has an atomic number of 13 and an atomic mass of 26.9815. It does not show allotropes and crystallizes in a wall-centered regular network of A1 type with a parameter of 0.40408 nm. The melting point of aluminum is 660.37°C and the boiling point is 2494°C. The density of aluminum is 2.6989 g/cm^3 at 20°C. Aluminum in the annealed condition is characterized by tensile strength 70–120 MPa, yield strength 20–40 MPa, elongation 30–45%, and narrowing 80–95%. Aluminum can be cold and hot worked plastic deformed. In the cold plastic deformed state with 60–80% degree of crushing, the tensile strength reaches 140–230 MPa, yield strength 120–180 MPa, hardness 40–60 HBW, with a reduced elongation 1.5–3%.

Aluminum is characterized by high electrical conductivity – 37.74 MS/m, constituting approx. 65% of the electrical conductivity of copper, and good thermal conductivity. The electrical conductivity is significantly reduced with increasing concentrations of impurities and impurities, mainly Fe and Si, as well as Cu, Zn, and Ti. These elements also reduce the plasticity but increase the strength properties. The Fe admixture hardly dissolves in aluminum, creating the brittle Al_3Fe phase. On the other hand, Si shows low solubility and does not form any phases, it is in the free form. In the case of the simultaneous presence of Fe and Si, the $Fe_3Si_2Al_{12}$ or $Fe_3Si_2Al_9$ intermetallic phases occur. Their separations, especially at the grain boundaries, significantly reduce the plastic properties of aluminum.

Aluminum is highly resistant to corrosion. In the air, it is covered with a thin layer of Al_2O_3, which protects against atmospheric corrosion, water, concentrated nitric acid, numerous organic acids, and hydrogen sulfide. On the other hand, reducing acids HCl and HF, seawater, vapors, and mercury ions accelerate the corrosion of aluminum. In order to improve corrosion resistance, aluminum may be subjected to anodic oxidation (so-called anodization), i.e. an electrolytic process of producing an oxide coating, e.g. in a 10% sulfuric acid solution, combined with different colors of the metal surface.

Aluminum is extracted from ores in a complex process and, obviously, unknown throughout most of human history [5–7]. On the other hand, hydrated potassium-aluminum sulphate (KAl (SO_4) 2 12 H_2O) has been known for millennia as alum, or more precisely aluminum-potassium alum, which occurs naturally as a mineral in the form of granular, infiltrative clusters. Raids and efflorescence. It occurs in the weathering zone of clay rocks, coals containing pyrite, marcasite and other sulfides. It is also formed in the process of weathering igneous rocks rich in feldspar and feldspar, incl. in Uzbekistan in the Kara-kum desert, in Italy near Vesuvius, in the USA in Alaska and in Russia in Kamchatka. People have known alum since the 5th century BC and it was widely used for staining as early as antiquity [8]. It was widely distributed in international trade relations in the Middle Ages due to its use in dyeing [8–10]. Aluminum was discovered in 1825 by Hans Christian Oersted [11–13], a Danish physicist. The results of his work were extended by Friedrich Wöhler [8, 13, 14], a German chemist. In 1854, both Robert Wilhelm Bunsen, a German chemist, and Henri Étienne Sainte-Claire Deville, produced aluminum for the first time by electrolysis. Electricity supply was ineffective at the time, so their methods could not be developed in industry. At that time, aluminum was extremely valuable, and therefore its price significantly exceeded that of gold. The situation in this respect continued until 1856, when Deville first began industrial production and not through the production of electrolytes [8], which he considered impractical at the time, so he switched to chemical methods in the same year. In 1884, world aluminum production reached 3.6 tons.

In 1886, Paul Héroult, French engineer and Charles Martin Hall, American engineer, independently developed a method of producing aluminum, which became the first large-scale production technology [8]. Their process is still in use today and is known as the Hall-Héroult process (**Figure 1**) [15, 16]. This process and the process developed by Carl Joseph Bayer, an Austrian chemist, determined the wide dissemination of aluminum and its availability to the public, starting from 1889 (**Figure 2**) [8].

Nowadays, the commercial production of aluminum oxide by refining from bauxite practically covers the Bayer process [17, 18] (**Figure 2**), which consists of:

1. Digestion, in which bauxite is ground and slurried to caustic soda (NaOH) which is then pumped into large pressure fermentors where the caustic soda reacts with aluminum oxide minerals to form soluble sodium oxide $Na[Al(OH)_4]$;

2. Clarification, in which a sodium oxide solution is processed in cyclones to remove coarse particles; the remaining fluid is processed in thickeners where the flocculates are added to the solid agglomerates which are removed with fabric filters; the residue (red mud) is washed and discarded and the clarified solution (containing $Na[Al(OH)_4]$) goes to the next step;

3. Precipitation – aluminum oxide nucleating agents are added to the clarified solution in order to facilitate the separation of larger agglomerated aluminum oxide crystals; product-size crystals are separated from the small crystals (recycled serve as nucleating agents) and washed to remove caustic residues; agglomerates are transferred to the next step.

4. Calcination – $Na[Al(OH)_4]$ agglomerates are placed in rotary kilns or stationary fluidized bed calciners at a temperature higher than 960°C, ensuring dehydration and obtaining commercial grade aluminum oxide.

Both these processes are still used in the production of aluminum and it is them that have made aluminum popular in industry and everyday life. The statue of Anteros, which was erected in 1893 at Piccadilly Circus in London, Great Britain, can be considered a symbol of the growing role of aluminum and is considered the first large work made of this material (**Figure 3**).

World aluminum production reached 6,800 metric tons in 1900. **Figure 4** shows changes in the production volume and prices of aluminum in the years 1885–2020.

In the late nineteenth and early twentieth centuries, aluminum alloys began to be used more and more. The basis of modern aluminum production are the Bayer and Hall-Héroult processes, further enhanced in 1920 by Carl Wilhelm Söderberg, a Swedish chemist and his cooperators. In the twentieth century, the range of aluminum applications has significantly increased, including both in engineering and in construction, to gain use in aviation as a strategically important material during World War I. For these reasons, the importance of aluminum during the Second World War was even greater, to exceed 1,000,000 metric tons for the first time in 1941. After the war, since 1945, aluminum consumption has been increasing annually by another 10% for almost three decades, gaining popularity in construction applications, electrical cables, basic foil and the aerospace industry. In 1954, aluminum, ahead of copper, became the most widely produced non-ferrous metal, with world production increasing to 2,810,000 tons per year. In the second half of the 20th century, aluminum found its way into transport and packaging, although its production was critically assessed for environmental reasons, and as a result aluminum recycling developed. For example, in the United States, in the period from 1970 to 1980, the level of aluminum recycling increased 3.5 times, and by 1990 even 7.5 times. At the same time, mining and processing costs decreased and aluminum production increased, exceeding 10,000,000 metric tons per year for the first time in 1971. This was accompanied by a fall in aluminum prices since the beginning of

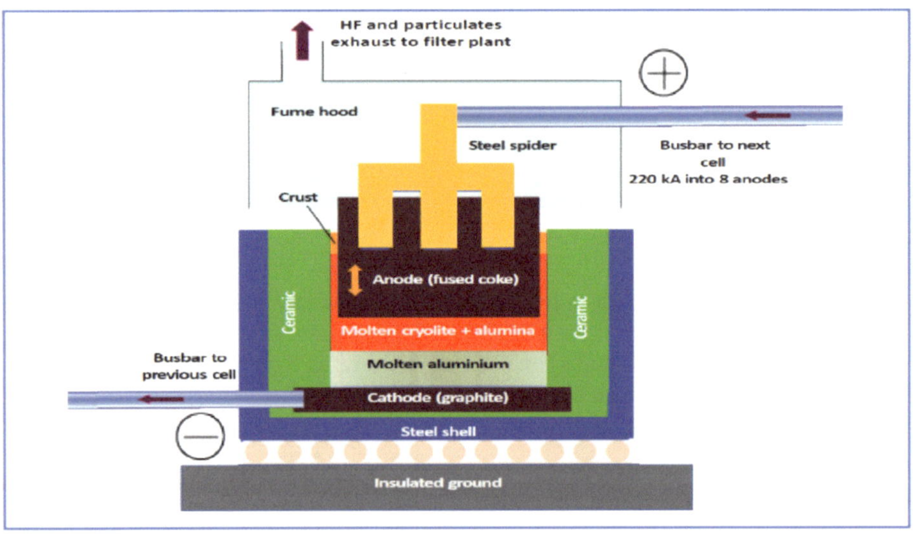

Figure 1.
The diagram of the Hall – Héroult cell for process for smelting aluminum through dissolving aluminum tained in molten cryolite, and electrolyzing the molten salt bath.

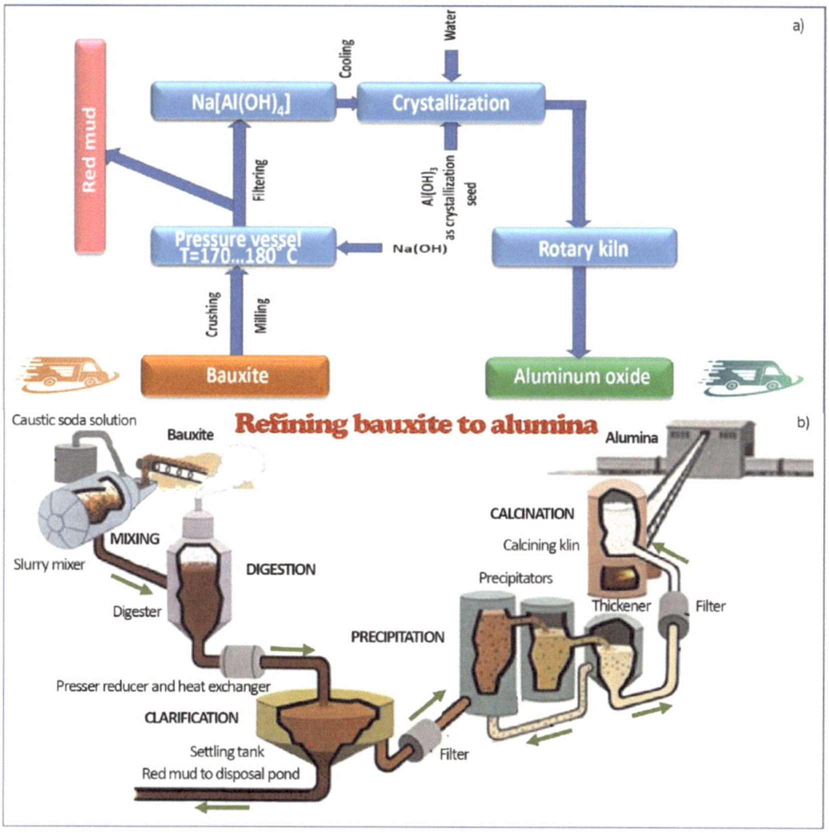

Figure 2.
The Bayer process of obtaining aluminum oxide from bauxite ores (a) the block diagram; (b) the example of the industrial application.

the 1970s. In 1973, that price was US$ 2,130 per metric ton (1998 US price level). The launch of the production of beverage cans prompted a significant increase in aluminum production in the early 1970s, despite the fact that in the 1970s and 1980s the costs of primary aluminum production increased. However, it contributed to an increase in the share of secondary aluminum in the overall production, due to its significantly increased recycling. In the 1970s, aluminum became a stock exchange commodity. In the 1970s, aluminum production from the United States, Japan and Western Europe was gradually allocated to Australia, Canada, the Middle East, Russia and China, where it was cheaper, e.g. in view of favorable tax regulations in these countries. Production costs fell in the 1980s and 1990s. In the first decade of the 21st century, the total share of the BRIC countries (Brazil, Russia, India and China) increased from 32.6% to 56.5% in primary production and from 21.4% to 47.8% in primary consumption. China amassed a particularly large share of world production thanks to abundant raw materials, cheap energy and government incentives, and its share in consumption increased from 2% in 1972 to 40% in 2010. In the United States, this share decreased to 11%. In the United States, Western Europe, and Japan, most aluminum is used in transport, engineering, construction, and packaging. World production continues to increase, surpassing 50,000,000 metric tons in 2013 and surpassing a record 58,500,000 metric tons per year in 2015. World production has now exceeded more than 62,500,000 metric tons per year [19]. Aluminum is produced in greater quantities than all other non-ferrous metals combined.

Figure 3.
Statue of Anteros, the Greek god of revered love, on Piccadilly Circus in London, UK: (a) view of Piccadilly circus; (b) detail of the monument with the figure of Anteros.

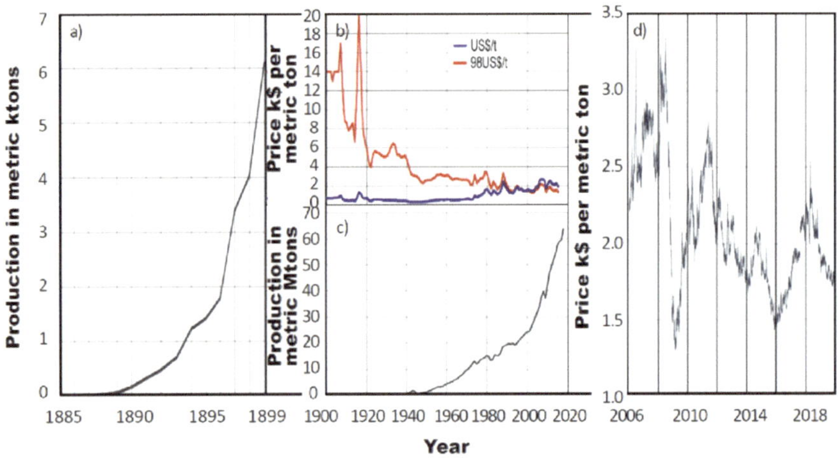

Figure 4.
Changes in the volume of aluminum production and prices in the years 1885–2020; (a) production volume in the years 1885–1900; (b) changes in aluminum prices in the years 1900–2020 – Blue line, the actual price in current values; red line – Price comparable in 1998; (c) production volume in the years 1900–2020; changes in aluminum prices in the period from 1 December 2005 to 24 November 2020 (own study based on statistical data).

The basic raw material for the production of aluminum is bauxite [20–22]. Pierre Berthier the French geologist discovered bauxite in Provence, France in 1821. Bauxite is a sedimentary rock which contains a high aluminum content as mixture of few minerals i.e. gibbsite ($Al(OH)_3$), boehmite (γ-$AlO(OH)$) and diaspore (α-$AlO(OH)$), goethite ($FeO(OH)$) and hematite (Fe_2O_3), clay mineral kaolinite ($Al_2Si_2O_5(OH)$) and also anatase (TiO_2) and ilmenite ($FeTiO_3$ or $FeO.TiO_2$). Australia is the largest producer of bauxite, followed by China. **Figure 5** shows the distribution of the main raw materials of bauxite in the world with the differences in their abundance. **Table 1** presents the production of bauxite and aluminum oxide in the leading countries in the world in 2018 and 2019 and the current state of bauxite raw material reserves [23].

Aluminum is produced in many grades [24] with varying degrees of purity from 99.999 to 99.0%, a few examples of which are given below:

- the purest one containing 99.995% Al is used in the production of chemical equipment and condenser films,

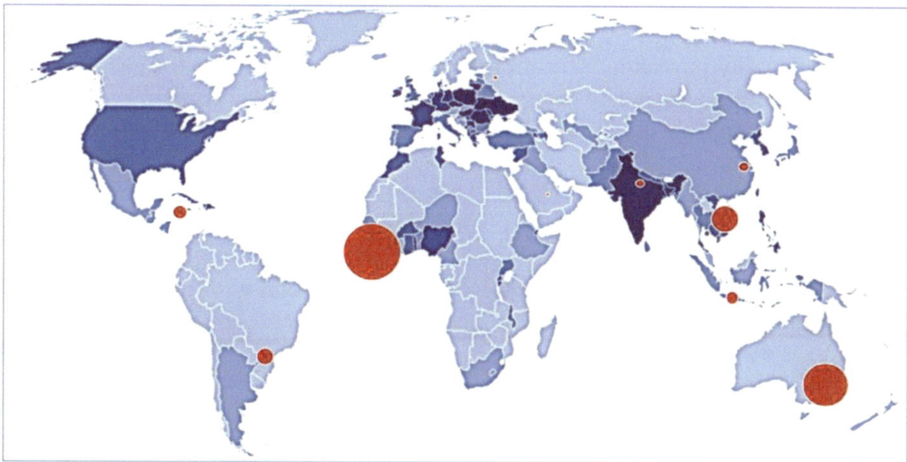

Figure 5.
Approximate distribution of bauxite reserves in the world according to data from 2019, with an indication of differences in their abundance (the diameter of the red circle corresponds to the resource size) (own study).

Country (alphabetical order)	Bauxite Reserves	Bauxite		Aluminum oxide	
		2018	2019	2018	2019
Australia	6,000,000	86,400	100,000	20,400	20,000
Brazil	2,600,000	29,000	29,000	8,100	8,900
Canada	—	—	—	1,570	1,500
China	1,000,000	79,000	75,000	72,500	73,000
Guinea	7,400,000	57,000	82,000	180	300
India	660,000	23,000	26,000	6,430	6,700
Indonesia	1,200,000	11,000	16,000	1,000	1,000
Jamaica	2,000,000	10,100	8,900	2,480	2,100
Malaysia	110,000	500	900	—	—
Russia	500,000	5,650	5,400	2,760	2,700
Saudi Arabia	200,000	3,890	4,100	1,770	1,800
Vietnam	3,700,000	4,100	4,500	1,310	1,300
United States	20,000	W	W	1,570	1,600
Other countries	5,000,000	17,000	15,000	11,400	12,000
World total (rounded)	30,000,000	327,000	370,000	131,000	130,000

Table 1.
Production of bauxite and aluminum oxide in the leading countries in the world in 2018 and 2019 and the current state of bauxite raw material reserves.

- containing 99.8% Al is used for foils, cable coatings and plating,
- containing 99.5% Al for electric wires,
- containing 99% Al for everyday products.

There are three different types of primary aluminum, corresponding to different production methods:

- commercially pure aluminum from the reduction of Al_2O_3 in the electrolyser, representing the majority of industrial production; it may contain up to 1% of impurities, and the purity very often exceeds 99.9%.
- refined aluminum produced by electrorefining commercially pure aluminum in the cell layer; its purity ranges from 99.9% to 99.99 + %.
- zone refined aluminum produced by zone refining and with the appropriate starting material and technique can produce metal with less than 1 ppm impurities.

This purity was obtained thanks to the zone refining of electrodeposited metal from organic baths.

Secondary aluminum is produced by remelting production scrap or obsolete equipment and is usually recovered as alloys. Even after remelting high-quality electrodes, the resulting "secondary pure aluminum" is less pure than the corresponding primary aluminum because contamination by other materials occurs.

Most of the impurities in primary aluminum come from raw materials (bauxite, soda, carbon). Therefore, their purity should be controlled, although it is also important to vary the chemical composition of individual shipments, even if they come from the same source, as they may differ in their composition, even by one order of magnitude.

Metallurgical aluminum grades (with limited purity) are used in the production of alloys and numerous everyday products, equipment for the food industry, for some electric cables, heat exchangers, in construction, for coatings of steel products, and in the form of foil – for food packaging. Refined aluminum (high purity) is used in electronics and electrical engineering and for the construction of special chemical apparatus.

1.2 General presentation of aluminum alloys

The relatively low strength properties of aluminum can be increased – even several times – by introducing alloying elements and heat treatment of the alloys [24–30]. **Table 2** gives the general classification of aluminum alloys according to their chemical composition.

Most generally – due to the method of manufacturing – aluminum alloys are divided into:

- for plastic deformation,
- cast.

Some of these alloys can be used both as cast and for plastic deformation.

Aluminum alloy series	Major alloy additives	Characteristic
1xxx	Lack (admixtures <1%)	Low strength properties and very high plasticity, making them susceptible to cold and hot forming; thanks to the high purity of the alloy, they are highly resistant to corrosion, and they are also characterized by high thermal and electrical conductivity
2xxx	Cu	Has an increased hardness compared to 1xxx, but the corrosion resistance is much lower
3xxx	Mn	Low strength, but susceptible to plastic deformation; they are very well weldable and corrosion resistant
4xxx	Si	High strength and corrosion resistance
5xxx	Mg	Corrosion resistance is even higher than for 4xxx
6xxx	Mg + Si	Corrosion resistant, shows good plasticity, medium mechanical strength
7xxx	Zn + Mg	Hard, high-strength aluminum alloys; they are characterized by the highest strength of all aluminum alloys, but also low corrosion resistance
8xxx	Others, no classification	Covers aluminum alloys which were not classified in the previous series

Table 2.
Classification of aluminum alloys by chemical composition.

The alloys for plastic deformation usually contain up to about 5% of alloying elements, most often Cu, Mg, Mn, sometimes also Si, Zn, Ni, Cr, Ti, or Li. Some of these alloys are used in a plastic worked or in recrystallization annealed state, and some are subjected to a precipitation hardening heat treatment. Aluminum alloys with a concentration of alloying elements greater than 5% can also be subjected to plastic deformation, under special conditions. Aluminum alloys can be shaped by various methods of plastic deformation and are supplied in the form of many metallurgical intermediates.

Cast aluminum alloys are mostly multi-component alloys with a high concentration – from 5 to 25% – of alloying elements, mainly Si, Cu, Mg, Zn, and Ni or their various combinations. Cast Al alloys are characterized by good fluidity and often low casting shrinkage. Alloys containing less than 5% alloying elements can also be used as cast. In general, aluminum alloys can be sand-molded, die-cast, pressure-cast, and lost wax methods although some of these methods are rarely used today.

1.3 General information on composite materials with aluminum alloys matrices

Advanced possibilities of shaping the properties of materials containing aluminum alloys are provided by the production of composites with an aluminum matrix (AMC), where the matrix most often used is not aluminum but aluminum alloys [31–45]. AMCs in relation to aluminum and its alloys show better physical and mechanical properties, including a very favorable strength-to-weight ratio, high strength and high modulus of elasticity, better fatigue strength, good ductility, low thermal expansion coefficient, good wear resistance and corrosion resistance and creep strength. Therefore, AMCs have found numerous applications, many times unrivaled in comparison with aluminum alloys, when high properties are required, e.g. in the automotive, aviation, manufacturing, armaments, energy and medicine industries [46–62].

Generally, a composite material, often simply referred to as a composite, is a combination of two or more materials (reinforcement members, fillers and binder constituting the matrix of the composite) differing in type or chemical composition on a macroscopic scale [3]. Components of composite materials retain their identity because they do not completely dissolve in each other and do not combine into other elements, but interact together. Typically, these components can be physically identified and exhibit interfacing surfaces.

Composite materials are usually artificially produced to obtain properties that cannot be obtained separately by any of the components present. Composite materials are used to ensure appropriate mechanical properties, but also electrical, thermal, tribological, various environmental and other properties. Composite materials most often contain fibers or phase particles and are stiffer and stronger than the continuous matrix phase. The many reinforcement elements also provide good thermal and electrical conductivity and a lower coefficient of thermal expansion and/or good wear resistance. However, there are exceptions that can also occur in composite materials, such as rubber-modified polymers, where the discontinuous phase is more compliant and more ductile than the polymer, resulting in increased impact strength. Similarly, steel wires can be used to reinforce gray cast iron in truck and tractor brake drums.

In general, composite materials can be classified on two separate levels. This classification, although it may seem a bit eclectic, is of practical importance. The first level of classification concerns the matrix material, and in this case the analysis is limited to metals and their alloys, in particular aluminum and its alloys. At the second level (**Figure 6**) the classification concerns the geometrical features of the reinforcing materials which is presented in the **Figure** 7 also.

Composite materials based on non-ferrous metals, including aluminum, may contain dispersoid particles with a diameter of 10–250 nm. These particles, usually metal oxides, distributed in the matrix, affect the blocking of the dislocation movement and even if they are not coherent with the matrix, they influence its strengthening. These particles cannot dissolve in the matrix, although a low solubility may improve the adhesion of the dispersoids and the matrix. As Al_2O_3 does not dissolve in Al, the Al_2O_3 – Al system effectively enhances the dispersive hardening of such a composite material. There are numerous examples of this type of composite

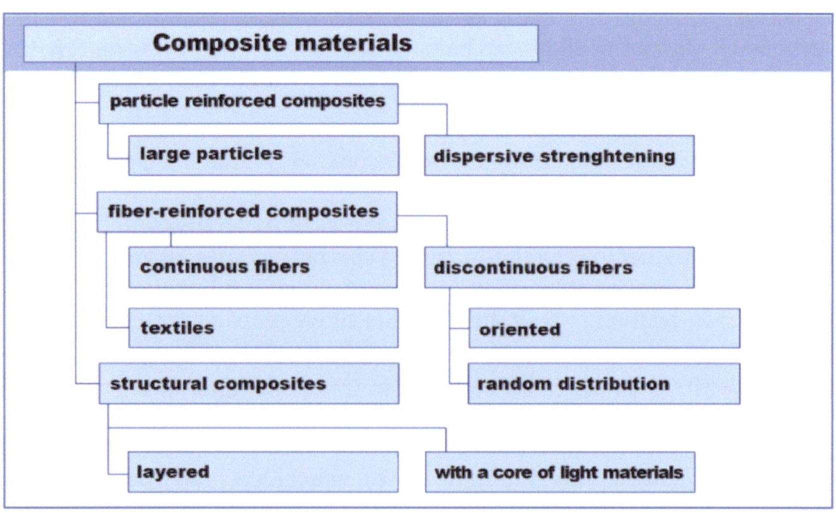

Figure 6.
Classification of composite materials according to the type and geometrical features of reinforcing elements.

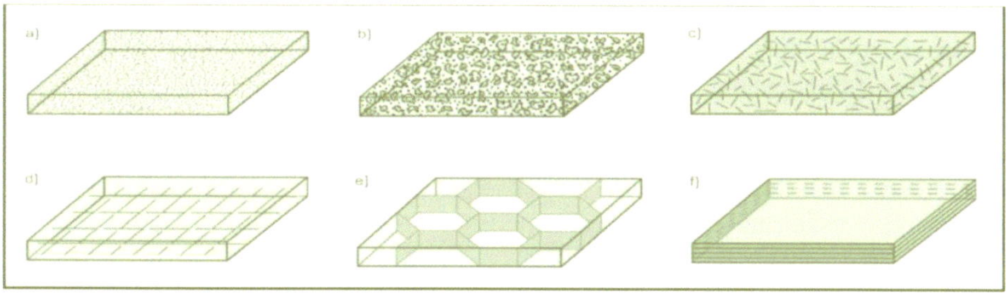

Figure 7.
Examples of different types of reinforcement elements for composite reinforced materials: (a) dispersive particles, (b) flakes, (c) discontinuous fibers, (d) continuous fibers in the form of a textile, (e) skeletally, (f) layered in laminates.

materials, including sintered aluminum powder SAP reinforced by Al_2O_3 in 14% and dispersion hardened nickel or iron alloys, oxide dispersion-strengthened ODS.

Powder metallurgy is a frequently used technology to produce metal matrix composite materials, and this distinct group of composite materials is labeled P/M MMCs (powder metallurgy metal – matrix composites) or P/M AMCs respectively for aluminum matrix.

In the case of composite materials with a non-ferrous matrix, including aluminum and its alloys, reinforced with large particles, due to the relatively large size of the reinforcing particles and the lack of their coherence with the matrix of non-ferrous metals, dispersive hardening cannot take place. These groups of composite materials, on the other hand, achieve a combination of functional properties resulting from the interaction of both the matrix and the particles of the second phase arranged therein, which often guarantee an increase in strength properties.

Fiber-reinforced composite materials are most often used to provide increased static and fatigue strength, including specific strength and stiffness, which is obtained by incorporating strong, stiff but usually brittle fibers into a soft but tough matrix. The matrix only transmits applied load to the fibers, which for the most part carry it. Composite materials of this group can show high strength properties both at room temperature and at elevated temperature.

Many reinforcement materials are used. Boron, metal and ceramic fibers are used to strengthen e.g. composite materials with a metal matrix, as well as with a composition corresponding to the intermetallic phases. The described fibers have gained a wide range of applications due to their very high utility values. However, it should be noted that their price varies by as much as 50,000 times. Glass fibers are used very commonly, although not in the case of composite materials with a non-ferrous metal matrix, because their relative specific cost (cost of 1 kg of material) ranges from 1 to 3. Aramid, carbon, ceramic oxide and an oxide fibers show very good performance properties, but at a high price, because the relative cost related to the cost of glass fibers ranges from about 15 for aramid and carbon fibers, although also not used in the case of composite materials with a non-ferrous metal matrix, to approx. 4500 for SiC fibers, which determines the limited use of these materials. In very special cases, however, sapphire fibers are used, where special reinforcing properties are required, as the relative price of this material exceeds 45,000.

Often, the fibers are stapled, their length does not exceed 10 mm, and they are arranged randomly in the matrix of composite materials, especially those with a polymer matrix.

Composite materials with a non-ferrous metal matrix including AMCs reinforced with continuous fibers are a class of materials with a wide range of applications due to their mechanical properties, wear resistance and thermal properties. It is possible to adapt their properties to requirements and provide increased strength, stiffness, thermal conductivity, abrasion resistance, fatigue strength and dimensional stability. These composite materials are non-flammable, do not evaporate under vacuum, and are resistant to organic liquids, including fuels and solvents. The matrix of these materials consists of monolithic alloys, usually light metals, including aluminum, but also heat-resistant superalloys. They are reinforced with boron, carbide, oxide and tungsten fibers in a continuous form, usually with a volume fraction of 10–70% in the composite material, produced in addition to materials with the same matrix reinforced with discontinuous fibers. Aluminum oxide particles are a cheap solution and are most often used when the metal matrix is cast, while the most common are silicon carbide and boron carbide particles as a reinforcement in the form of discontinuous but also continuous fibers, especially in aircraft structures (**Table 3**).

Reinforcing fiber material	Manufacturing process	Application examples
Boron	Hot pressing of the fiber system between the metal foils	Tubular struts of the central part of the space shuttle hull, cold heat dissipator (diffuser) in the housings of electronic microchips of multi-layer plates, neutron radiation shielding material (e.g. in spent nuclear fuel storage facilities or containers for transporting such fuel), fan blades in turbo-jet engines, aircraft wing plating, aircraft landing gear components, bicycle frames, golf clubs; Due to the degradation of boron fibers above 480°C, high-temperature applications are impossible and manufacturing using casting or high temperature low pressure pressing
SiC	Investment casting, low pressure hot forming, superplastic forming with diffusion welding, isostatic hot pressing HIP, green tape production by winding fibers on films covering a rotating drum and their initial bonding to the film using a polymer resin, removed then in the volatilization (evaporation) process or by plasma spraying and subsequent pressure diffusion welding of several such cut tapes stacked at a temperature close to the melting point and pressure consolidation in an autoclave in a vacuum metal container, as well as in closed forms to give the required final shapes	Construction of aircraft elements, wings up to 3 m long, elements of portable bridges, small pressure vessel cylinders, bullet stabilizers, missile casings
Graphite	Vacuum diffusion welding of raw metal (aluminum) strips with longitudinally arranged fibers, diffusion rolling welding Rapi-Press, continuous pressing, direct metal infiltration process, casting	Very high strength components and stiffness, e.g. the mast of the Hubble cosmic telescope, thin-walled pipes of small and large diameters (up to 2 m), with complex shapes, produced with the near-net-shape technology
Al_2O_3	Casting	Pistons for internal combustion engines (Toyota)

Table 3.
General characteristics of composite materials with aluminum matrix reinforced by continuous fibers.

Various types of ceramics, e.g., Al_2O_3, SiC and B_4C, are widely used to reinforce aluminum matrix composites. They show very good properties, such as high strength, high hardness, abrasion resistance.

Composite materials in metal-to-metal and metal-ceramic systems are produced using liquid and solid-phase technologies. Liquid-phase technologies are based on direct (in situ) methods, during which the fibrous structure of the composite is obtained by solidifying the alloy with unidirectional heat dissipation, and on indirect (ex situ) methods using foundry technologies, plastic processing and powder engineering.

The basic manufacturing processes for AMC on an industrial scale can be divided into the following groups:

- Liquid state processes, including stir casting, extrusion casting, injection molding and in situ (reactive) process, ultrasonic assisted casting, vacuum infiltration, pressureless infiltration and dispersion methods;

- Solid-state processes involving powder mixing followed by consolidation (powder metallurgy), friction by shot-assisted high-energy powder grinding, ion bonding by diffusion and vapor deposition techniques

- Liquid–solid processes including composite casting, semi-solid molding.

Methods of manufacturing composite materials with an aluminum alloys matrix with continuous reinforcement			
With the participation of the liquid phase	Gravitational • vacuum • pressure	In solid state	Diffusion welding
			Hot rolling welding
			Hot extrusion welding
	Bonding		Bonding high-energy shaping
	Hot pressing with the participation of the liquid phase		
	Plasma spraying		
Methods of manufacturing composite materials based on aluminum alloys with discontinuous reinforcement			
With the participation of the liquid phase	Methods of mixing metal in liquid state • by mechanical methods • electromagnetic methods • gas injection	In solid state	Powder engineering methods
Thixotropic manufacturing in the liquid–solid state			
Spray methods			
In situ generation methods: • reactions in the system: liquid–liquid, liquid–gas, liquid–solid, • reactions in salt mixtures, • direct oxidation and nitriding, • reactive spray molding, • self-developing high-temperature synthesis, • reactive plasma synthesis.			

Table 4.
Classification of methods of manufacturing composite materials with a matrix of aluminum alloys with continuous and discontinuous reinforcement.

Method	Range of shape and size	Range of volume fraction	Damage to reinforcement	Cost
Stir casting	Wide range of shapes, larger size up to 500 kg	Up to 0.3	No damage	Least expensive
Squeeze casting	Limited by preform shape up to 2 cm height	Up to 0.5	Severe damage	Moderate expensive
Spray casting	Limited shape, large shape	0.3–0.7		Expensive
Powder engineering	Wide range, restricted size		Reinforcement fracture	Expensive

Table 5.
Comparative analysis of different technique of composites manufacturing.

A detailed classification can also be made, based on the type of reinforcement (continuous and discontinuous) (**Table 4**).

Table 5 compares additionally some of the methods mentioned.

During the last quarter of a century, the processes of mechanical alloying, known for a long time, have been used in the production of composite materials reinforced with particles, causing improvement in the structure and dispersion hardening of these materials and even distribution of hardening particles in the matrix [33, 47, 63–65]. The resulting nanocrystalline or submicrocrystalline structure improves the properties of newly developed composite materials [34, 37, 47, 53, 63, 65].

2. Description of the materials taken into account and the technologies used for their manufacturing

2.1 The range of matrices of aluminum alloy matrix composite materials presented in this chapter

This chapter presents a generalization of the results of numerous own studies and provides a detailed overview of the structure and properties and manufacturing technology of several AMCs selected for presentation due to their advanced manufacturing methods and unexpected properties and the resulting wide application possibilities.

Among these materials, there are basically three groups of such materials presented in this chapter, examined within the research projects [66–74] of the Author of this chapter and doctoral dissertations [75–80] carried out under the supervision of the Author of this chapter. The details were published earlier [81–112]. The description in this chapter covers the main groups of composite materials presented in **Table 6**, as follows:

1. composite materials manufactured by gas-pressure infiltration with liquid aluminum alloys,

2. nanocomposite materials manufactured by mechanical synthesis of a mixture of aluminum alloy powder with HNTs or MWCNTs nanotubes and the press consolidation,

3. composite surface layers on the substrate of aluminum alloys as a result of laser feather by introducing carbides powder.

No.	Matrix	Reinforcement	A form of reinforcement	Technology of manufacturing	Literature
1	AlSi12	Sintered Al_2O_3 powder with carbon fibers	Skeleton structure	Gas-pressure infiltration	[73–75]
2.	AlSi12	Sintered HNTs halloysite nanotubes with carbon fibers	Skeleton structure	Gas-pressure infiltration	[68, 70, 74, 78]
3	AlSi12	Selective laser sintering of titanium powder	Skeleton structure	Gas-pressure infiltration	[66, 70, 74, 80]
4	AlSi7Mg0.3	Selective laser sintering of titanium powder	Skeleton structure	Gas-pressure infiltration	[66, 70, 74, 80]
5	AlMg1SiCu	Halloysite HNTs nanotubes	Reinforcing by dispersive particles	Mechanical synthesis & press consolidation	[68, 76]
6	AlMg1SiCu	Multi-wall carbon nanotubes MWCNTs	Reinforcing by dispersive particles	Mechanical synthesis & press consolidation	[67, 79]
7	AlMg3	WC	Powder particles in remelted zone	Laser feather by introducing powder in remelted zone	[69, 70, 72, 77]
8	AlMg3	SiC	Powder particles in remelted zone	Laser feather by introducing powder in remelted zone	[69, 70, 72, 77]
9	AlMg5	WC	Powder particles in remelted zone	Laser feather by introducing powder in remelted zone	[69, 70, 72, 77]
10	AlMg5	SiC	Powder particles in remelted zone	Laser feather by introducing powder in remelted zone	[69, 70, 72, 77]
11	AlMg9	WC	Powder particles in remelted zone	Laser feather by introducing powder in remelted zone	[69, 70, 72, 77]
12	AlMg9	SiC	Powder particles in remelted zone	Laser feather by introducing powder in remelted zone	[69, 70, 72, 77]

Table 6.
General characteristics of the discussed composite materials.

In the above-mentioned composite materials, aluminum alloys listed in **Table 7** were used as the matrix.

2.2 Manufacturing technologies of porous skeletons as a reinforcement of pressure-infiltrated aluminum alloy matrix composite materials

In the case of the 1st composite material in question, porous Al_2O_3 ceramic skeleton was produced by sintering Alcoa Industrial Chemicals CL 2500 powder (**Table 8**) with the addition of 30, 40, or 50% by weight Sigrafil C10 M250 UNS carbon fibers from SGL Carbon Group (**Figure 8**) with the properties given in **Table 9**.

Al_2O_3 powder was wet-milled in a ball mill for 5 minutes to break up powder agglomerates. In order to eliminate mutual electrostatic interactions and prevent agglomeration of carbon fibers, Dolapix CE 64 was used, introduced into the Al_2O_3

Aluminum alloys	Elements concentration, wt. %							
	Si	Fe	Cu	Mn	Mg	Ti	Other	Al
AlSi12	10.0–12.0	Max 0.3	Max 0.03	0.2–0.4	0.25–0.32	0.08–0.15	—	rest
AlSi7Mg0.3	6.5–7.5	Max 0.15	Max 0.05	Max 0.10	0.25–0.45	0.08–0.15	V max 0.03	
AlMg1SiCu	0.6	0.47	0.22	0.11	0.95	0.006	Cr 0.26; Zn 0.015	
AlMg3	0.07	0.07	0.01	0	2.86	0.01	—	
AlMg5	0.08	0.07	0.01	0	5.55	0.01	—	
AlMg9	1.32	0.1	0.003	0.43	10.1	0.008	Zn 0.006	

Table 7.
Chemical composition of the aluminum alloys used.

Density, g/cm^3	Particle diameter D50, μm	Share of other phases, wt%						
		Al_2O_3	Na_2O	Fe_2O_3	SiO_2	CaO	B_2O_3	other
3.98	1.80	99.80	0.05	0.02	0.01	0.01	0.01	0.10

Table 8.
Properties Al_2O_3 Alcoa CL 2500 powder.

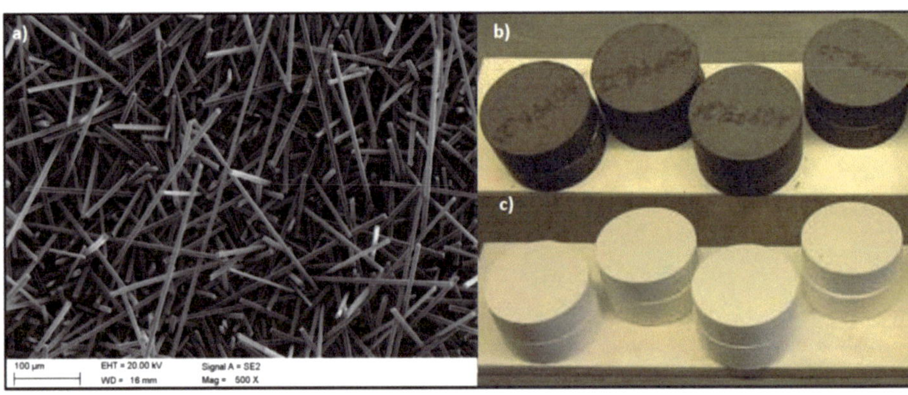

Figure 8.
(a) Morphology of Sigrafil C10 M250 UNS carbon fibers (SEM); (b, c) powder mix of aluminum oxide and carbon fibers (b) after pressing, (c) after sintering (macro).

suspension. To facilitate subsequent compression, 1% water-soluble polyvinyl alcohol Moviol 18–8 was added to the mixture. The powder was then dried by freezing and sublimating water under reduced pressure and sieved through a 0.25 mm sieve. The polyvinyl alcohol was activated after the powder was moistened with distilled water and left for 24 h in a polymer foil bag. Then the powders were pressed uniaxially in a die with a diameter of 30 mm on a manual laboratory press "Nelke" and in a rectangular die with dimensions of 65x46 mm on a hydraulic plate press Fontune TP 400 at a pressure 100MPa, for 15 seconds. The pressed preforms were sintered in Gero tubular furnace in an air atmosphere with a flow of 20 l/min, with slow heating up to 800°C at a rate of 20°C/h, heating for 10 hours for complete thermal degradation of carbon fibers, heating up to 1500°C at a rate of 300°C/h,

Diameter, μm	Average length, μm	Density, g/cm^3	Strength, GPa	Young's modulus, GPa	Carbon content, wt. %
8	135	1.75	2.5	26	>95

Table 9.
Properties of Sigrafil C10 M250 UNS carbon fibers.

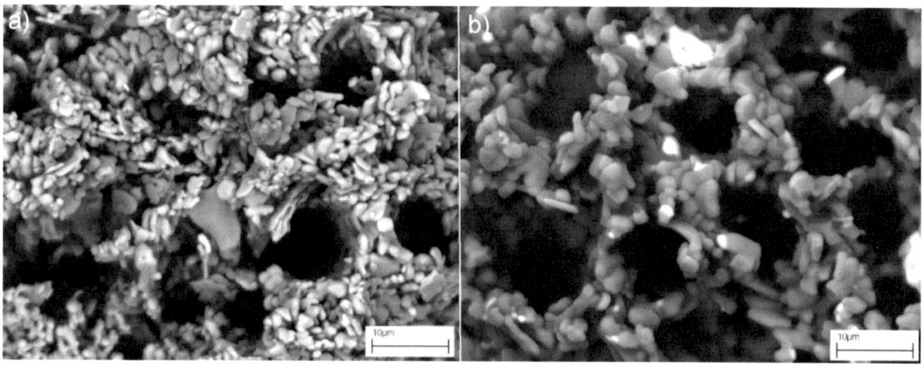

Figure 9.
The fracture's structure of the ceramic skeleton produced by sintering Al_2O_3 powder with (a) 30%; (b) 50% carbon fiber content.

sintering for 2 h and cooling with the furnace. The sintered skeletons show larger pores after the carbon fiber degradation and of much smaller dimensions around the individual ceramic particles that were intentionally not densified to the maximum extent (**Figure 9**). As a result, the created ceramic skeleton is characterized by a system of interconnected pores, ensuring high permeability of 2.48–10.56 $m^2 \cdot 10^{-13}$ and facilitates infiltration with liquid aluminum alloy.

Appropriate Al_2O_3 wettability by the liquid aluminum alloy was ensured by nickel coating the pores of the ceramic skeletons by using Atotech's Futuron technology (**Table 10**) used to make metallic coatings on polymeric materials to create a tin and palladium layer on the surface.

The solution of Noviganth Activator AK II ensures the exchange of tin on the surface with metallic palladium, after which it is possible to apply a layer of the Ni-P alloy by the chemical electroless method. The internal surfaces of the ceramic skeletons were coated using a solution pumping device which used a spiral cooler in the shape of a copper coil to ensure the required temperature. The sintered porous skeletons were glued with the two-component adhesive "UHU plus endfest 300" to the aluminum rings, which could be infiltrated with the liquid AlSi12 alloy after heating in a furnace at 800°C. The sintered ceramic skeleton in a matrix coated with graphite and heated to 450°C, was infiltrated by the AlSi12 liquid alloy at 800°C using a punch in a Fontune TP 400 hydraulic plate press with a pressing speed of 17 mm/s and a maximum press pressure of 100MPa for 120 s. They were removed from the mold and cooled with a stream of compressed air.

In the second case of the composite materials analyzed in this chapter, the ceramic porous skeleton, which is the reinforcement of the composite, was manufactured with the use of halloysite nanotubes (HNTs) by their mechanical milling, press consolidation, and sintering. HNTs are provided by NaturalNano (USA), and their phase composition is shown in **Table 11**. The nanotubes are obtained from halloysite, which is a clay mineral of volcanic origin, which is

Treatment	Reagent	Temperature, °C	Time, min
Surface activation	Futuron Activator	Room temperature	1
	Noviganth Activator AK II	40–45	3
Applying Ni-P coating by chemical electroless method	Noviganth PA Chemical Nickel	55	1.5

Table 10.
The course of coating Al_2O_3 ceramics with Ni-P alloy.

No.	Phase	Share, wt.%		
		Halloysite nanotubes	Halloysite nanotubes heated at 1100°C	Halloysite nanotubes heated at 1500°C
1	SiO_2	45.63	53.85	53.92
2	Al_2O_3	37.93	44.86	45.1
3	Fe_2O_3*	0.46	0.58	0.59
4	TiO_2	0.11	0.16	0.13
5	CaO	0.01	0.02	0.01
6	MgO	0.07	0.09	0.1
7	K_2O	0.01	0.02	0.02
8	Na_2O	0.03	0.18	0.09
9	P_2O_3	0.35	0.42	0.41
10	Loss on the kiln at 1025°C	15.58	0.24	0.04

**Total Fe converted to Fe_2O_3.*

Table 11.
The phase composition of halloysite nanotubes provided by NaturalNano (USA).

characterized by high porosity, large specific surface, high ion exchange and easy chemical and mechanical processing. Halloysite is a mineral [$Al_2Si_2O_5$ $(OH)_4{\cdot}H_2O$] that occurs in three deposits in the world, including in Poland. At least 30% of naturally occurring resources are in the form of nanotubes in the shape of polyhedron, hollow inside as cylindrical objects of the diameter 40–200 nm and length of up to 1–2 μm. After properly insulation HNTs can be used in nanotechnology. The morphology and particle size distribution of the powders used are shown in **Figure 10**.

For the manufacturing of porous ceramic skeletons, a mixture of halloysite powder with a mass fraction of 30, 40, 50, 60, or 70% by weight, respectively, was mechanically dry milled with Sigrafil CIO M250 UNS carbon fibers from SGL Carbon Group, analogous to that used in the previous case. Additionally, the MA7050 micronized amide wax powder with a mass fraction of 1% was used as a lubricant, also reducing the adhesion between the powder material and the mill and milling media, as well as leveling the friction between the die walls and powder particles. The technology of manufacturing the porous skeleton was very similar to that described in the 1st case. The milling of the mixture of the above-mentioned components was performed with the use of a Fritsch Pulverisette 6 centrifugal ball mill for 15 min.

The fragmented and homogeneous powder mixture obtained in this way was formed by uniaxial cold pressing in closed dies, respectively, of circular (φ30 mm) or rectangular (12 × 35 mm) cross-section for 15 s on a LabEcon 600 Fontijne Grotnes hydraulic plate press. Preliminary tests allowed to set the pressing pressure as 50, 75, and 100 MPa. The degradation and sintering cycle of the moldings in the PRS 75 W high-temperature furnace by Czylok in the air atmosphere with the flow of 5 l/h was analogous to that in the first case, but the sintering temperature, determined on the basis of the analysis of the structure and mechanical and functional properties, is 1500°C for 1 h heating and an optimal share of HNTs of 70%. In this way, the structure of sintered porous mullite is obtained (**Figure 11**), which is subjected to pressure infiltration in the device shown schematically in **Figure 12** with the AlSi12 alloy.

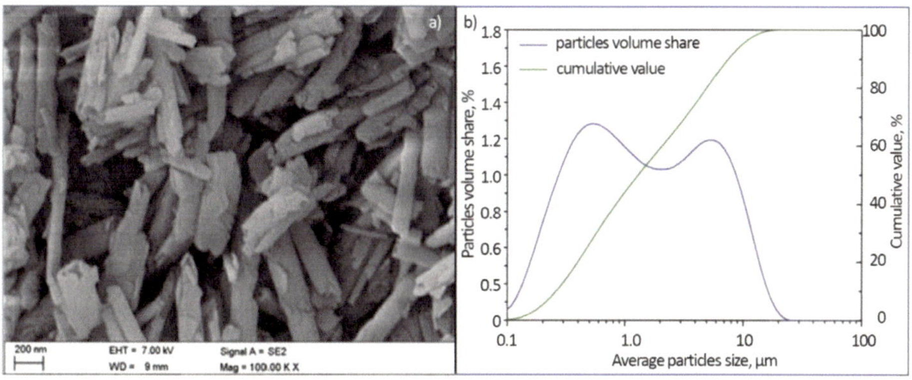

Figure 10.
Morphology of halloysite nanotubes (a) and particle size distribution (b).

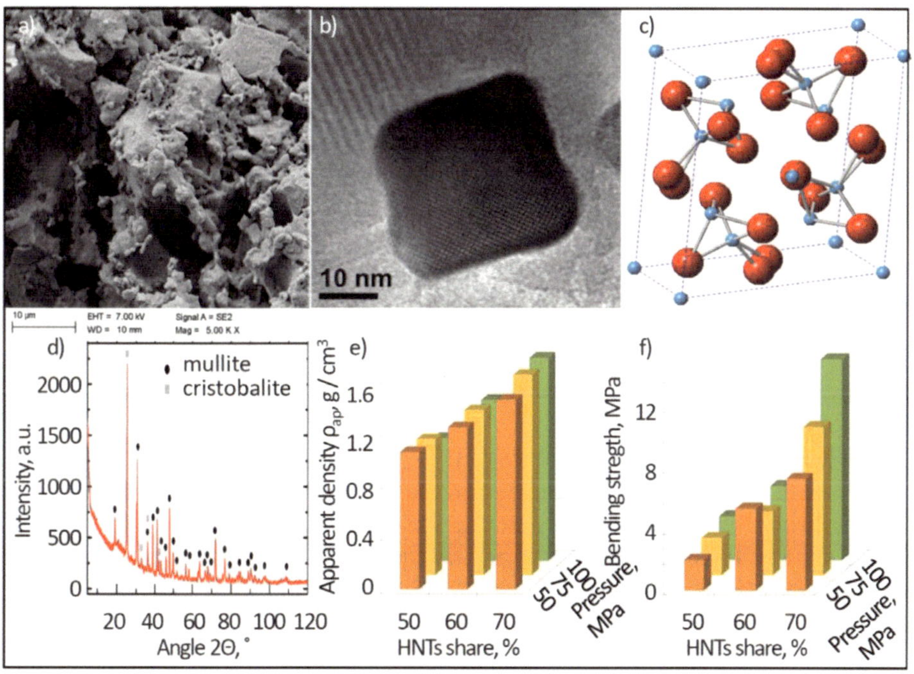

Figure 11.
Structure of the sintered skeletons (a) mullite with cristobalite nano-areas as confirmed by X-ray analysis (d); (b) nanoparticle of mullite (HRTEM); (c) mullite structure. (f) elementary crystallographic cell of mullite from fig. (b); (e), (f) influence of HNTs share and pressure on the apparent density (e) and bending strength of skeleton (f).

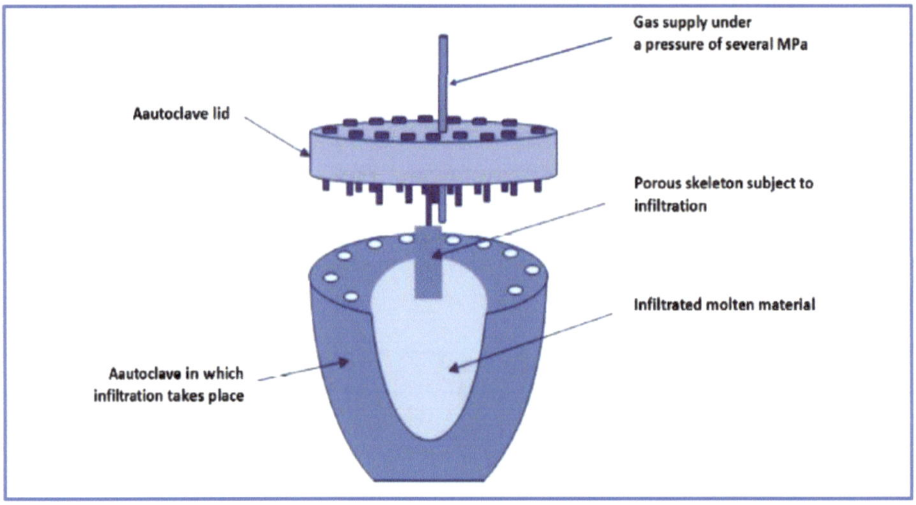

Figure 12.
Diagram of the infiltration device.

In the third and fourth cases of composite materials, selective laser sintering SLS was used for the manufacturing of a porous skeleton, divided into two main stages: design and manufacturing this skeleton (**Figure 13**).

In stage I, a given element is designed. The available specialized three-dimensional software allows for almost complete control and repeatability of the size and geometrical features of the designed elements, and also enables full control of the conditions of the manufacturing process. The design result is a three-dimensional computer-aided design CAD model in stereolithography STL format, showing the surface of the element by means of a triangle mesh. The smaller triangles are, the more precise the mapping of the surface is. By successively dividing the element into layers of a specific thickness, the optimal conditions for the manufacturing process are established, including laser power, scanning speed, layer thickness, distance between successive remelting paths, and the diameter of the laser beam. Software for Manufacturing Applications 3D Marcarm Engineering AutoFab, integrated with the Renishaw AM 125 machine, used for selective laser sintering, was used. The three-dimensional skeleton model was designed by CAD using unit cells of a defined structure and size and by subsequently multiplicating such unit cell to design a repetitive structure skeleton in which the pore size and wall thickness are fully designed by the designer, characterized by completely open pores.

On the basis of numerous technological experiments, the selection of "hexagon cross" unit cells (**Figure 14**) was made and the conditions for the manufacturing by selective laser sintering of porous titanium skeletons with a selected pore size of 250 μm (controlled unit cell dimensions 500 μm) and 350 μm (dimensions unit cell 600 μm). Experimentally, the relationship between the design arrangements and the actual pore dimensions was established (**Figure 15**).

After giving the model the appropriate size and structure, it is divided into layers with a predetermined thickness, with the number of layers corresponding to the number of layers of powder that will sinter until obtaining the accuracy of the model production depends on the laser power used: for low power 50 W it is ±20 μm in the XY plane, and for high power 200 W ± 100 μm in the XY plane with a layer thickness of 20 to 50 μm. Manufacturing conditions are transferred to the

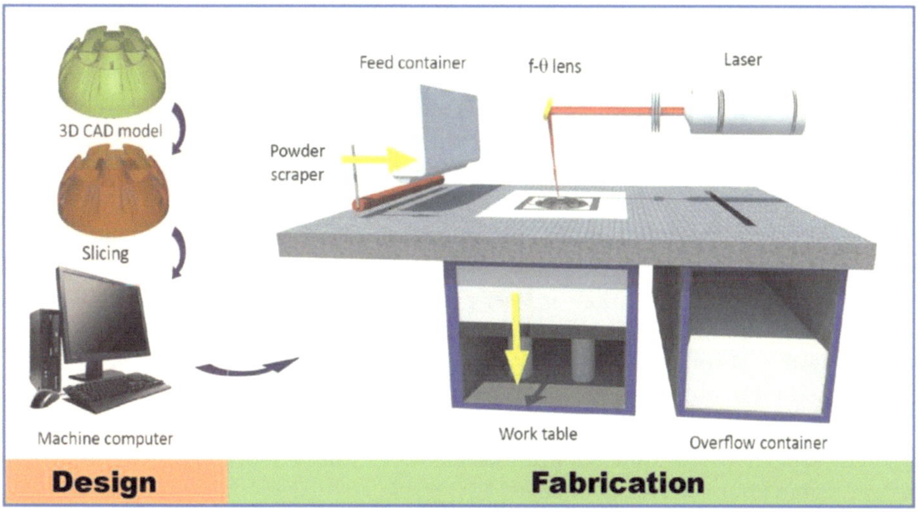

Figure 13.
Schematic diagram of the selective laser sintering technology.

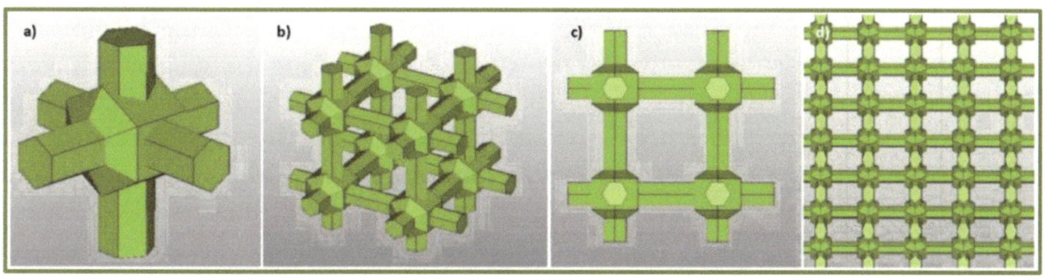

Figure 14.
(a-c) Basic hexagon cross unit cell, (d) computer model structure image showing the arrangement of unit cells at an angle of 45° to the x axis of the coordinate system.

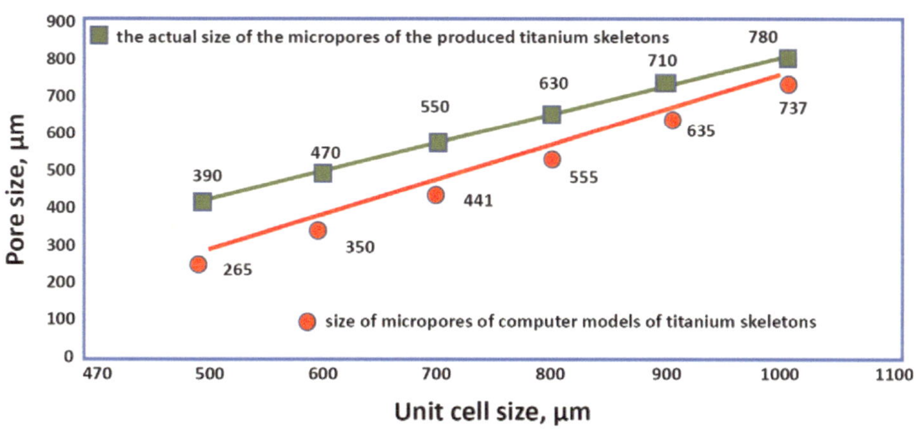

Figure 15.
Graph of the dependence of the pore size designed with the use of AutoFab software and titanium micro-skeletons produced by the selective laser sintering method on the size of the base unit cell.

designed model in STL format to the machine software for selective laser sintering. The most favorable properties of titanium skeletons with the given pore sizes are provided by the arrangement of unit cells at an angle of 45° to the x axis of the coordinate system (**Figure 11d**).

In the second stage, the previously designed element is manufactured by the method of selective laser sintering, layer by layer, until the final product is obtained. Selective laser sintering is a complex thermophysical process, the course of which depends on the type of material, the characteristics of the laser and the environment, and the thickness of the powder layer, as previously stated, as well as the diameter of the laser spot, the distance between the laser spots and the distance between the laser paths (**Figure 16**).

In the case of porous materials, the laser power of 60 W was selected and, in addition to the II laser path, the diameter of the spot was smaller than the distance between the laser spots, marked as laser path I. The AM 125 system uses a YFL fiber laser with ytterbium-doped active material and a maximum power of 200 W. The AM 125 system is equipped with a vacuum chamber, which allows the use of a unique method of emptying the working chamber of the device in such a way that all gases are first pumped out of the working chamber, and then an inert gas such as argon or nitrogen is introduced, providing an operating environment in which the oxygen concentration does not exceed 100 ppm.

The technology of selective laser sintering can be treated as a modern technology of powder engineering, where the process of building an element begins with the distribution of the powder layer on a working table with position adjustable in relation to the z-axis. This layer acts as a substrate for the created object. The laser beam is guided along the surface of the powder in accordance with the previously entered and appropriately configured information concerning the successive layers of the cross-section of the spatial image of the object. Next, the working table with the powder is lowered by the height (layer thickness) set by the user, another thin layer of powder is spread, where the grains sintering, which takes place by surface melting of another layer of metal powder grains with the previously sintered layers of the manufactured element. Whether the powder is sintered or fused depends on the manufacturing conditions. The process involves sintering with a liquid phase. Successive layers of the cross-section are sintered together. The process is repeated until a coherent element is obtained. After the construction is completed and the temperature is lowered, the element is removed from the powder bed and subjected to a finishing treatment, depending on the application, e.g. sandblasting or grinding. The AM 125 device has a powder container with automatic valves delivering an additional portion of the powder during the entire process, and its excess after the distribution of the next layer in the working chamber is drained to the container, which can be easily and quickly dismantled and placed at the powder sieving or

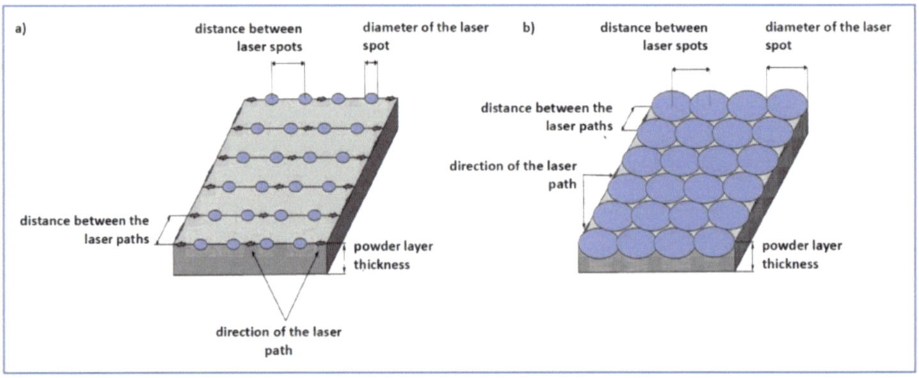

Figure 16.
Characteristics of a laser beam with a spot diameter: (a) smaller than the distance between the laser spots, (b) greater than the distance between the laser spots.

selection station thanks to which it is possible to reuse the powder in the next process.

For the production of porous titanium skeletons, titanium powder was used with the composition presented in **Table 12** with the purity grade 2 and the grain size up to 45 μm with a spherical shape. The powder used for the tests also has a reduced oxygen concentration down to 0.14% (in titanium powders the average oxygen concentration is about 0.5%).

The technology of selective laser sintering allows the production of microporous titanium skeletons with completely open pores, with the full possibility of controlling the shape and size of these pores, and also allows the production of elements of any designed shape, which makes it very competitive in relation to other material manufacturing technologies of porous materials.

The mechanical properties of selectively laser-sintered titanium micro-skeletons depend on the spatial orientation with respect to the main axes of the applied external stress, the assumed pore size, the laser power, the size of the laser spot, and the path of the laser spot. The best mechanical properties of 36 MPa (tensile strength), 97 MPa (bending strength), and 125 MPa (compressive strength) are obtained after selective laser sintering according to spatial orientation 45° in relation to the x-axis, with a laser spot width of 50 μm, laser power 60 W, an assumed porosity of 50–60%, corresponding to a pore size of 250 μm, for the course of a laser spot for which the distance between the laser spots and the laser remelting paths in relation to the diameter of the laser spot is equal to or smaller than the diameter of the laser spot. The lowest mechanical properties of 8 MPa (tensile strength), 15 MPa (flexural strength), 14 MPa (compressive strength), respectively, are ensured by the 45° orientation in relation to the y axis and 45° in relation to the x-axis for the same pore size ~250 μm. The smaller assumed pore size influences the greater the strength, and it varies from 8 MPa (tensile strength), 17 MPa (bending strength), 12 MPa (compressive strength) to 18 MPa, 43 MPa, 37 MPa, respectively, with a reduction the pore size from 450 μm to 250 μm. As the pore size decreases, the diameter of the laser spot also decreases from 90 to 50 μm, respectively, which helps to increase the strength. With the increase of the laser power from 50 W to 60 W, the mechanical properties increase, changing from 28 MPa (tensile strength), 72 MPa (bending strength), 74 MPa (compressive strength) to 36 MPa, 97 MPa, 125 MPa respectively. Using the same technological conditions for selective laser sintering, changing the ratio of the diameter of the laser spot to the value of the distance between the spots and paths of laser remelting from ≥1 to <1 results in more than double the strength of porous titanium.

The titanium porous elements after selective laser sintering were subjected to preliminary cleaning in isopropanol solution using an ultrasonic cleaner. After the excess powder was removed from the pores of the titanium skeleton, the specimens were digested with aqua regia for 1 hour using an ultrasonic bath to remove excess powder loosely attached to the titanium skeleton that had not been removed in the pre-treatment. Composite engineering materials, the model of which structure is shown in **Figure 14a**, were produced by pressure infiltration of microporous

Mass concentration of elements, %							
C	O	N	H	Fe	Other		Ti
					Each	Together	
0.01	0.14	0.01	0.004	0.03	<0.01	<0.4	rest

Table 12.
Chemical composition of titanium powder.

titanium skeletons produced by the selective laser sintering technology at a temperature of 800°C under a pressure of 2–3 MPa for 2 minutes in a manner analogous to the second case, using casting matrix material aluminum alloys, respectively AlSi12 in third case and AlSi7Mg0.3 in fourth case (**Figure 17**).

2.3 Manufacturing technologies by mechanical synthesis and subsequent plastic consolidation of aluminum alloy matrix composite materials

In cases 5 and 6, the composite materials were manufactured by mechanical synthesis and subsequent plastic consolidation of the mixture of aluminum alloy AlMg1SiCu powders as a matrix and reinforcement, respectively for variant 5 halloysite nanotubes HNTs as in variant 2 in the volume fraction of 5 to 15% and in the variant 6 multi-wall carbon nanotubes MWCNTs in the volume fraction of 0.5 to 5% (**Figure 18**).

By using the mechanical alloying method, it is possible to produce nanostructured composite materials with the controlled particle size of the reinforcing phase and their uniform distribution and resulting in increased mechanical properties. Air-atomized aluminum alloy powder with nominal particle size <63 μm was obtained from ECKA, Austria. Multiwalled carbon nanotubes MWCNTs with the properties given in **Table 13** were supplied by Cheap Tubes (USA).

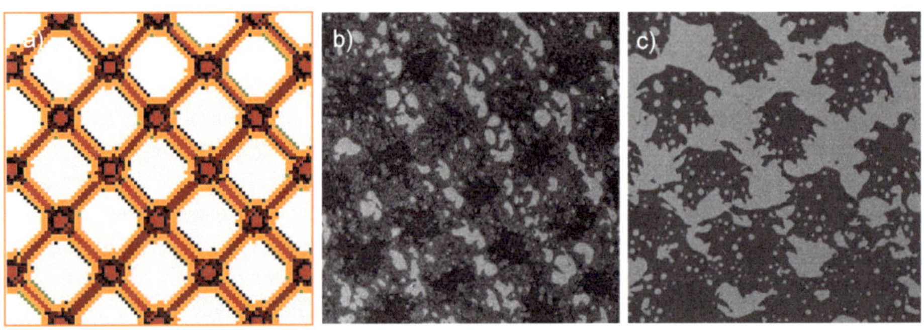

Figure 17.
Structure diagram of engineering composite materials with reinforcement of a titanium skeleton laser selectively sintered and pressure infiltrated with casting aluminum alloys; (a) diagram; (b) with AlSi12 matrix; (c) with AlSi7Mg0.3 matrix (b, c: SEM).

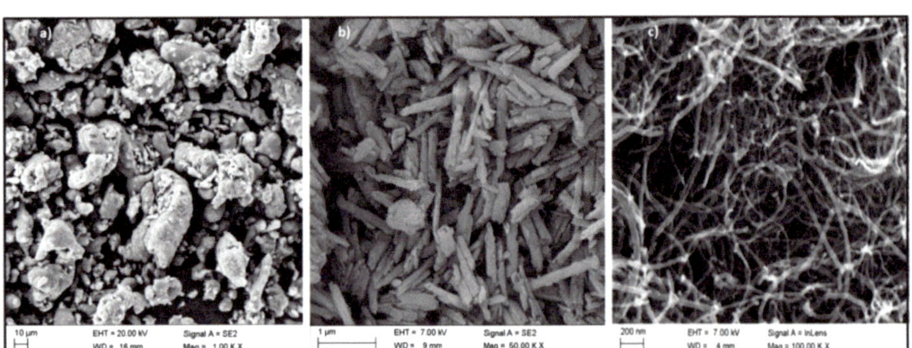

Figure 18.
Structure comparison in the delivered state; (a) AlMg1SiCu aluminum alloy powder; (b) halloysite nanotubes HNTs; (c) multi-wall carbon nanotubes MWCNTs (SEM).

As before, the lubricant MA7050 was used for all powders as a process control agent PCA reducing the adhesion between the feed and mill material with a mass fraction of 1%. The four stations planetary ball mill Pulverisette 5 by the Fritsch Company was used for mechanical synthesis using high energy mechanical milling for 0.5 to 20 hours at 200 rpm. Detailed conditions are given in **Table 14**.

In the case 5th, halloysite nanotubes had to be annealed prior to high-energy milling. A series of structural changes takes place in the halloysite, starting from dehydroxylation causing a metastable state, during heating of halloysite nanotubes at a temperature of ~100°C, associated with the 1st endothermic reaction consisting in the release of adsorption water (about 2% by weight), through the transformation into a regular spinel phase, in the temperature range of 400–600°C (peak at 548°C), when the 2nd endothermic reaction occurs, associated with hydroxylation (OH- ions in the octaendric layer are exchanged for O_2- ions (approx. 13% by mass), ending with the formation of mullite at a temperature close to 950°C as a result of an exothermic reaction associated with the reconstruction of the halloysite structure and the formation of the Si-Al spinel phase as a transition form [113–115] (15% by mass) [115]. These phenomena are confirmed by the thermogravimetric analysis (**Figure 19**) performed with the use of a corundum crucible in the range it is 20–1000°C in the air atmosphere to which 0.5018 g of halloysite nanotubes was subjected. The halloysite nanotubes used for the tests were therefore annealed at the temperature of 100, 500 or 700°C (HNT_{100}, HNT_{500} and HNT_{700}).

No.	Properties	Dimension	Value
1	Outer diameter	nm	20–30
2	Inner diameter	nm	5–10
3	Ash	wt. %	<1,5
4	Purity	wt. %	>95
5	Length	μm	10–30
6	Inner diameter	m^2/g	110
7	Electrical conductivity	S/cm	>100
8	Bulk density	g/cm^3	0.28
9	True density	g/cm^3	~2.1

Table 13.
Properties of multi-wall carbon nanotubes.

No.	Description	Reinforcement type	
		HNTs	MWCNTs
1	Reinforcement's concentration	5; 10; 15 wt.%	0.5; 2; 3.5; 5 wt.%
2	Time of milling	0.5; 1; 1.5; 2; 2.5; 3; 3.5; 4; 5; 6 h	0.5; 1; 1.5; 2; 3; 4; 5; 10; 15; 20 h
3	Ball material	AISI 420 quenched stainless steel	
4	Ball diameter	20 mm	
5	Ball-to-powder weight ratio	20:1	

Table 14.
Milling process parameters.

As a result of high-energy mechanical milling, crushed and permanently bonded composite powders were manufactured. Composite powders obtained as a result of high-energy mechanical milling were then cold pressed in a mold with a seat diameter of 26 mm and pressure of 300 MPa, and then pressed at a temperature of 460–500°C using a graphite suspension in oil as a lubricant, without degassing, in a cover made of AlMg1SiCu alloy. Co-extrusion of cold-pressed powders of rod-shaped composite materials consisted of placing the molded part in a thick-walled sleeve closed on one side by a punch, and on the other – by a die with a forming hole. At elevated temperature, under the pressure of the punch, the material was extruded through the die opening, obtaining a rod with a circular cross-section and a diameter of 8 mm in the case of using HNTs or 6 mm in the case of using MWCNTs, respectively.

2.4 Manufacturing technologies of the composite layers on the substrates of aluminum alloy

All other cases 7–12 concern composite layers of several millimeters obtained on the surface of casting aluminum alloys containing 3–10% magnesium. The desired composite layer of the tested aluminum alloys was obtained as a result of smelting the surface of the elements with the high-power HPDL ROFIN DL020 diode laser and fusing ceramic powders directly into the liquid metal pool using a gravity feeder (**Figure 20**). This laser is characterized by a very high beam power density, up to 107 W/cm^2, which means that the thermal impact on the surface of the element is limited, and thus causes only slight stresses and thermal deformations. The use of a diode laser for feathering ceramic powders into the surface layer of Al-Mg alloys allows for a predictable and repeatable way to obtain a composite layer on the surface of the analyzed aluminum alloys.

In order to minimize the power of the laser beam, shorten the time of the beam's impact on the surface and reduce the absorption of surface radiation, e.g. by grinding, anodizing, or tarnishing with the use of etching chemicals. The use of a flux, e.g. lithium chloride, is preferred. If this flux is not used, it would be necessary to reduce the scanning speed and increase the power of the laser beam, which results in increased linear energy and can lead to structure's changes in a larger volume of the material, making the technology less predictable and more difficult to control than with laser feathering using flux on surfaces of aluminum-magnesium alloys.

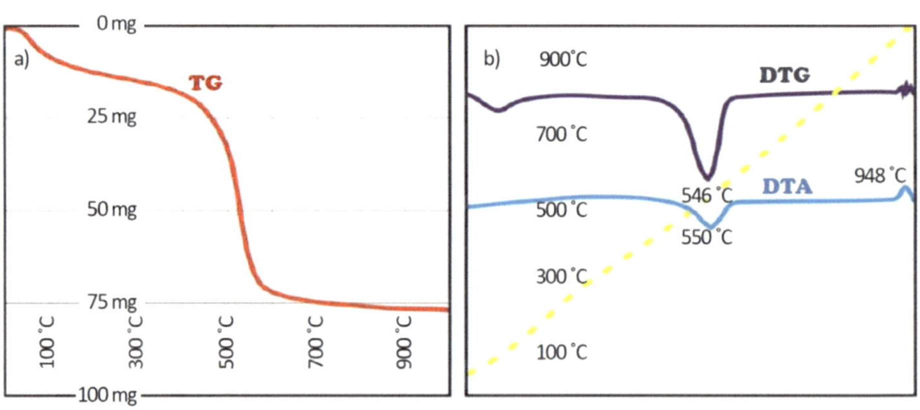

Figure 19.
Thermogravimetric curve of TG (a) and DTG and DTA (b) curves of halloysite nanotubes and the course of temperature changes during the tests (yellow dashed line).

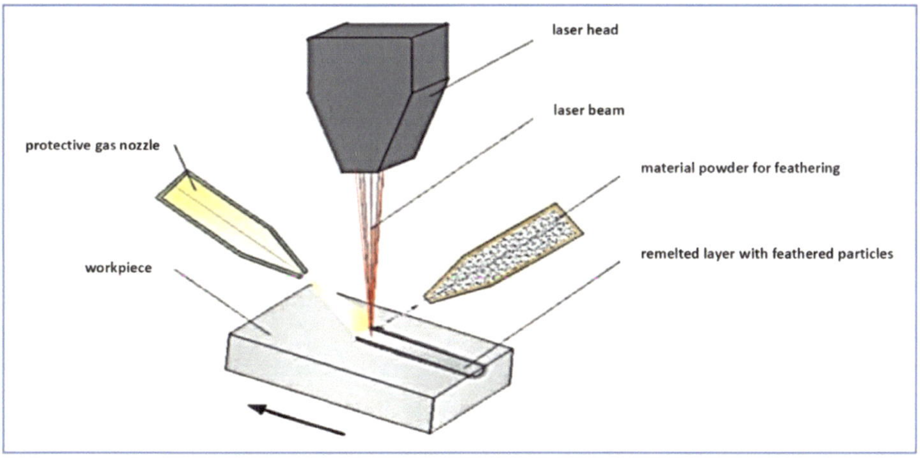

Figure 20.
Diagram of laser feathering of ceramic powder particles.

The purpose of selecting and optimizing the technological conditions of laser feathering is to produce a surface durable composite layer of aluminum with embedded carbides, characterized by a high coefficient of friction without the need for additional heat and surface treatment. The surface of the elements before laser feathering was mechanically processed by grinding with sandpaper with a gradation of 60 μm and then covered with a lithium chloride-based flux under the trade name "AluFlux". Lithium chloride powder was mixed with anhydrous ethyl alcohol in a ratio by volume of 1:5, resulting in a liquid suspension which was uniformly covered over the entire surface of the laser-processed part. The surface prepared in this way was dried for 30 minutes in a furnace at 50°C in order to evaporate the alcohol and moisture from the element surface. WC/W_2C and SiC powders were selected for laser surface treatment, taking into account the wettability of their powders (**Figure 21**). The conditions of the laser surface treatment, including the scanning speed and power of the laser beam, the method of powder feeding, and the shape of the laser spot, were determined on the basis of numerical calculations of the Finite Elements Method FEM computer simulation. In the numerical model, the laser beam power ranging from 0.5 to 2.5 kW, the beam scanning speed ranging from 0.2 to 0.8 mm/s, and the method of feeding the feathered powder through a paste consisting of a mixture of water glass and powder, milled on the surface of two or

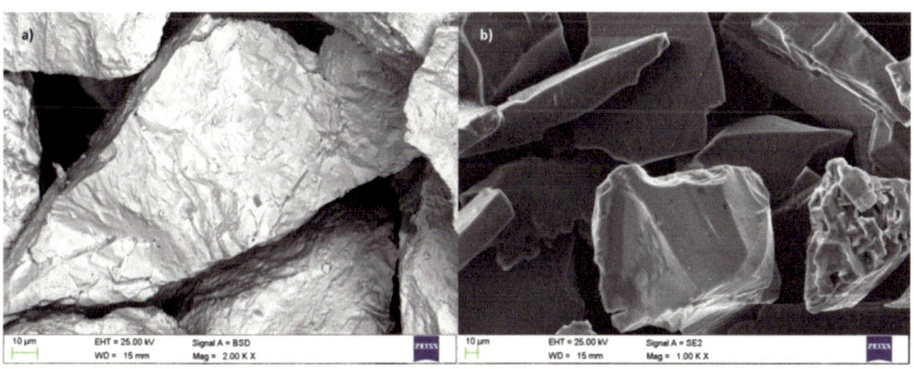

Figure 21.
Feathering powder of a) WC/W_2C tungsten carbide; b) SiC silicon carbide.

No.	Technical characteristic	Value	
1	Laser beam power, kW	1.8; 2.0; 2.2	
2	Laser beam scanning speed, m/min	0.5	
3	Spot size, mm	1.8 x 6.8	
4	Protective gas	Argon	
5	Protective gas flow, l/s	20	
6	Alloying powder	SiC	WC
7	Powder particle size, μm	45–180	45–180
8	The amount of powder fed, g/min	1	8
9	Theoretical powder density, g/cm^3	3.21	15.7
10	Melting point, °C	2730	2870

Table 15.
Technical conditions of the laser feathering process and general characteristics of the ceramic powders used.

three grooves with a depth of 0.5 and 1 mm and a gravity feeder with a speed depending on the density of the fed powder from 0.5 to 12 g/min. **Table 15** presents the decided experimental technical conditions of the laser feathering process and basic information on the ceramic powders used. The method of feeding the feathering particles into the area of the molten metal pool by means of a gravity feeder at a speed of 1 g/min for SiC and 8 g/min for WC/W_2C was selected. After the laser surface treatment, no other thermal or thermo-mechanical treatments were performed.

3. Structure and properties of the composite materials with aluminum alloys matrices taken into account in this chapter

3.1 Structure and properties of the composite materials with reinforcement in form of skeletons infiltrated with aluminum alloys matrices

In the first case of composite materials obtained by infiltration of porous skeletons, it is them that largely determine the properties. Pressure infiltration of porous skeletons with liquid metal alloys is increasingly used in the production of metal-matrix composite materials [116–140]. The benefits of using this technology include near-net-shape mapping and high surface quality, the possibility of local product strengthening and the use of a wide range of matrix and reinforcement materials, high efficiency also at mass production scale with relatively low production costs [87, 88, 141]. The interest in using aluminum oxide as the backbone of this group of materials has increased significantly in the current two decades [88, 141–145]. In the case of composite materials with reinforcement in the form of skeletons and an infiltrated matrix of aluminum alloys, a metal-ceramic connection also plays an important role due to the lack of wettability of aluminum oxide by aluminum alloys [82–84, 87, 88, 141–143, 146–154]. It has been improved by applying Ni-P coatings on Al_2O_3. A small proportion of pores with small dimensions ensures that after infiltration a density close to theoretical, which increases with the increase of the Al_2O_3 proportion. It was confirmed that composite materials with aluminum alloy matrix can be manufactured by pressure infiltration of porous ceramic skeletons made of a mixture of carbon fibers and Al_2O_3 particles, and the distribution of the reinforcing phase in the matrix is uniform. The infiltration is complete, and the fracture is mixed because the matrix is plastically deformed around the reinforcing phase, while at the metal-ceramic interface, the fracture is brittle (**Figure 22**). The Ni-P layer has a continuous structure with a thickness that does not clog the pores of the sintered ceramic skeletons.

The hardness of composite materials reinforced with porous skeletons made of Al_2O_3 particles increases with the increase in the proportion of the ceramic phase, obtaining 48.88 and 51.74 HRA, respectively, with a 30% share of reinforcement covered by Ni-P, and without it, which is more than 2.5 times increase in relation to the matrix with a hardness of 19.19 HRA (**Figure 23**). The tensile strength of the composite material reaches its maximum value with 25% reinforcement and amounts to 299.07 MPa and 320.14 MPa, respectively, in the case of the material with the reinforcement not covered or covered with Ni-P respectively. It proves the desirability of covering the internal surfaces of porous frameworks with a Ni-P layer in order to improve Al_2O_3 wettability by the liquid aluminum alloy and to improve the metal-ceramic bond.

The abrasion resistance of these composite materials increases with the increase in the proportion of the ceramic phase, and composite materials are characterized by much lower wear than their matrix, and the smallest of them is the material with a 25% ceramic phase (**Figure 23c**). Potentiodynamic studies confirm that the

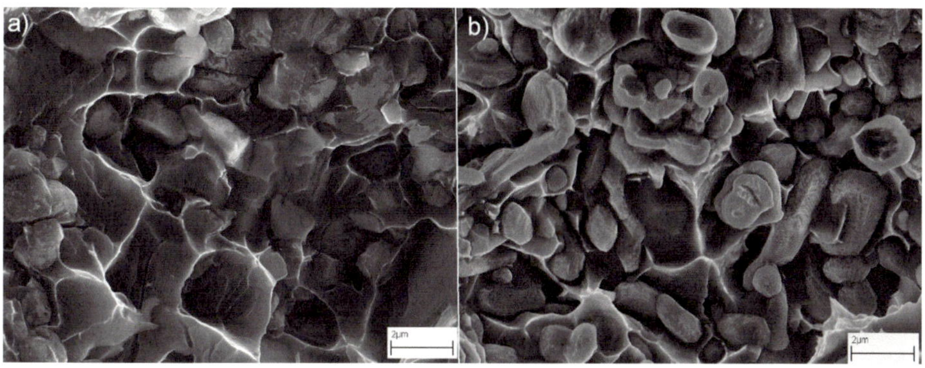

Figure 22.
The fracture's structure of the composite material a) without b) with an N-Pi layer on the Al_2O_3 skeleton.

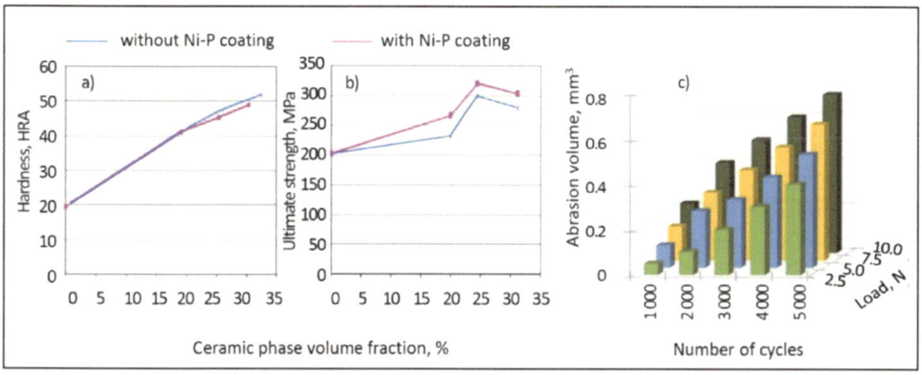

Figure 23.
Dependence of (a) hardness; (b) tensile strength from the proportion of Al_2O_3 reinforcement in the form of a porous sintered skeleton; (c) wear resistance with 30% Al_2O_3 reinforcement.

corrosion resistance of composite materials reinforced with ceramic skeletons in this way increases with the increase of the reinforcing phase. The presence of the Ni-P coating applied to the surface of the ceramic, however, promotes corrosion acceleration compared to the cases where this coating is absent, although the corrosion resistance also improves in these cases along with the increase in the proportion of the ceramic phase.

Composite materials reinforced with porous ceramic skeletons obtained from Al_2O_3 particles or fibers consisting in their pressure infiltration with liquid AlSi12 alloy constituting the matrix provides the required structure, as well as mechanical properties and wear resistance much more favorable than the aluminum alloy constituting the matrix.

In the second case, composite materials are manufacturing using liquid AlSi12 alloy for infiltration of the skeleton from sintered mineral nanotubes obtained from halloysite. Halloysite, like other secondary layered silicates, such as kaolinite or montmorillonite, and in general aluminosilicates or aluminum silicates, can be used to produce alternative reinforcement of composite materials with a matrix of light metals, especially aluminum, reinforced with ceramic particles [155–168]. These minerals can be used to produce mullite $3Al_2O_3{\cdot}2SiO_2$, among others by sintering, melting, or mechanical or chemical synthesis [169–173]. The mullite has good mechanical and physicochemical properties, meeting the requirements for porous skeletons as reinforcement for composite materials [174–177]. The porous skeletons produced in this way are an alternative to preforms commonly used for the production of composites based on aluminum alloys by infiltration methods, which are ceramic skeletons or foams formed from particles or short fibers, with high porosity 60–90% [119, 163, 177–189]. They replace the particles of oxides, carbides, nitrides, graphite, carbon, boron, glass, Al_2O_3 and SiC fibers, intermetallic phases used in their production; fly ash microspheres and microgranules; Al_2O_3, SiC, SiO_2 nanotubes; multi-walled carbon nanotubes, graphene and most often SiC and Al_2O_3 fibers. For economic reasons, it is beneficial to replace Al_2O_3 fibers with particles, which also increases the strength of infiltrated materials by approx. 35%, despite a slight reduction in abrasion resistance [84, 86, 87, 106, 148]. Halloysite in the form of NHTs can be an alternative for both economic and mechanical properties reasons. Among the sintered, porous ceramic skeleton, it is most preferable to sinter HNTs at 1500°C with the share of 70% HNTs, while the differentiation of the pressing pressure in the range of 50 to 100 MPa is of little importance (compare **Figure 19**). All composite materials analyzed here are characterized by an infiltrated structure with complete filling of the pores of the ceramic skeleton with the matrix alloy and good adhesion at the phase boundary of the ceramic reinforcement and the metallic matrix (**Figure 24**).

Composite materials manufactured according to the technology described as the second case are characterized by a structure of evenly interpenetrating phases with a contrasting structure. The matrix material is an α-solution with numerous precipitates of the coarse-crystalline β phase and fine precipitates of the iron-rich phase, while the reinforcement is an evenly distributed, continuous mineral phase in the form of irregularly shaped mullite. The strengthening of the above-mentioned composites in the form of fine-grained mullite and cristobalite occurs as a result of sintering mineral nanotubes and the accompanying series of structural changes. The gas-pressure infiltration of these ceramic skeletons leads to the diffusion dissolution of the cristobalite formed during sintering in the liquid aluminum alloy. As a result, the excess of silicon in the liquid solution crystallizes in the form of coarse precipitates, analogous to hypereutectoid Al-Si alloys. The relatively short infiltration time does not affect the dissolution of mullite in the liquid solution. The mechanical and functional properties of the discussed composite materials with

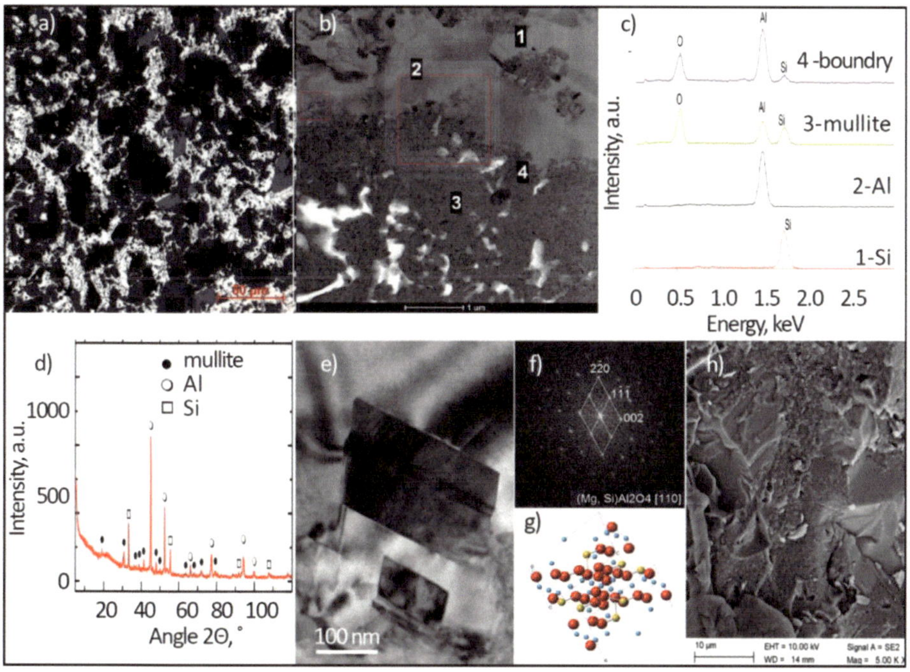

Figure 24.
Structure (a) of composite material with Al12Si matrix with reinforcement of sintered HNTs in 70% share; (a) light microscope; (b) structure of intermediate zone between the reinforcement and the matrix with marked points of analysis by EDS method (SEM); (c) results of the local EDS analysis; (d) X-ray analysis of the surface of the composite material; (e) thin foil structure in the intermediate zone between ceramic and metal (HRTEM); (f) solution of the FFT of the nano diffraction from the area (e); (g) $MgAl_2O_4$ unit cell model; (h) fracture morphology of the composite material with 70% HNTs after the bending test (SEM).

HNTs reinforcement are the best in the case of sintering ceramic skeletons at 1500°C and at 70% HNTs (**Figure 25**).

The mentioned technology enables the manufacturing of composite materials with enhanced mechanical properties and wears resistance, significantly exceeding the functional properties of the alloy used as a matrix, providing the prospect of their implementation on structural elements, e.g. in the aviation and automotive industries.

The third and fourth cases of composite materials concern the use of microporous titanium skeletons manufactured by selective laser sintering with an adjustable pore size and their controlled spatial orientation as reinforcement, manufactured by pressure infiltration with liquid alloys AlSi12 or AlSi7Mg0.3, respectively.

Additive technologies have found a very wide application in numerous industries, including military, automotive, aviation, machine, household goods, but also in medicine, both for the production of solid and porous elements [128, 190–218]. Porous materials produced with this technology can be used as skeletons for the production of infiltrated composite materials with a matrix of aluminum alloys. The structure of these composite materials with aluminum alloys matrix and titanium skeleton reinforcement manufactured by additive technology does not contain any local voids and pores. It is made of a matrix corresponding to the chemical composition of the alloys used, respectively, AlSi12 or AlSi7Mg0.3 (**Figure 26**) inside the pores of a titanium skeleton with a size of 250–350 μm, which is a reinforcement with a shoulder thickness of 200–300 μm depending on the sintering conditions. The matrix and reinforcement composition was confirmed by X-ray diffraction (**Figure 26i-k**). The results of the X-ray diffraction tests of the structure AlSi12/Ti

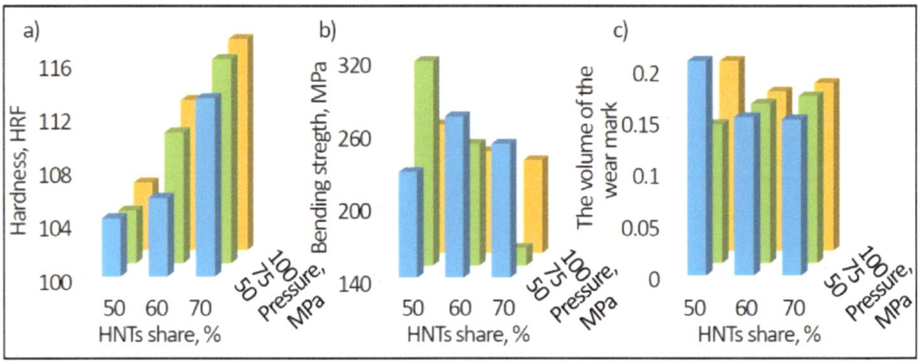

Figure 25.
The relationship of (a) hardness, (b) bending strength, (c) the volume of the wear mark on the share of HNTs and compaction pressure.

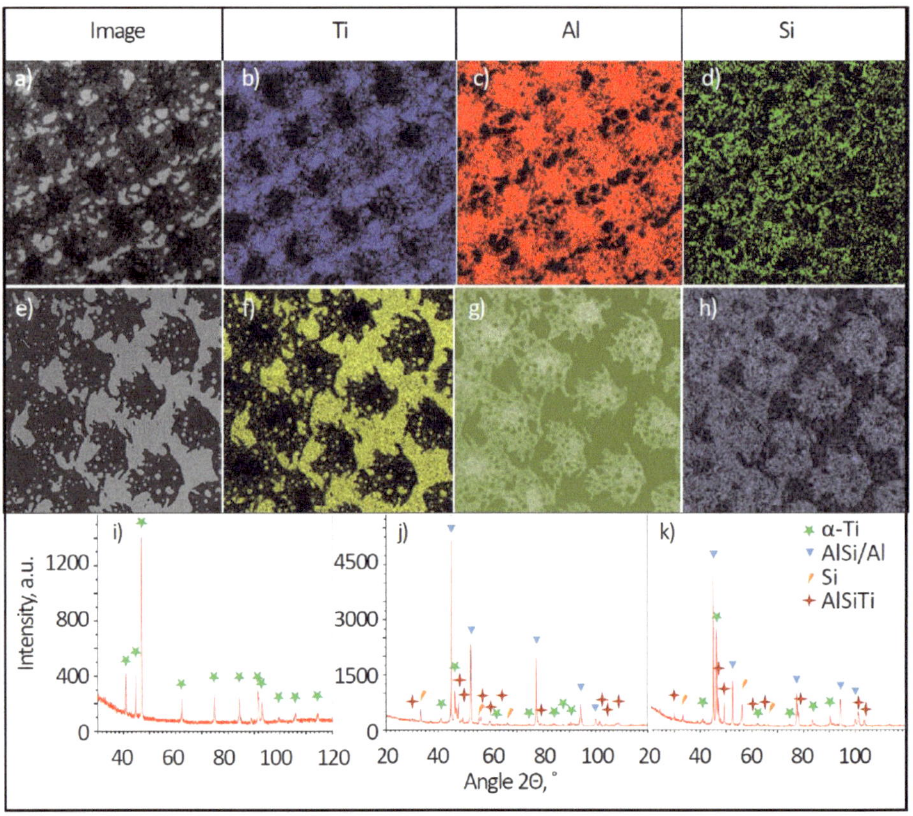

Figure 26.
The structure of the surface layer of the composite material with the reinforcement in the form of a titanium skeleton manufactured by selective laser sintering SLS and a matrix made of aluminum alloy: (a-d) AlSi12, (e-h) AlSi7Mg0.3; (a, e) SEM; (b-d, f-h) surface distribution of elements: (b, f) Ti, (c, g) Al, (d, h) Si; (i-k) X-ray diffraction from the (e) image; (i) Ti skeleton; (j, k) composite materials with the matrix (j) AlSi12; (k) AlSi7Mg0.3.

(**Figure 26j**) and AlSi7Mg0.3/Ti (**Figure 26k**) composite materials show that during infiltration, phases with the participation of individual elements of the composite material, such as Al, Si, may be formed as well as Ti, AlSi and the AlSiTi phase. The AlSiTi phase is formed in the area between the matrix material and the reinforcement (**Figure 27**), and its chemical composition differs from that of the reinforcement and the matrix, as Al, Ti, and Si were identified in the qualitative and quantitative EDS method. In the matrix material, Mg, Cu, Ni can also be identified, the concentration of which in the matrix is less than 1%.

The mechanical properties of pressure-infiltrated composite materials with a matrix of selected aluminum AlSi12 or AlSi7Mg0.3 casting alloys (**Figure 28**) depend on the spatial orientation with respect to the main axes of the titanium micro-skeleton produced by selective laser sintering, the size of the pores in the micro-skeleton and the sintering conditions, especially laser power and the size of the laser spot. The bending strength is 280–430 MPa, and the compressive strength is 480–600 MPa. The most advantageous mechanical properties: bending strength of 430 MPa and compressive strength of 600 MPa, these composite materials obtain after laser sintering of the skeleton with a laser power of 60 W, with a laser spot

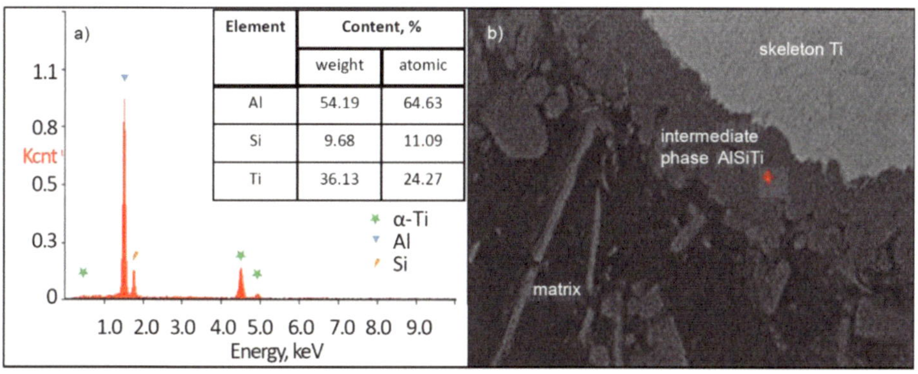

Figure 27.
The structure of the AlSi7Mg0.3/Ti composite material (SEM) (b) and the scattered X-ray energy diagram and the results of the quantitative analysis of the chemical composition of the AlSiTi phase of the place marked with a red cross (a).

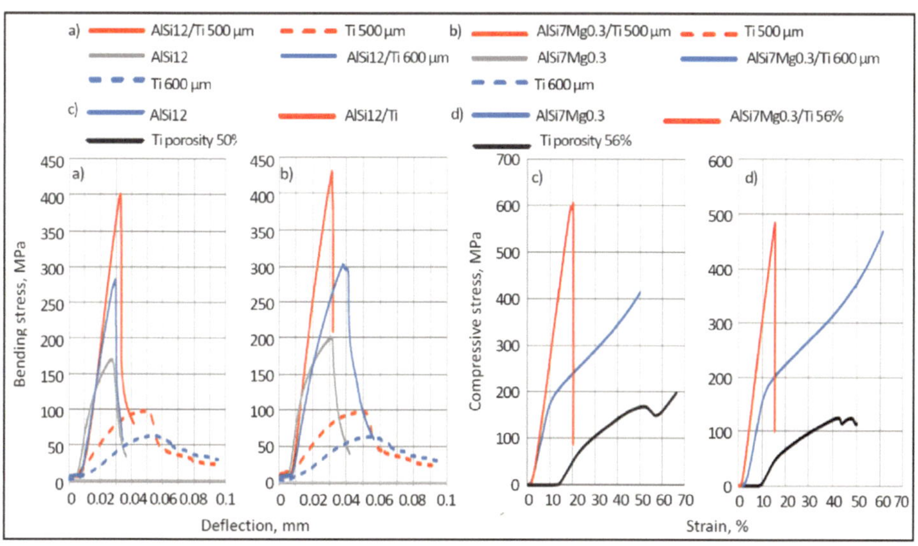

Figure 28.
Diagrams of the dependence of bending stress on deflection (a, b) and compressive stress on deformation (c, d) for the AlSi12 alloy, titanium skeleton and AlSi12/Ti composite material (a, c) and AlSi7Mg0.3 alloy, titanium skeleton and composite material AlSi7Mg0.3/Ti (b, d) with pore sizes of 500 or 600 μm (a, b) or with a porosity of 50% (c) and 56% (d), respectively.

with a diameter of 50 μm, the distance between the laser spots and between laser remelting paths equal or smaller than the diameter of the laser spot when the spatial orientation is 45° to the x-axis and if the pore size is 200–250 μm with a porosity of approx. 50–60%.

Therefore, reinforcing the aluminum alloy with porous titanium skeletons has a positive effect on the mechanical properties of the analyzed composite materials. The AlSi12 alloy is characterized by a bending strength of about 170 MPa, while in the case of the AlSi12/Ti composite material, the maximum bending strength was obtained using a titanium skeleton with a porosity of 56%, obtaining a bending strength of about 400 MPa, and with a titanium skeleton with a porosity of 66%. strength of about 280 MPa was obtained. The strength of the AlSi12/Ti composite material compared to the strength of solid titanium (1682 MPa) is more than four times lower, however, compared to the strength of the AlSi12 alloy itself, it increases twice. The use of a titanium skeleton with lower porosity, i.e. a skeleton of greater strength, in the composite material with the AlSi12 alloy matrix, increases the bending strength by about 230 MPa compared to the matrix material, while when the skeleton with lower strength and higher porosity is used, the increase in bending strength is not so significant at about 110 MPa compared to the matrix material. In the case of a composite material with the AlSi7Mg0.3 matrix with the use of titanium skeletons, the increase in bending strength compared to the matrix material is approximately 100 and 130 MPa, respectively. When the same titanium skeletons and different matrix materials are used, the matrix bending strength has a dominant effect on the bending strength of composite materials, while if the same matrix materials are used, the strength of the composite material is mainly determined by the strength of the titanium matrix.

The comparison of the fracture structure of the aluminum alloys constituting the matrix of the composite materials discussed with the fracture structure of the composite material after bending tests shows significant differences. The breakthroughs of aluminum alloys are flat, with visible planes along which the material cracked, while in the case of composite materials, places characteristic of titanium are distinguished (**Figure 29**). The studies of the breakthrough of the AlSi12/Ti and AlSi7Mg0.3/Ti composite materials indicate that the combination of titanium and aluminum alloys is durable and does not break when the bending force is applied. The structure of the breakthroughs after the compression test is significantly different from the breakthroughs after bending. After the compression test, such a clear separation of the matrix material from the reinforcement material cannot be noticed, which is so clearly visible in the structure of breakthroughs after bending.

Micro-skeletal composite materials, manufactured by pressure infiltration with selected aluminum casting alloys AlSi12 or AlSi7Mg0.3 to give the final shape and geometric form, dedicated, among others for the automotive, machine-building, and aerospace industries, as well as for medical and dental applications.

3.2 Structure and properties of the aluminum alloy matrix composite materials manufactured by mechanical synthesis and subsequent plastic consolidation

In cases 5 and 6 for mechanical synthesis of the composite materials in a planetary mill (the so-called high-energy process) of AlMg1SiCu alloy powder and respectively nanotubes obtained from halloysite or MWCNTs carbon nanotubes, was used to produce the composite materials. In the initial stage of mechanical synthesis, the originally spherical particles are deformed, flattened, and have a lamellar shape. After 5 hours of mechanical milling, it was found out that most of the particles were equiaxed with no flake shaped particles present. Changes in the

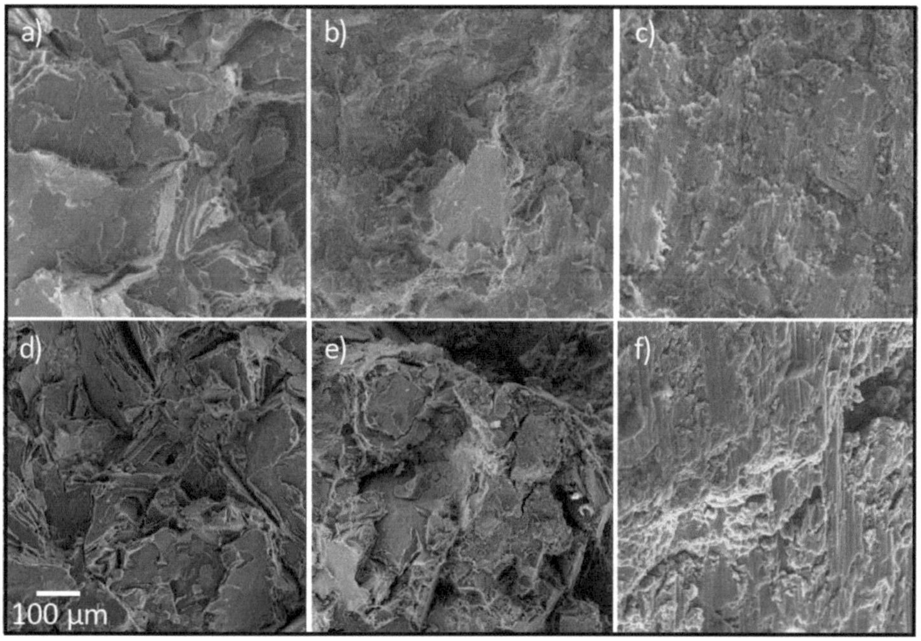

Figure 29.
Breakthrough structure (a) AlSi12 alloy; (d) AlSi7Mg0.3 alloy; and composite materials based on (b, c) AlSi12 alloy; (e, f) an AlSi7Mg0.3 alloy; after tests; (a, b, d, e) of static bending; (c, f) static compression.

shape and size of the milled powders are caused by the welding of individual particles as a result of collisions with the milling media or the walls of the mill. The flattened particle conglomerates formed as a result of welding become significantly hardened, hard, and therefore prone to cracking. As the milling progresses, these particles become fragmented and reassembled, which ultimately affects the random orientation of the boundaries of the particles to be welded. The fact that the process has reached a steady-state is provided by the relatively equiaxial shape of the milled powder particles. In contrast to the milled powder of the AlMg1SiCu alloy, in the case of the powder of the composite material, it was found out that the deformed particles joined tightly, creating a morphologically homogeneous structure, free of pores and discontinuities. Despite the twice longer milling time, the powder of the AlMg1SiCu alloy consists of fine particles, which to a large extent form porous conglomerates with irregular, non-aligned shapes and sizes varying in a wide range from 10 to 200 μm. Extending the milling time results in a homogeneous distribution of the reinforcing particles, and it is also possible to produce powders with equiaxed particles. The influence of the presence of reinforcement subject to fracture under the conditions of mechanical alloying of the ductile aluminum matrix is given in the diagram of the successive stages during mechanical alloying (**Figure 30**). The examples of the corresponding structure for the case respectively 5 and 6 are given in **Figure 31**. Examples of correctly and incorrectly made rods in the stage of plastic consolidation are shown in **Figure 32**.

In the case of 5th, the composite materials with the AlMg1SiCu matrix are reinforced with nanotubes obtained from halloysite. The use of HNTs particles reduces the bulk density of the mixture with the matrix powder of the AlMg1SiCu alloy even by more than 50%, and this effect becomes present with the increase in the proportion of the reinforcing phase. Reinforcement material particles are characterized by a larger specific surface, irregular, often oblong shape, and form porous conglomerates, thus characterized by a much lower bulk density compared to globular AlMg1SiCu particles. In the case of mechanical milling of a plastic

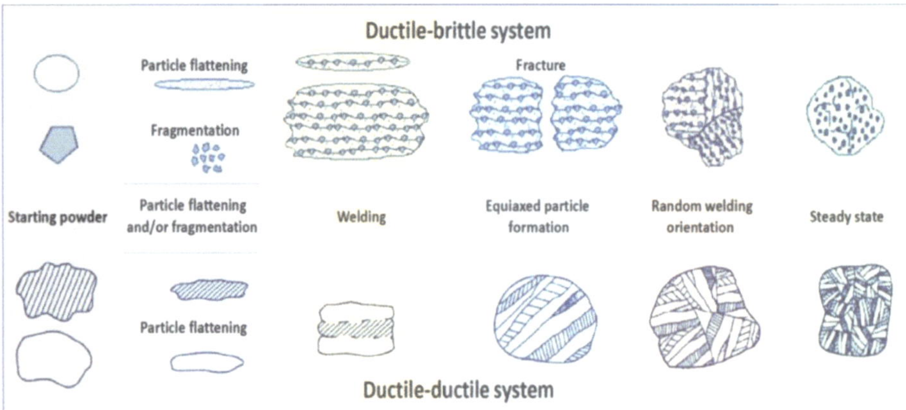

Figure 30.
Diagram of the successive stages of the ductile-brittle/ductile-ductile systems during mechanical alloying.

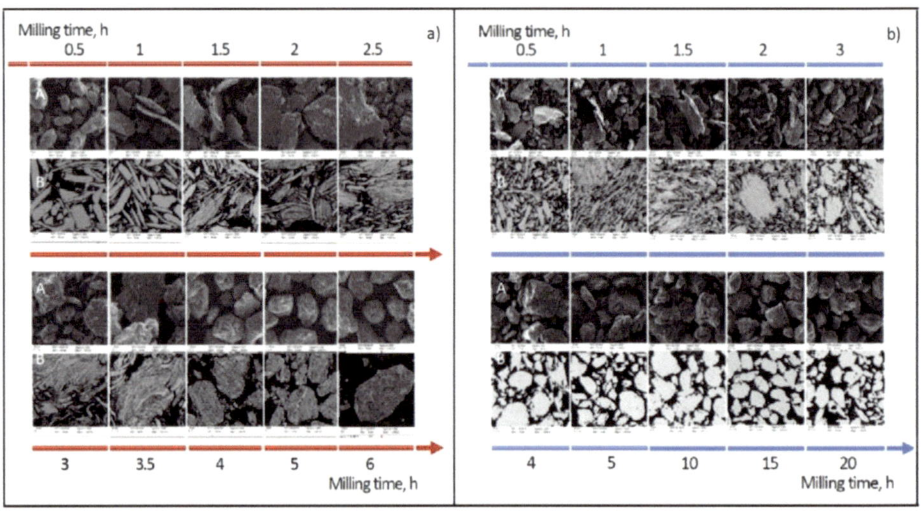

Figure 31.
Morphology (A), structure (B) after high-energy mechanical milling of a powder mixture of AlMg1SiCu alloy with (a) 15% of halloysite nanotubes (red) or (b) 2% MWCNTs carbon nanotubes (blue) respectively in the range of (a) 0.5–6 h or (b) 0.5–20 h.

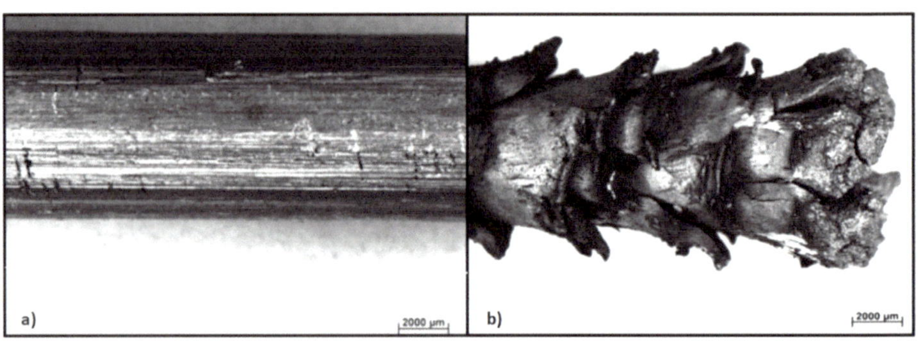

Figure 32.
Exemplary view of bars containing AlMg1SiCu alloy with (a) 5% and (b) 2% of multi-wall carbon nanotubes MWCNTs extruded at 460°C (a) and 500°C (b).

powder of an aluminum alloy with a brittle reinforcing phase in the form of halloysite nanotubes with a mass fraction of 15%, the increase of the bulk density takes place in 2 times shorter time than in the case of matrix powder. As a result of high energy milling, an equiaxial shape of the particles is obtained and an even distribution of fine reinforcing particles confirming that the process has reached a steady state. Further milling does not significantly affect the obtained structure and properties of the powders. In the initial stage, the particle size distribution is characterized by two peaks (**Figure 33**), because the tested powders are a mixture of particles with a significant difference in size (the aluminum alloy powder has a median of 62.30 μm, and halloysite nanotubes – 2.58 μm). The particle size of the powders undergoes continuous, non-linear changes during milling. As the milling time increases, the deformed particles are joined together and the peak on the particle size distribution curve becomes wider at the base, and then with the increase of the milling time and the associated increase in the proportion of large particles resulting from repeated welding, an asymmetric deviation is achieved. Due to the breakage of large particles, the asymmetry that has arisen gradually disappears. Eventually, the process reaches a steady state, corresponding to a balance between the powder particle aggregation and fragmentation mechanisms, which is represented by symmetrical or a narrow particle size distribution curve (**Figure 33**).

Mechanical synthesis, in addition to the effect on the morphology and structure of the manufactured composite powders, also affects their phase composition, and increasing the milling time also affects the generation of a large number of defects and results in the fragmentation of the crystal structure [219–221]. Already after several minutes of milling, low-angle reflections from mineral halloysite crystals disappear, leaving only reflections identified as α-Al on the diffraction patterns. Structural changes occur during annealing of halloysite [169, 222–226]. The interplanar distance of all halloysite powders is 0.72 nm, which proves permanent dehydration [113, 114]. Annealing of halloysite nanotubes at the temperature of 500–700°C causes disorder of their crystal structure, which with increasing temperature causes amorphization of halloysite as a result of disordering the packet structure [227–230]. Mechanical high energy milling improves the distribution of HNT_{100} reinforcement particles in the AlMg1SiCu matrix, more advantageously than when using HNT_{500} and HNT_{700}. With a short milling time, this ensures a homogeneous structure of the nanocomposite material with an even distribution of reinforcing particles without their agglomeration after consolidation. The consolidation of powders consists in their successive cold pressing and hot extrusion, resulting in a lack of discontinuities of the structure in the form of pores or cracks and obtaining a material with a very fine structure (**Figure 34a**), devoid of agglomerates of reinforcing particles. In the produced material, there are grains with a size of several nanometers (**Figure 34b-d**) in the structure of the solid

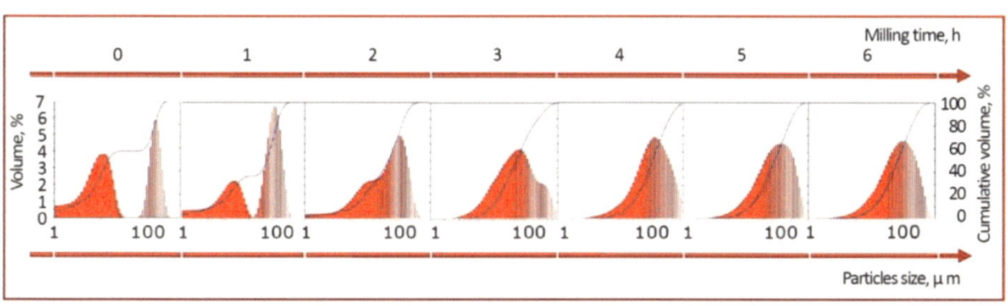

Figure 33.
Relationship between the size of the mixture of AlMg1SiCu particles and HNT_{100} halloysite nanotubes in a share of 10% from the time of high-energy milling.

solution of the AlMg1SiCu alloy in a metastable state (**Figure 34e-h**). There are also many regular, parallel, rhomboidal regions not exceeding a few nanometers (**Figure 34e**), which most likely constitute the Guinier–Preston zones with the concentration of Cu atoms [1, 231–233].

Based on the results of the X-ray phase analysis, the presence of only the matrix phase in the analyzed composite materials, i.e. solid aluminum solution (**Figure 35**).

In addition to the nanometric grains of the Alα solution (**Figure 36a**) in the matrix, there are precipitations of the Al_2O_5Si phase (**Figure 36b**) probably formed during mechanical milling and the intermetallic primary phases $AlFe_3$ and Al_4 (Fe, Cr, Mn) $Si_{0.74}$ (**Figure 36c**) [234].

The high degree of plastic deformation caused by mechanical milling, nanometric fragmentation of the structure, and dispersion hardening with particles

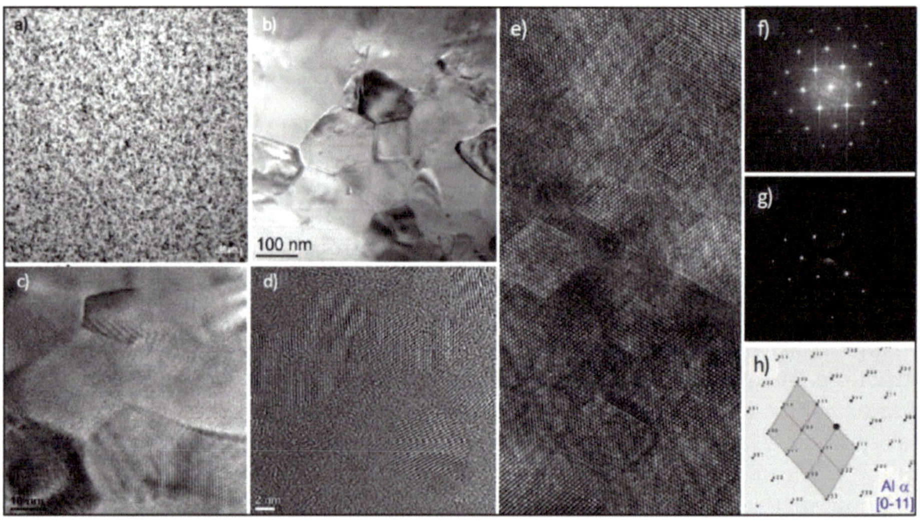

Figure 34.
The structure of the nanocomposite with AlMg1SiCu matrix with the share of 15% HNTs: (a) uniform structure (light microscope); (b-d) very fine grains matrix structure (HRTEM); (e) rhomboidal nanostructured regions in the matrix (HRTEM); (f) Fourier transform; (g) nano diffraction; (h) solution of the diffraction pattern from Figure e.

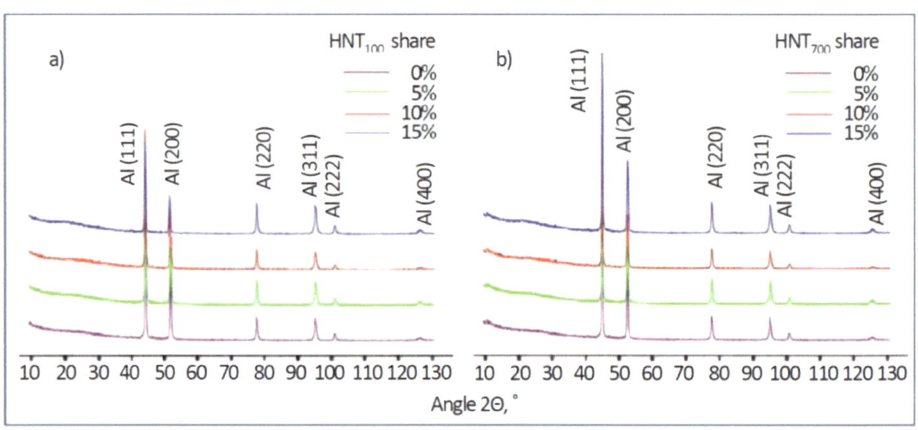

Figure 35.
Results of X-ray diffraction analysis using a Cu lamp of AlMg1SiCu alloy powders and high energetic milled nanocomposite materials with a matrix of this alloy with the participation of 5, 10, or 15% strengthening with halloysite nanotubes, respectively (a) HNT_{100}; (b) HN_{700}.

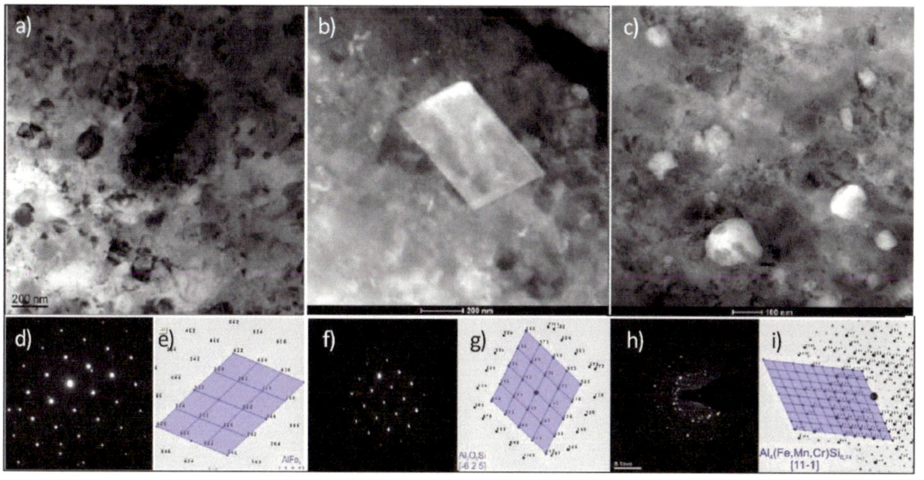

Figure 36.
The structure of nanocomposite with AlMg1SiCu matrix with 15% HNTs (TEM): (a) nanometric grains of Alα solution; (b) Al_2O_5Si phase precipitations probably during mechanical milling; (c) intermetallic primary phases Al_4 (Fe, Cr, Mn) $Si_{0.74}$; (d, f, h) diffraction patterns; (e, g, i) solutions of diffraction patterns.

of halloysite and oxides results in an almost 3-fold increase in the microhardness of powders than before milling and despite its reduction by about 30% after plastic consolidation, the produced nanocomposite materials show microhardness 110–150 $HV_{0.1}$ (**Figure 37a**), higher than the Mg1SiCu alloy. The introduction of HNTs particles into the Mg1SiCu alloy matrix using high-energy milling causes more than a twofold increase in the yield point under compression, because the yield point for compression with HNT_{100} nanotubes is 450–495 MPa at the test temperature of 25° C (**Figure 37b**), and at the test temperature of 225°C is 265–280 MPa depending on HNTs share. When using nanotubes HNT_{500} and HNT_{700}, respectively, it is slightly smaller (**Figure 37c**). **Figure 37f-j** show the course of compression diagrams at various temperatures of 25–225°C. The tested nanocomposite materials show a characteristic "dimple fracture" [235–238], sometimes also referred to as a ductile foveal fracture [239] (**Figure 38**). Nanocomposite materials reinforced with HNTs nanotubes are more resistant to frictional wear, as measured by changes in the surface layer related to permanent deformation, mainly weight loss [1, 240, 241], therefore they are characterized by a lower abrasion trace volume than the AlMg1SiCu alloy (**Figure 37d,e**). Nanocomposite materials reinforced with HNT_{100} halloysite nanotubes with a mass fraction of 15% are characterized by over 180% higher yield strength under compression, over 200% higher microhardness, and nearly 250% higher friction wear resistance compared to the AlMg1SiCu alloy, which is the matrix of the composite material.

In the case of 6th composite materials with AlMg1SiCu matrix reinforced with MWCNTs, the main problem during manufacturing is the homogeneous distribution of carbon nanotubes in the matrix material. Of the several methods of dispersing nanotubes in a metal matrix, such as mechanical milling, mixing in a turbulent mixer or mixing with an ultrasonic homogenizer, only high-energy mechanical milling leads to an homogeneous distribution of the reinforcement in the matrix material, without agglomerating the MWCNTs on the surface [242–250]. Carbon nanotubes can be placed between particles of a gas-atomized or spheroidal cold-alloyed aluminum alloy. After mechanical milling on the surface of the ground particles, no carbon nanotubes are present on the surface of any of the powders ground for 5 or more hours [95, 242, 244, 246, 251–255]. With the increased share of MWCNTs, the particle size decreases (**Figure 39**).

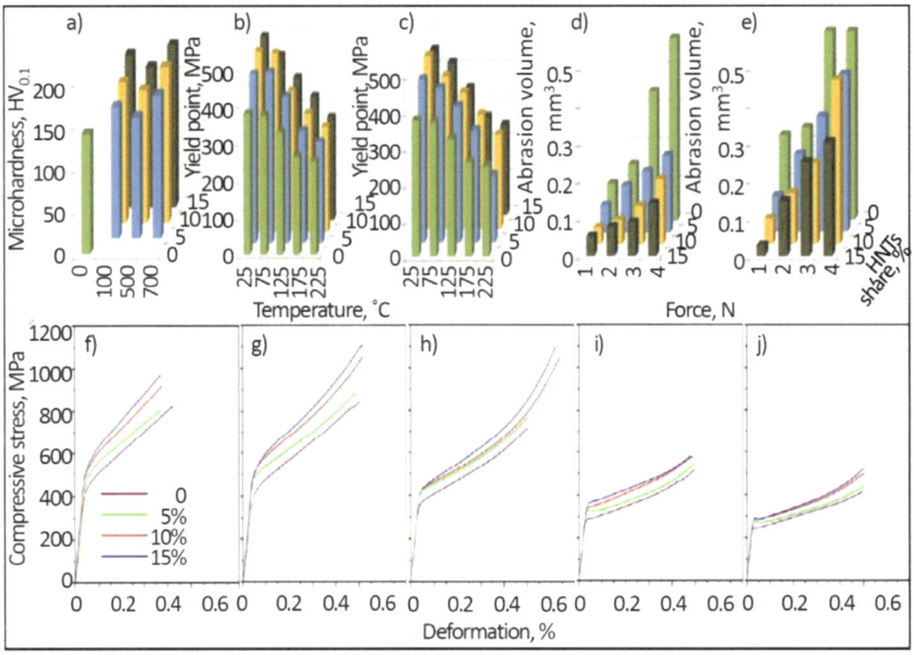

Figure 37.
Characteristics of the properties of a nanocomposite material with AlMg1SiCu matrix reinforced by HNTs in the proportion of 0–15%: (a) the influence of the HNTs annealing temperature on the microhardness; (b, c) influence of test temperature on compression yield point of nanotubes (b) HNT_{100}; (c) HNT_{700}; (d, e) effect of a load in the range 1–4 N on abrasion volume after post-measurement; (d) 1000 cycles; (e) 2000 cycles; (f-j) compressive stress curves as a function of deformation at temperature; (f) 25°C; (g) 75°C; (h) 125°C; (i) 175°C; (j) 225°C.

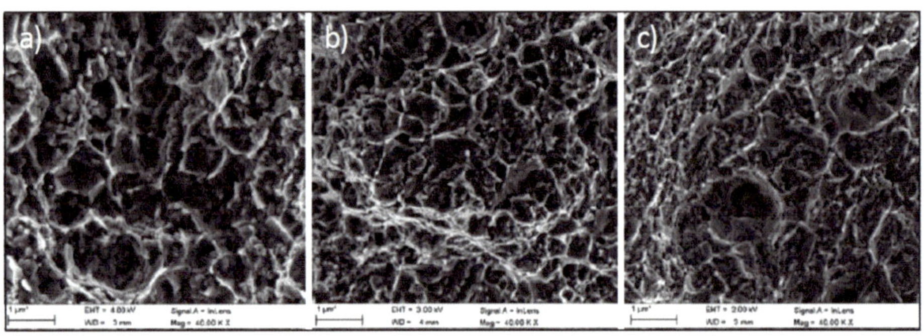

Figure 38.
The breakthrough of samples as dimple fracture style after the static compression test of nanocomposite materials with the participation of halloysite nanotubes (a) 5%; (b) 10%; (c) 15%.

As a result of mechanical milling, the order of the structure of carbon nanotubes and the interaction between the reinforcement and the matrix are degraded as a result of breaking the sp2 bonds. The results of the Raman analysis confirm the defect of the structure and the amorphization of carbon nanotubes (**Figure 40**), indicating a reduction in the intensity of the G and G' bands as well as an increase in the intensity of the D band, which is confirmed by the ID/IG and IG'/IG intensity factors ratios (**Figure 40**).

Carbon nanotubes during high-energy mechanical milling were dispersed and placed between deformed aluminum alloy particles, which increases the density of

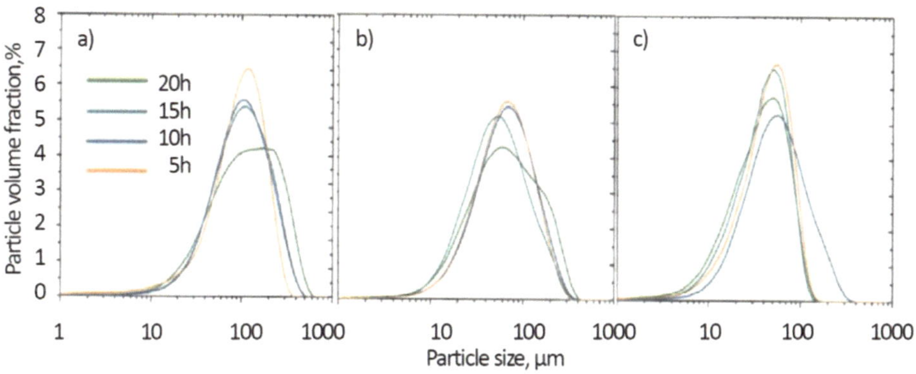

Figure 39.
The particle size distribution of the powder after high-energy mechanical milling for 5, 10, 15, and 20 h, respectively, depending on the share of MWCNTs nanotubes: (a) 0%; (b) 2%; (c) 5%.

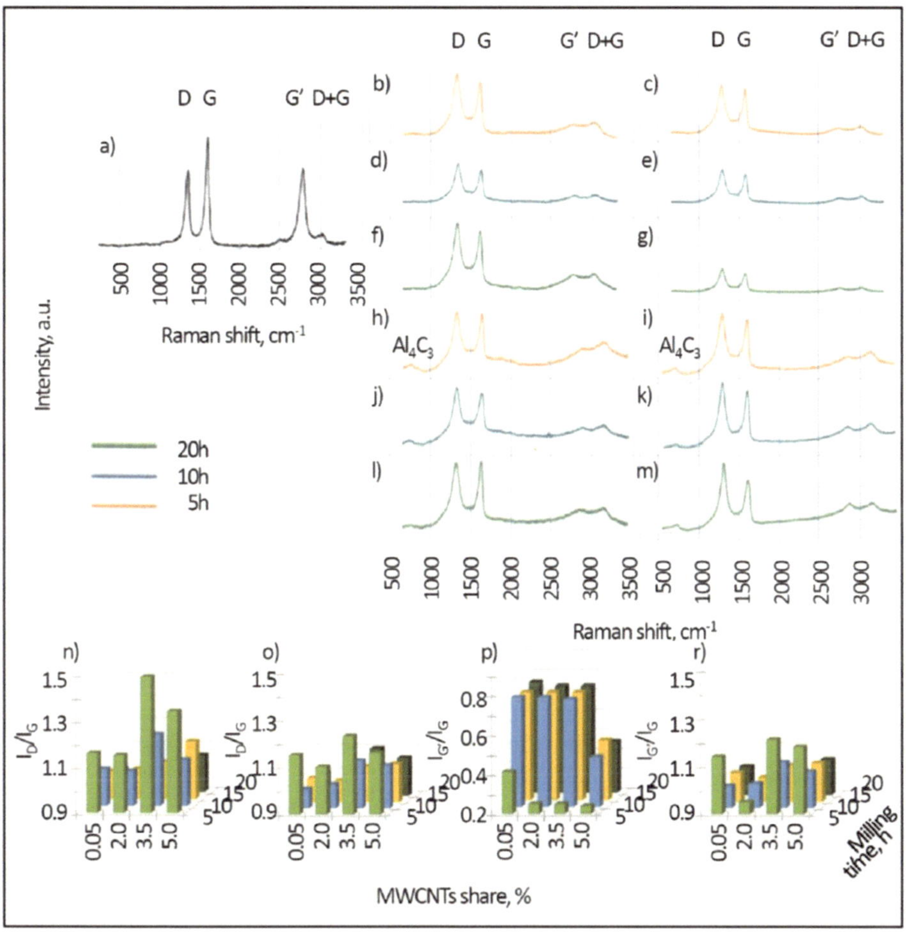

Figure 40.
Results of (a-m) Raman analysis and (n-r) distribution of intensity factors ratios (n, o) I_D/I_G and (p, r) $I_{G''}/I_G$: a) MWCNTs nanotubes; (b-g, n, p) powder obtained from high-energy mechanical milling and (h-m, o, r) after plastic consolidation, with milling time respectively for (b, c, h, i) 5 h; (d, e, j, k) 10 h; (f, g, l, m) 20 h, depending on the share of MWCNTs nanotubes: (b, d, f, h, j, l) 2%; (c, e, g, i, k, m) 5% and (n-r) depending on MWCNTs share and milling time 5, 10, 15 and 20 h.

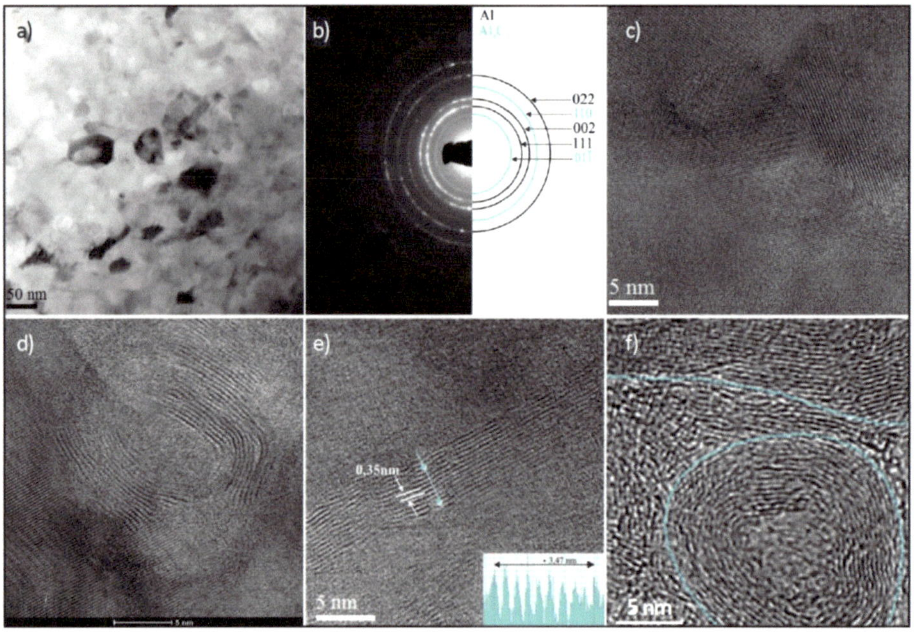

Figure 41.
The results of the thin foils investigations in the high-resolution transmission electron microscope HRTEM of the nanocomposite with the share of 5% by volume of MWCNTs after plastic consolidation; (a) structure; (b) the SAED diffraction pattern and the solution of the diffraction pattern from fig, a; (c) structure of nanograins; (d, e) nano-areas of the nanocomposites inherited from carbon nanotubes; (f) the structure of carbon nanotubes remaining after the HCl digestion of the material after high-energy mechanical milling for 10 h.

powders of nanocomposite materials (**Figure 41a, b**). The structure of the nanocomposite material with the matrix of the AlMg1SiCu aluminum alloy reinforced with MWCNTs is homogeneous (**Figure 41c**) with symmetrical, approximately spherical grains, with a size of about 50–70 nm. Carbon nanotubes show homogeneous dispersion and are strongly deformed (**Figure 41d,e**), which was revealed after complete removal of the matrix by etching with HCl (**Figure 41f**). However, literature reports [244, 251, 255–263] concerning the degree of degradation of carbon nanotubes are not unequivocal.

However, extending the milling time from 5 to 20 hours causes increased degradation of the structure of carbon nanotubes. It is accompanied by the amorphization of nanotubes, which is associated with a reduction in the intensity of the G 'bands (**Figure 40n-r**). It has been found, however, that the structure of carbon nanotubes may even be completely destroyed. Regardless of the share of MWCNTs above 0.5% and the mechanical milling the time the manufactured materials show a brittle fracture (**Figure 42**). The structure of the fractures also shows the fine grain structure of the material.

Defective nanotubes are much more reactive, so during the next consolidation involving plastic deformation at elevated temperature, as a result of the interaction of carbon nanotubes with the AlMg1SiCu alloy matrix, in places where there are nanotube defects and amorphous areas on their surface, probably intercalation of aluminum atoms occurs between the carbon layers in MWCNTs. It is the direct cause of the following precipitation of Al_4C_3 carbides (**Figure 43**) with a diameter of approx. 5–10 nm and a length of 20–30 nm. The precipitation of these carbides was also confirmed by X-ray analysis (**Figure 44**). Information on the Al_4C_3 carbides precipitation is also available in the literature in many cases of composite materials involving carbon nanotubes [264–274]. However, neither X-ray structural

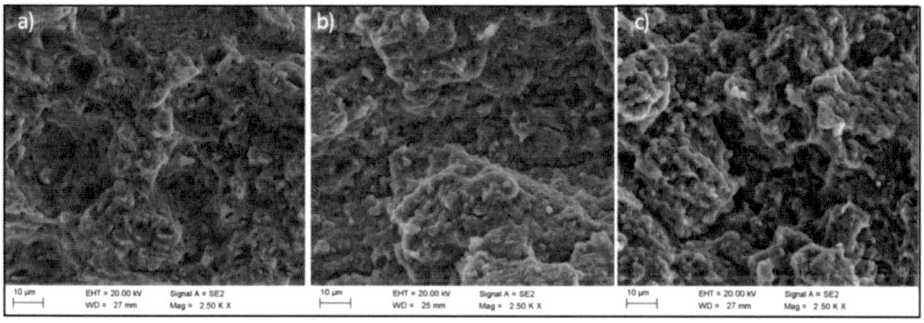

Figure 42.
Breakthrough structure of a nanocomposite material manufactured with (a, b) 2% (c) 5% by volume MWCNTs after high-energy mechanical milling for (a) 10 h; (b, c) 20 h (SEM).

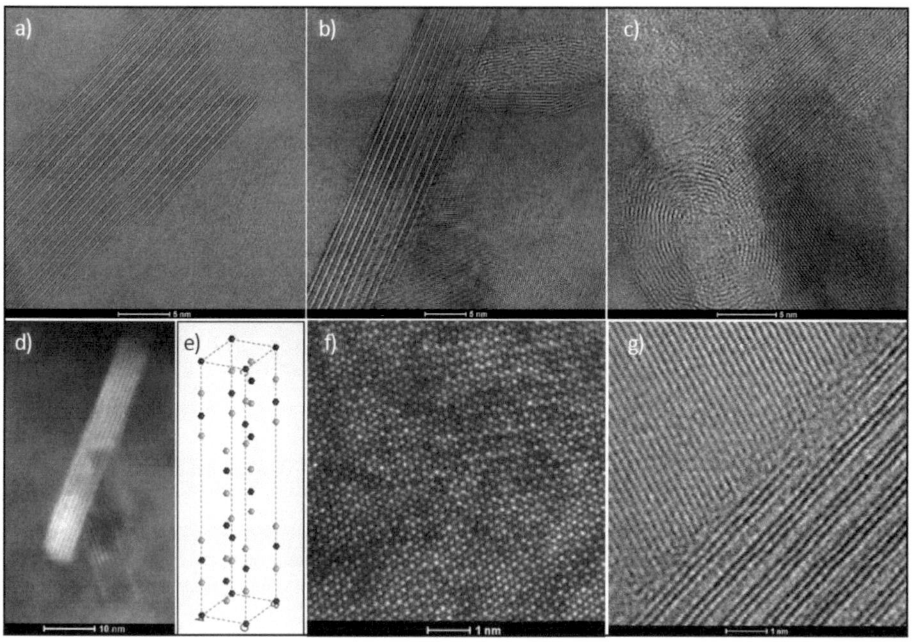

Figure 43.
Structure of thin foils of a nanocomposite material reinforced with 5% by volume MWCNTs (STEM-BF); (a) longitudinal section of the Al_4C_3 carbide in an AlMg1SiCu matrix; (b) MWCNT nanotube in longitudinal section and Al_4C_3 carbide in longitudinal section in an AlMg1SiCu matrix; (c) MWCNT nanotube in cross-section and Al_4C_3 carbide in longitudinal section in an AlMg1SiCu matrix; (d) Al_4C_3 carbide in an AlMg1SiCu matrix; (e) scheme of the distribution of Al (blue) and carbon (black) atoms in the Al_4C_3 carbide crystallographic cell; (f) Al_4C_3 carbide in cross-section – [001] direction; (g) Al_4C_3 carbide in longitudinal section and an AlMg1SiCu matrix.

analysis nor literature data indicate that Al_4C_3 is secreted directly as a result of mechanical synthesis by milling for up to 50 h [275].

The share of MWCNTs and the manufacturing conditions of nanocomposite materials have a relatively small effect on their density 2.56–2.57 g/cm^3 (**Figure 45a**), in contrast to the microhardness (**Figure 45b**). The manufactured nanocomposite materials reinforced with carbon nanotubes are characterized by increased microhardness by up to 300% (**Figure 45b**), greater ultimate tensile strength by over 200% (**Figure 45c, e-g**) and five times higher frictional wear resistance to the unground AlMg1SiCu alloy, because with the increase of share of MWCNTs, the volume of the wear mark decreases (**Figure 45d**). The coefficient of

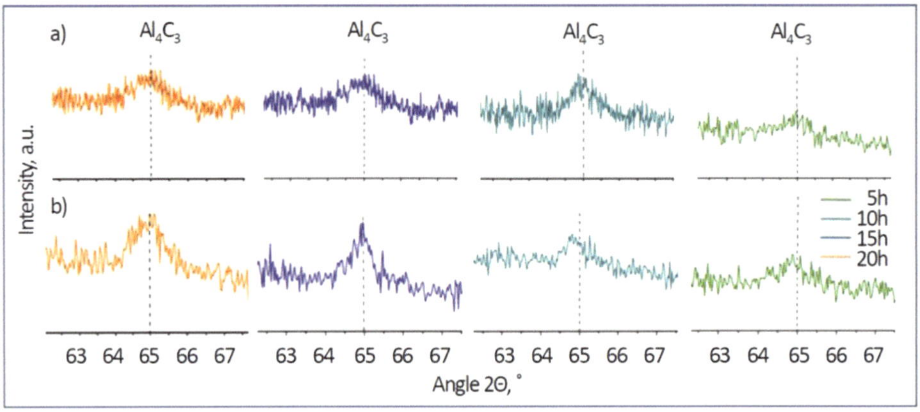

Figure 44.
Fragments of X-ray diffraction patterns confirming the precipitation of Al_4C_3 carbides in nanocomposite materials with AlMg1SiCu matrix reinforced by 2% (a) and 5% (b) MWCNTs carbon nanotubes as a result of plastic consolidation after prior high-energy mechanical milling for 5, 10, 15 or 20 h, respectively.

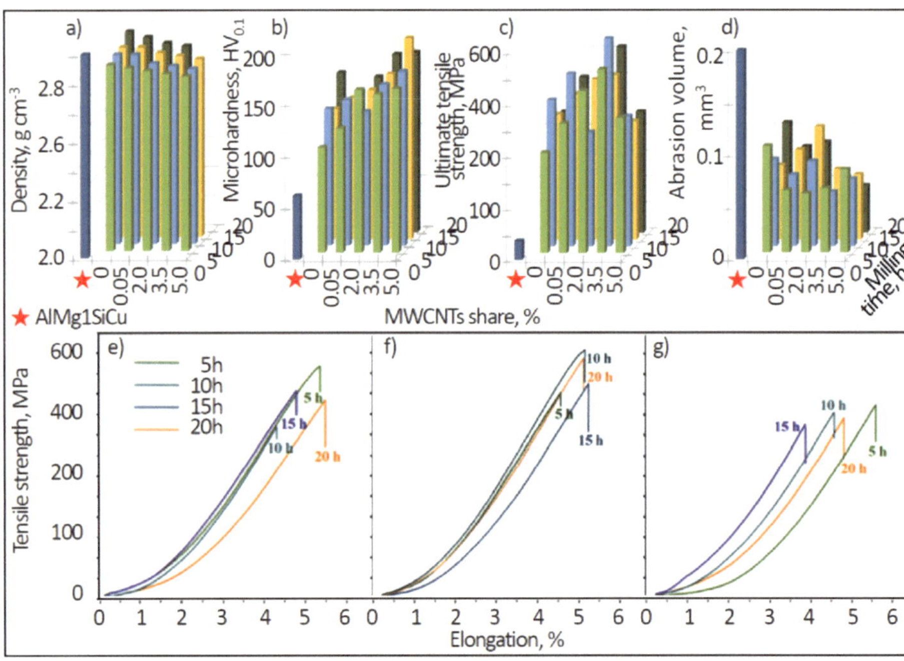

Figure 45.
Comparison of (a-d) properties of nanocomposite material obtained by consecutive high-energy mechanical milling and plastic consolidation depending on MWCNTs share 0.05; 2.0; 3.5 and 5.0% and milling time 5,10,15 and 20 h respectively comparing to AlMg1SiCu; (a) density; (b) microhardness HV 0.1; (c) ultimate tensile strength; (d) abrasion track volume; (e-g) the course changes the strength as a function of elongation during a static tensile test of a nanocomposite material manufactured after high-energy mechanical milling for 5, 10, 15, and 20 h with MWCNTs share (e) 2%; (f) 3.5%; (g) 5%.

friction of nanocomposite materials is not only lower than the coefficient of friction of materials made of the AlMg1SiCu alloy without MWCNTs, but also occurs in a much smaller range, which is also confirmed by the literature [276].

The reason for such a significant strengthening of nanocomposite materials with AlMg1SiCu alloy matrix, reinforced with MWCNTs carbon nanotubes, is dispersion nanostructured precipitates of Al_4C_3 carbide. Without the precipitation of these carbides, these positive effects would not exist. However, note that this carbide is highly hygroscopic and completely hydrolyzes in an aqueous medium to form aluminum hydroxide and gaseous hydrocarbons such as methane or acetylene, according to the reaction:

$$Al_4C_3 + 12\ H_2O \rightarrow 3\ CH_4 + 4\ Al(OH)_3 \tag{1}$$

Similarly, the reaction with hydrochloric acid also causes decomposition of aluminum carbide according to the reaction:

$$Al_4C_3 + 12\ HCl \rightarrow 3\ CH_4 + 4\ AlCl_3 \tag{2}$$

It leads to a deterioration of the corrosion resistance of the composite materials and is a common factor contributing to mechanical failure. Even a small share of aluminum carbide destroys composite materials with the participation of aluminum and nano additives of allotropic forms of carbon in aggressive environmental conditions, and even in atmospheric conditions, causing uneven pitting corrosion. It limits the practical use of such composite materials [277]. Al resistance to electrochemical corrosion improves with the formation of Al_4C_3 on the Al surface, which prevents the dissolution of the metal in the chloride-based solution [278, 279]. The practical use of such nanocomposite materials will therefore be possible only in cases where their surface will be tightly covered with appropriate, durable coatings that effectively prevent contact of the material surface with water and other aggressive environmental media that can be applied. It is obvious that the technological process of applying the coatings cannot dissolve Al_4C_3 carbides. This requires further detailed studies as well as the eventual application of other coatings described as preventing corrosion of various alloys and composite materials based on aluminum or its alloys [280] including also the polymeric ones [281–284] and maybe it could be also possible to use the atomic layers deposition ALD technology while ensuring the required adhesion of such layers [285–288].

3.3 Structure and properties of the composite layers on the substrates of aluminum alloy

In all other cases, 7th–12th, the result of the work carried out is to obtain a hard and abrasion-resistant composite surface layer by introducing tungsten carbide or silicon carbide powder, respectively, through the interaction of a laser beam on the surfaces of aluminum AlMg3, AlMg5, AlMg9 alloys. Due to the use of laser treatment, the surface of the aluminum material is characterized by a relatively low density, receives better mechanical and tribological properties. Laser technologies are now widely used in surface treatment not only in the treatment of aluminum alloys, but many other alloys [289–317].

The share of feathering carbides WC/W_2C or SiC in the surface layer of the AlMg alloys depends on the conditions of the laser treatment. The largest share of the feathered WC/W_2C powder occurs for the AlMg9 alloy when a laser beam power is 1.8 kW. The surface layer manufactured in this case is characterized by a uniform distribution of the embedded particles in the matrix at a depth of up to 150 μm (**Figure 46a**), analogous to the laser treatment of the AlMg5 alloy surface (**Figure 46i, j**). Increasing the power of the laser beam causes an increase in the depth of remelting zone (**Figure 47a**), which results in greater dispersion of the feathered particles in the remelted layer, and the accompanying increase in the intensity of convection movements in the liquid pool affects the displacement of a part of the powder to a depth of about 1.5 mm both in the case of AlMg9 alloy

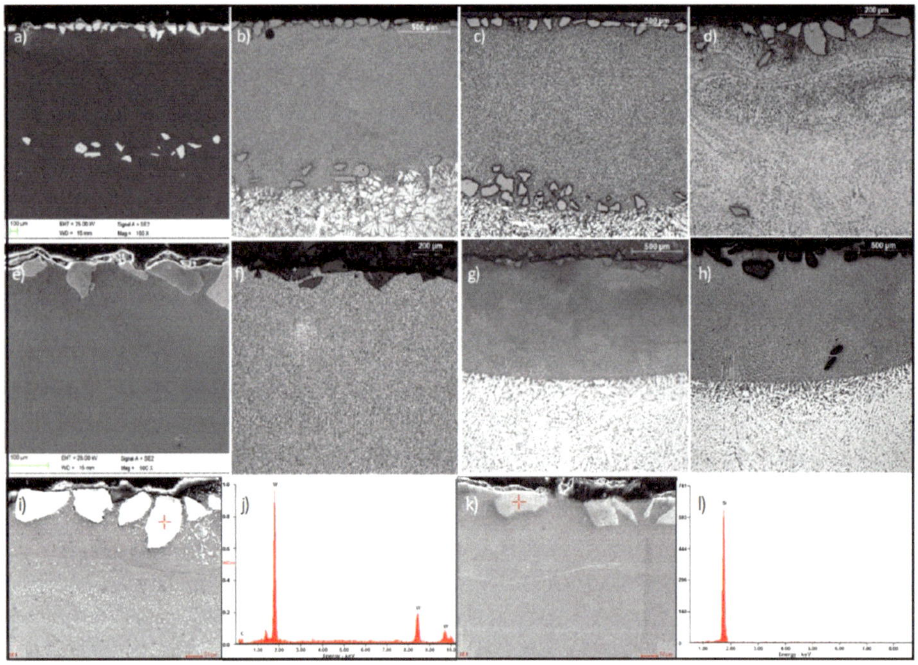

Figure 46.
Structure after laser feathering with (a-d, i, j) WC/W$_2$C carbides; (e-h, k, l) SiC carbides; (a-c, e, f) AlMg9 alloy; (d, g, i, j) an AlMg3 alloy; (h, k, l) AlMg5 alloy; using a laser beam with power: (a, b, f, g, h) 1.8 kW; (e) 2.0 kW; (c, d, g, i-l) 2.2 kW; (a, e, i-l) SEM; (j, l) SEM-EDS; (b-d, f-h) light microscope.

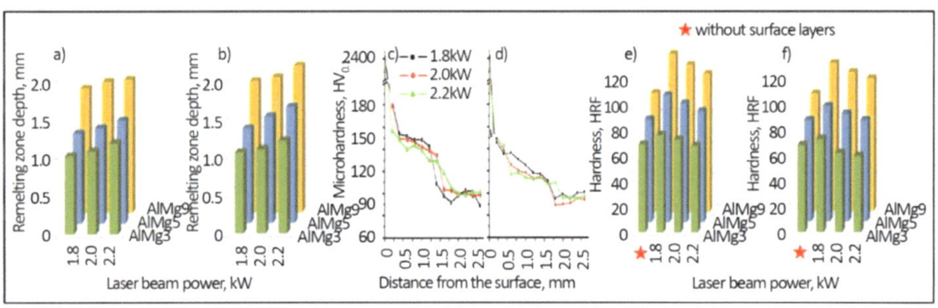

Figure 47.
The effect of laser beam power of AlMg3, AlMg5, and AlMg9 alloys when feathering carbides (a, c, e) WC/W$_2$C; (b, d, f) SiC; on (a, b) depth of the remelting zone; (c, d) course of hardness changes depending on the distance from the surface of the AlMg9 alloy; (e, f) the surface hardness versus the hardness of non-laser treated alloys.

(**Figure 46c, d**) and AlMg5 alloy (**Figure 46i, j**). The AlMg3 alloy after laser feathering of tungsten carbide with 1.8 kW laser beam power into the surface of the AlMg3 alloy results in an even distribution of carbides both at a depth of up to 160 μm and at the bottom of the layer. On the other hand, increasing the power of the laser beam causes gaps between the individual particles at a distance of up to 100 μm (**Figure 46d**).

The smallest proportion of the feathered powder occurs for the AlMg3 alloy after laser feathering of SiC carbide powder with a laser beam power of 2.2 kW (**Figure 46e**). As a result of laser feathering with a beam power of 2.0 and 2.2 kW, the carbides are evenly distributed in the near-surface part of the layer at a depth of up to 100 μm (**Figures 46e,g** and **47b**). The best results for laser feathering of silicon carbide are for the AlMg5 alloy (**Figure 46k,l**). Composite layers manufactured with a laser beam power of 2.0 and 1.8 kW contain evenly distributed embedded powder particles in the matrix of an aluminum alloy, including AlMg9 (**Figure 46e,f**), with the largest share of feathered SiC powder in the case of alloy fusion AlMg5 with 1.8 kW laser beam power (**Figure 46h**). Single carbides with an average diameter of 100 μm occur in the central part of remelting zone also at the maximum power of the laser beam of 2.2 kW (**Figure 46k, l**).

Structure studies with the use of transmission electron microscopy and X-ray phase analysis showed the presence of intermetallic phases and $AlMg_2$, Al_3Mg_2, Mg_5Al_{18}, Mg_2Si precipitates.

Among the AlMg3, AlMg5 and AlMg9 alloys, the AlMg9 alloy has the highest hardness (**Figure 47f**). The mechanical and tribological properties of the manufactured composite layers containing respectively WC/W_2C or SiC carbides on the above-mentioned alloys largely depend on the power of the laser beam, because too high power adversely affects the properties of these layers (**Figure 47f**). As a result of laser feathering of carbide powder, the greatest increase in hardness occurs for the AlMg9 alloy using a 1.8 kW laser beam (**Figure 47e, f**). In the case of the two other alloys, the highest hardness occurs at this laser power, and increasing the laser power causes a slight reduction in the hardness of the surface layer (**Figure 47e, f**). The highest hardness of the layer is provided by laser feathering of the WC/W_2C powder into the AlMg9 alloy matrix with a laser beam power of 1.8 kW (**Figure 47e**). The smallest increase in hardness is obtained when laser feathering SiC powder into the AlMg3 alloy matrix with a laser beam power of 2.2 kW (**Figure 47f**). Increasing the power of the laser beam causes an increase in the hardness with respect to the Al-Mg alloy substrate at a greater depth (**Figure 47c, d**). The maximum value of hardness is lower than for treatment made with a lower power of the laser beam (**Figure 47c, d**). The introduction of WC/W_2C into the Al-Mg alloys matrix enables obtaining a higher hardness of the layers than in the case of laser feathering of SiC powder (**Figure 47c-f**).

Ball-on-disk abrasion resistance tests of the obtained composite layers as well as aluminum alloy without laser treatment showed a significant increase in abrasive wear resistance of surfaces obtained as a result of laser feathering of hard powders of WC/W_2C and SiC carbides, respectively, into the remelted matrix of AlMg3 alloys, AlMg5 and AlMg9. The highest abrasive wear resistance among all alloys and laser feathering variants is obtained after laser feathering SiC powder into the AlMg9 alloy matrix with a laser beam power of 1.8 kW. The highest wear resistance among these alloys as a result of laser feathering of WC/W_2C tungsten carbide powder is demonstrated by AlMg9 alloy after feathering with a laser beam power of 1.8 kW.

The developed technology can find numerous applications, including in the mechanical engineering, automotive and aviation industries.

4. Final remarks on the importance of advanced composite materials based on aluminum alloys in the development of Industry

4.1 The general importance of engineering materials and the technology of their manufacturing and processing

Already a dozen or so years ago, the concept of a half-life of detailed knowledge was introduced by analogy to the half-life of radioactivity [318]. After this period, characteristic for a given area of knowledge, about 50% of detailed information in this area will be up-to-date, and the rest will become outdated due to new scientific and technological discoveries. It is a measure of increasing the level of general knowledge of engineering staff, enabling them to be quickly transferred to the practice of engineering design and product manufacturing as well as a wide application area. It is undeniable that in the area of engineering materials and the technology of their manufacturing and processing, the half-life of knowledge has been greatly shortened over the course of history and now reaches only a few years (**Figure 48**).

The development of natural materials, and with time predominantly engineering materials, has become the basic determinant of the development of material culture and human civilization in general, largely determining the level and quality of life of societies achieved. It should be realized that to manufacture any product that is accepted on the market, it is necessary to use properly selected engineering materials. Initially, for thousands of years, as described in publications, including own publications [319–337], but also a hundred years ago, the selection of materials was carried out by trial and error, which is considered the Materials 1.0 stage (**Figure 48**). To date, even in 80% of cases, materials are systematically tested according to the Materials 2.0 protocol [319, 320, 324, 333, 337, 338], after the concept is developed and its subsequent verification, and then the prototype is tested in laboratory conditions and only then tested and validated in real conditions. The impact of environmental conditions should be investigated and the product life cycle designed. In the case of the Materials 3.0 protocol, the procedure is similar, except that the use of IT tools at the concept development stage is characteristic of this stage. To design the target functionality, computational and simulation methods are used, e.g. using the principles of quantum physics and chemistry, methods of artificial intelligence, and databases. Nowadays, opportunities have opened up for the use of cyber-physical systems and big data, as well as advanced methods of artificial intelligence and machine learning, including artificial neural networks and fuzzy logic, and more and more often smart systems cooperating with people, which is the essence of the Materials 4.0 approach, which is now increasingly being implemented. It is accompanied by the application of materials on-demand, when not more than two decades ago, the principle of having to choose materials only from the assortment offered by material manufacturers was in force. The task set for the designer can now be formulated in the form of the question: "what can be used to create a product interesting for the customer on the market?" On the other hand, the approach that answers the question: "What can be

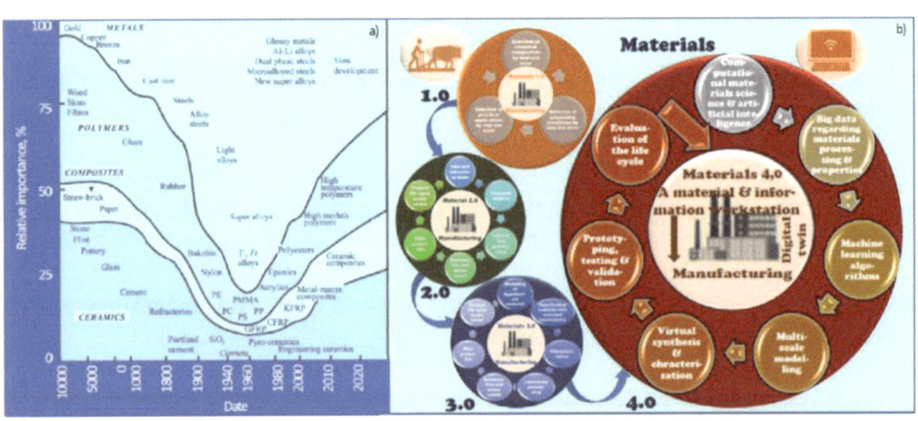

Figure 48.
Changes of: (a) the importance of individual groups of engineering materials throughout history; (b) the concept of selection of engineering materials for products from the trial and error method (materials 1.0) to the methodology using cyber-physical systems and advanced information technologies (Materials 4.0).

manufactured from a material that is currently available or already known to us?" It is anachronistic. Numerous grades from among much more than 100,000 currently known and available engineering materials give new innovative potential in the implementation of products. It is impossible to correctly select engineering material for any practical application without the use of modern and highly developed Computer-Aided Design CAD tools and other Computational Material Science methods. There are two aspects to this approach. On the one hand, the material properties of each product are designed by the constructor and an appropriate engineering material should be produced that meets the requirements of a given application solution, and on the other hand, the availability of one of the ready materials having a set of properties according to the criteria should be analyzed, using the available material databases. The problem is extremely important considering that an average small car has at least 15,000 parts and a passenger aircraft may have 4.5 million or more elements. Of course, in each case, the right decision must be made regarding the selection of engineering material and the appropriate technological process. All engineering materials are in each case equivalent from the point of view of engineering design, provided that they meet a set of required criteria, and the basis for multi-criteria optimization is a set of the best functional and technological properties with the lowest possible manufacturing and processing costs, as well as the use of the material and the product manufactured of it, taking into account ecological aspects and the product life cycle. Product engineering design is essential and is not an activity extracted from the overall product manufacturing process, but an important element in satisfying market needs and at the same time is dependent on them.

Product engineering design is composed of three elements that cannot be separated and cannot exist autonomously (**Figure 49**) [3, 318–320, 324, 339–351], which include:

- structural design, ensuring the shape and geometric features of the product,
- material's design, ensuring the required physicochemical and technological properties as well as product life cycle,

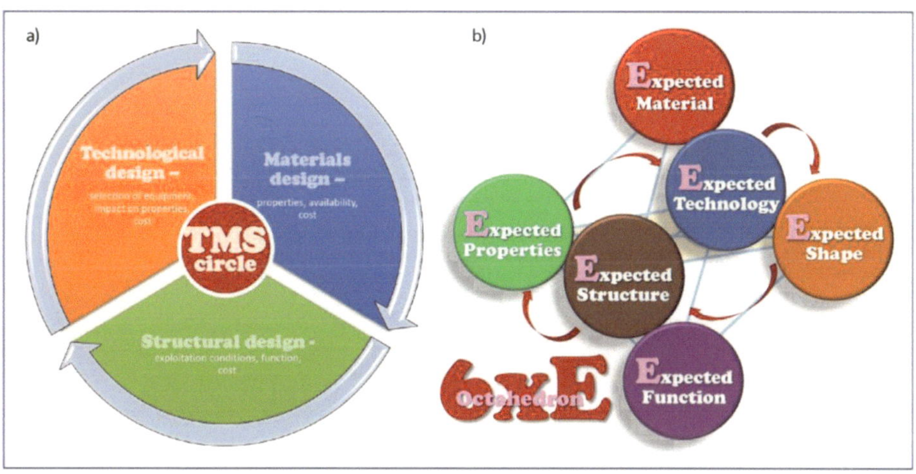

Figure 49.
Diagram of relationships between structural, material, and technological design as components of product engineering design (a) and the 6xE octahedron of expectations resulting from the material science paradigm and materials science (b).

- technological design, ensuring the manufacturing of products with the required properties, with the highest possible efficiency, the highest possible degree of automation and computer aiding, and the lowest manufacturing costs.

The functional properties of products, are including mechanical, physicochemical, thermal, magnetic, electrical, and optical properties, depending on the chemical composition and the cross-sectional structure and very often also the surface structure of the materials and the conditions of use of the products and their components. These conditions include also operate at high or low temperatures, in conditions of cyclic loads, as well as in an aggressive gaseous or liquid medium. Materials science deals with the assessment of the impact of the structure on the electronic, crystalline, micro, and macroscopic scale on the properties of materials, while material engineering covers the relationships between the structure, technological processes, and functional properties of materials used for numerous products and complex application systems, significantly determining the quality of life of modern societies. The materials science paradigm corresponds to an octahedron illustrating the 6xE rule (**Figure 49b**) [3, 318–320, 339, 340]. Expected operational functions of the product are ensured by designing the expected material, processed using the expected technology, in order to give the expected geometric characteristics and shape of the product, enabling the expected structure and deciding to obtain a set of expected properties of the designed product. The above-mentioned approach provides people with continuous access to products and consumer goods that directly affect the level and quality of life, climate, quality and potential of health protection, education, information exchange, and other aspects recognized by the United Nations as Sustainable Development Goals (**Figure 50**). It presents a deeply humanistic mission of the engineering community in order to activate the development of societies. For this reason, the real development of materials engineering, including nanotechnology and surface engineering, and the accompanying research, are an important determinant of present and future prosperity and high quality of life. Sustainable development of this area is therefore of great interest to societies.

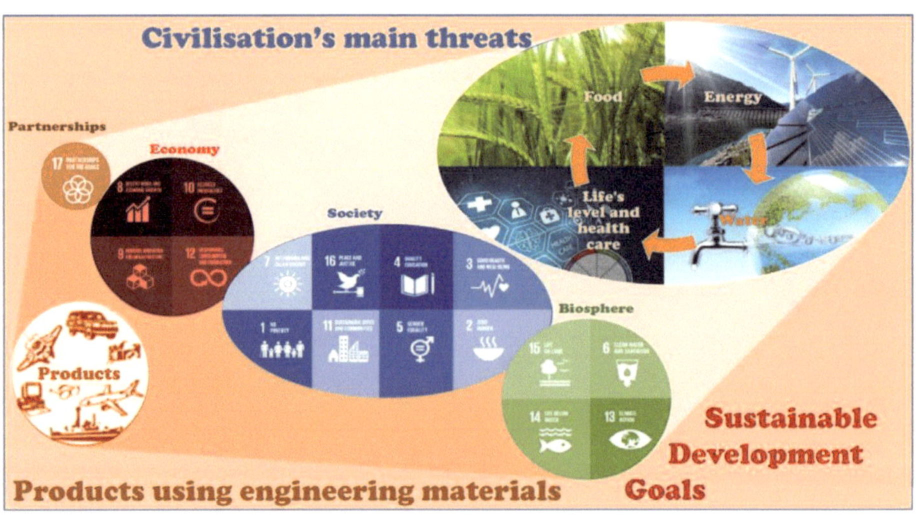

Figure 50.
Civilization's main threats and sustainable development goals enabling counteraction them and their relationship to the development of engineering materials used on products.

4.2 The importance of engineering materials and technological processes in the context of the current stage of Industry 4.0 of the industrial revolution

All products of interest to customers, without exception, are made of engineering materials and cannot be built without them. The processes of using engineering materials and other resources such as energy, capital, and people to make products available using different machines and different technological processes organized according to a well-designed plan are called manufacturing. The development of manufacturing processes is closely related to the development of the material culture of societies, which was reflected in the Japanese approach of the Society from groups of people practicing hunting and harmoniously coexisting with nature, which was considered Society 1.0 to information and achieving the goals of Society 5.0 of the smart sustainable development, oriented to people, their well-being and high quality of life, and ensuring economic development, which goals are achieved by connecting cyberspace with the real world (**Figure 51**) [319, 320, 322, 352–357].

In Europe, a few years earlier, the German concept of Industry 4.0 was developed, already popular in many countries around the world, and also presented in Author's own works, describing the stages of the industrial revolution (**Figure 52**) [195, 293, 319, 320, 322, 323, 332–337, 349, 350, 358–414].

The introduction of steam engines at the end of the 18th century, considered Industry 1.0, marked the beginning of the industrial revolution. The current stage

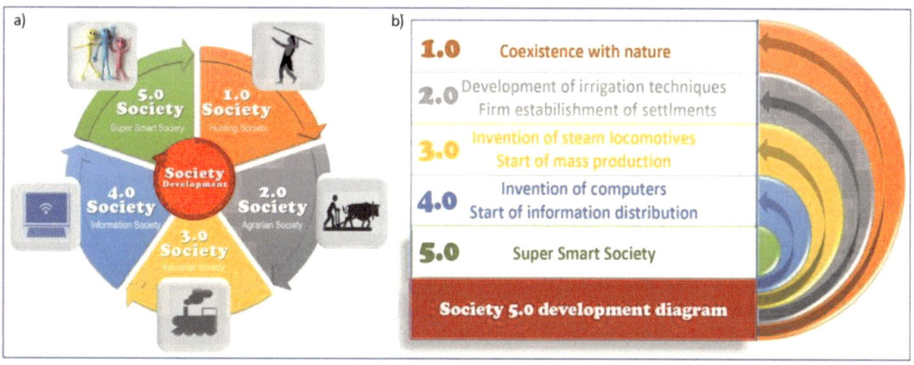

Figure 51.
The concept of development from Society 1.0 to Society 5.0 (a) and the main events determining the implementation of each of the stages (b).

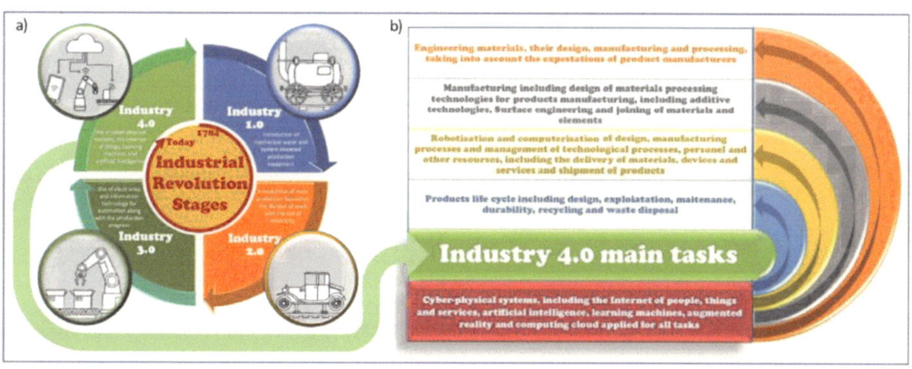

Figure 52.
The concept and tasks of Industry 4.0 (a) and main tasks in the Industry 4.0 stage (b).

of Industry 4.0 includes smart factories, supplied by smart suppliers, producing smart products. Physical production processes are monitored by manufacturing systems, making smart decisions, based on real-time communication with people, machines, and sensors, and based on experiments performed on the so-called "Digital twins" in virtual reality using the simulation of real conditions of operation and maintenance of manufactured products using cyber-physical systems (CPS), Internet of Things (IoT) and cloud computing, with the use of large data sets. In the classic Industry 4.0 model [319, 320, 322, 361, 367–369, 372–375, 378–382, 387–391, 393, 395, 402, 415, 416] it was recognized that this stage of using the nine basic technologies described in the literature was achieved (**Figure 53**) [293, 319, 320, 323, 345–349, 360–415].

It turned out that this model is one-sided and requires a significant augmentation, mainly by taking into account engineering materials, technological manufacturing processes and technological machines, and not only complex cyber-physical systems and very advanced tools and information systems (**Figure 54**).

The most important thing is that manufacturing possible only in reality, thanks to the physical system and on the condition of using real materials, usually engineering ones. No product can be manufactured without materials. Therefore, they should be absolutely included in the current Industry 4.0 model, because obviously they are and were necessary for every stage of the industrial revolution from Industry 1.0 to Industry 4.0. The simplified approach in the classic Industry 4.0 model creates the erroneous impression that the progress concerns only technologies covering monitoring, control, coordination, and integration of information and communication technologies that make up cyber-physical systems (**Figure 53**), without the need to make real progress in the field of technological machines, manufacturing technologies and necessary engineering materials to manufacture any product. It is obviously not true and therefore an adequate model must take into account all these aspects [319, 320, 322, 332–336, 365, 417]. A far-reaching simplification is also reducing the technological issues only to additive manufacturing, as it was done in the classic Industry 4.0 model. Additive manufacturing cannot be treated as ruthlessly competing with many known and used for many years, and sometimes even millennia, technologies for manufacturing products, such as foundry, machining, plastic working, heat treatment, surface engineering, joining and welding, and assembly [319, 320, 322, 332–337, 345–349, 364, 365]. The enormous technological progress that is taking place now and in the near future will not

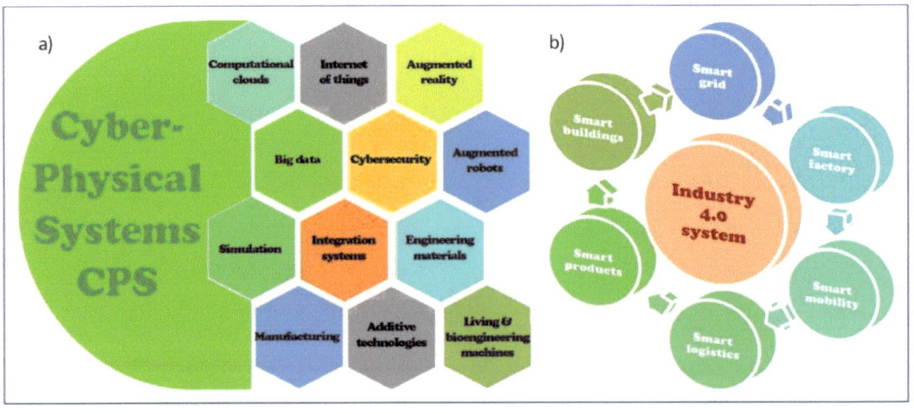

Figure 53.
The crucial components of the current industry 4.0 model: (a) augmented set of leading technologies included in cyber-physical systems; (b) smart systems included in the Industry 4.0 model.

eliminate these technologies, so it is necessary to include them in the Industry 4.0 model. It should be noted that the systematic modernization and progress in the field of these technologies concerns their mechanization, robotization, and above all computer-aided engineering design and manufacturing within cyber-physical systems. It is obvious that the classic model of Industry 4.0 in fact concerns only the computerization of technological processes, therefore, instead of it, it was necessary to introduce the appropriate holistic augmented model of Industry 4.0 [319, 320, 418] leaving cyber-physical systems as one of the elements technology platform of this new model (**Figure 53**).

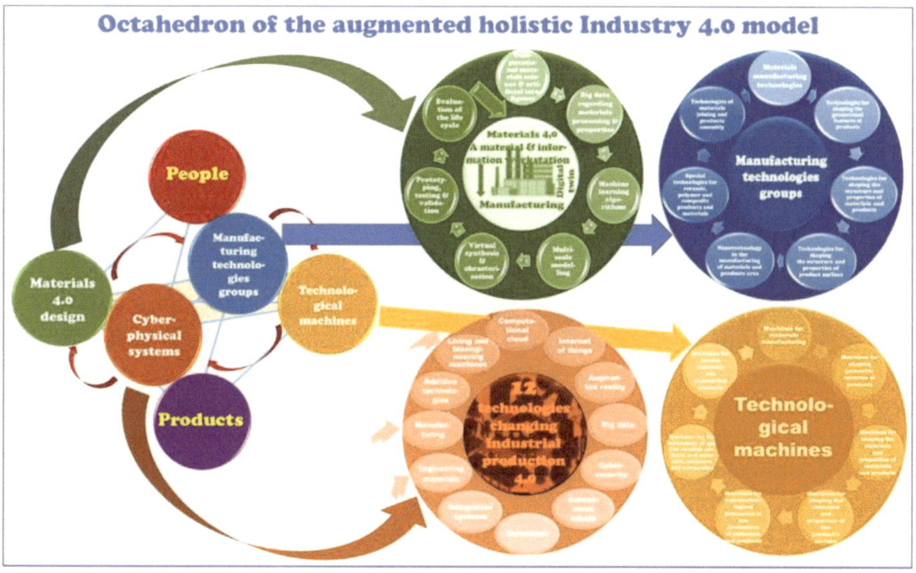

Figure 54.
The holistic augmented model of Industry 4.0 taking into account advanced engineering materials and technological processes and also the machines and production devices used, as well as twelve technologies related to cyber-physical systems conditioning the development of production in the most modern phase of the industrial revolution.

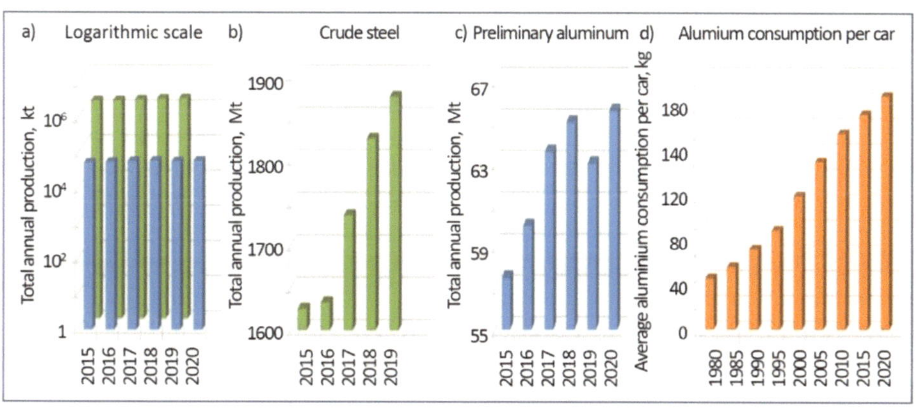

Figure 55.
Comparison of the annual total world production in 2015–2020 years of (a, b) crude steel (a, c) preliminary aluminum (a – logarithmic scale); (d) changes of the average aluminum consumption per car in 1980–2020 years.

4.3 The importance of advanced composite materials with an aluminum alloy matrix in the context of the current stage of Industry 4.0 of the industrial revolution

The growing and more sophisticated requirements set by the conditions of the current stage of the Industrial Revolution, conventionally designated as Industry 4.0, also pose challenges to designers and manufacturers of engineering materials. They apply equally to the implementation of new engineering materials, technological processes of their manufacturing and processing, advanced methods of their design, as well as the application of nanotechnology and surface engineering in this area. Among engineering materials, world steel production remains extremely high (**Figure 55**), despite reports of a decline in production in some countries due to the SARS-CoV-2 coronavirus pandemic.

The Worldsteel Association published crude steel production data in August 2020. Compared to August 2019, world production was 0.6% higher, despite the pandemic. The lowest production level, amounting to 137 million tons, was recorded in April 2020. Since then, production has been increasing and in August it amounted to 156.2 million tons. However, there has been a reshuffle in the market. Production declines were recorded in the United States (−24.4%), Japan (−20.6%), Germany (−13.4%), Russia (−4.6%), India (−4.4%) and South Korea (−1.8%). From the beginning of the year, production decreased by 19% in North America, by 18.6% in the European Union, by 15.6% in South America, and by 4.5% in the former Soviet Union countries. Production in the dominant country, i.e. China, despite the pandemic, does not decrease, but even increases. China, as the largest steel producer in the world, produced 94.8 billion tons in August, 8.4% more than in the previous year. Comparing total production in the first eight months of 2020 and 2019, China grew by 3.7%. Increases in steel production also occurred in Vietnam (32.9% compared to August 2019 and 11.5% in total since the beginning of the year), in Turkey (22.9% compared to August 2019 and 0.6% in total since the beginning of the year), in Iran (14.6%) and Brazil (6.5%).

Compared to steel, aluminum alloys are significantly lighter, and at low temperatures are characterized by higher impact strength. Aluminum alloys are the preferred structural materials used, among others, for the manufacturing of cars, in the aerospace industry and mining machines, in the processing industry for various high-performance elements for various applications, due to their relatively low weight and favorable set of properties. Ecological considerations require the reduction of the total weight of cars, which indicates the need to produce many new components from aluminum alloys (**Figure 55**). The use of lightweight materials for the manufacturing of car components allows for compliance with increasingly stringent air pollution regulations. It is obvious that a car with a lower weight allows for lower and more economical fuel consumption, which clearly reduces the emission of exhaust gases, mainly carbon dioxide, into the atmosphere.

Annual global aluminum production is expanding (**Figure 55**) at a level almost 30 times lower than that of steel, although the aluminum market may also have changed due to the COVID-19 pandemic. In 2015–2020 years, the total aluminum production amounted to 374,840 tons, and thanks to the increase in production in China in 2020, the total world production increased by a few percent compared to 2018, when it was higher than in 2019. The dynamics of aluminum production in 2015–2020 is shown in **Figure 54**. For example, in 2017, primary aluminum was produced in 41 countries. The main producers were China (55%), Russia (6%), India (5%), Canada (5%) and the United Arab Emirates (4%) in order of decreasing aluminum production. The share of China in the global aluminum production is growing year by year.

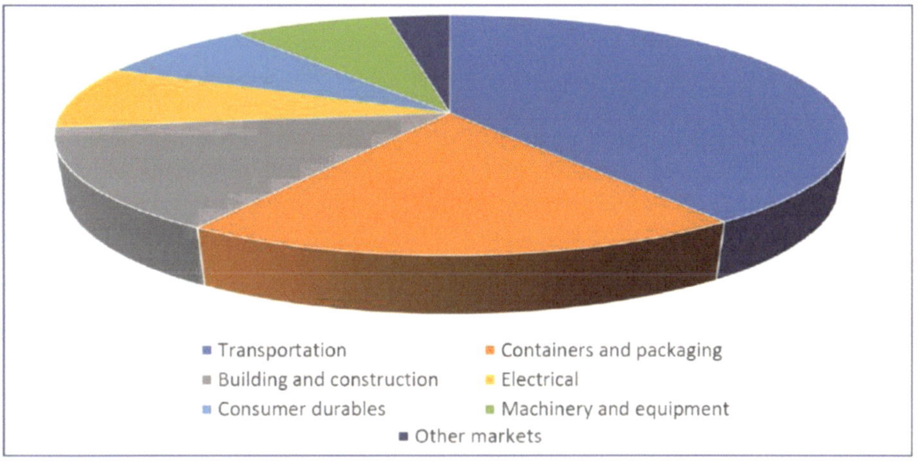

Figure 56.
Main areas of application of aluminum and its alloys in the US and Canada in 2017 (compiled according to data from the US Geological Survey and Aluminum Association 2017, available on July 15, 2018).

The end-use rate of aluminum and its alloys in various application areas has been established for many years. Aluminum has gained a good position on the world market due to its significant share in increasing the fuel efficiency of vehicles and in relation to its use in ecological construction products and ecological packaging. An example are the main areas of application of aluminum and its alloys in the United States and Canada (**Figure 56**).

According to the International Trade Center (ITC), the value of the aluminum market and its products in 2018 amounted to USD 194.32 billion [419]. It was estimated that the volume of the aluminum market in that year was 79 million tons, and it is estimated that aluminum consumption will increase with compound annual growth rate CAGR of 4.2% in the forecast period until 2025 [420]. Aluminum is widely used for packaging and construction, and is also used in machinery and equipment, and in the electrical and consumer durables industries. The main area of application of aluminum is the automotive industry. Since aluminum is a light metal, it is used mainly in car bodies and engines parts, what is an essential factor in reducing car fuel consumption. It also applies to trucks, as, for example, the Ford plant launched the production of an all-aluminum F-150 trak's body in 2015 [421], which reduced the weight of the truck by more than 300 kg, resulting in a significant improvement in fuel efficiency and reduced total operating costs. International agreements on fuel economy are influencing for increase the use of aluminum in the automotive industry around the world. The expected increase in the production first of all of electric vehicles, and also semi-autonomous and autonomous vehicles as well as self-driving cars and trucks will further strengthen the aluminum automotive market. This segment is estimated to grow at a CAGR of 4.3% by 2025. Aluminum is now widely used in smartphones, tablets, laptops, flat screen TVs, monitors and other consumer electronics products. According to estimates provided by the Global System for Mobile communications Association (GSMA), the number of mobile phone users worldwide will increase from 5.1 billion in 2018 to 5.8 billion in 2025. Therefore, the global smartphone and smart TV market will be subject to a CAGR growth of 8–10% in the forecast period until 2025, which will affect the increase in demand on the aluminum market. Aluminum is the second most used metal in construction after steel, and according to the World Bank, in 2016, the gross domestic product GDP of construction was 25.4%. Aluminum is widely used in windows, curtain walls, roofing and cladding, sunshades, solar panels, railings, shelves and other temporary structures. The growing construction activity around the world creates more and more opportunities for the aluminum market. However, there are differences in aluminum consumption by end-use category in high (HIC), low (LIC) and middle income countries (MIC). In low- and middle-income countries, aluminum is mainly used for the production of electrical systems and in the construction industry. In high-income countries the main areas of application are products in the transportation sector, including cars, commercial aircraft and railway stock.

It is estimated that the aluminum market will grow in CAGR above 3% in the forecast period until 2025, although earlier data even indicated an increase in aluminum consumption up to 120 Mt. with an average growth rate of about 4.1% per year. The global aluminum market is expected to grow by 6.5% CAGR to reach USD 235.8 billion in 2025. This means that the annual production of bauxite should increase to approx. 570 Mt., and of aluminum oxide to approx. 230 Mt. However, a diversified increase in the use of aluminum is expected in different geographic areas (**Figure 57**). The main driver of the global aluminum market is the growing construction activity in the Asia-Pacific region. In North America, the growing number of construction projects and growing investments in the United States and Canada are driving the market under study. This upward trend is likely to continue, depending on the financing of larger projects. On the other hand, the slowdown in the global automotive industry and the unfavorable conditions resulting from the COVID-19 outbreak are holding back market development, and earlier forecasts are obsolete. Therefore, it is estimated that the construction industry will dominate the global aluminum market in the coming years. The Asia-Pacific region will be the fastest growing in the world in this situation, due to rising consumption in China, India and Japan, with a growing population, rising middle-class income and urbanization. In fact, the Asia-Pacific region will generally dominate the global aluminum market. The electronics manufacturing market in this region will develop rapidly in the coming years due to the numerous Original Equipment Manufacturer OEMs, low cost and availability of raw materials and cheap labor, as well as cooperation with numerous entities in Germany, France and the United States in the field of production services and assembly. China is expected to be the world's largest

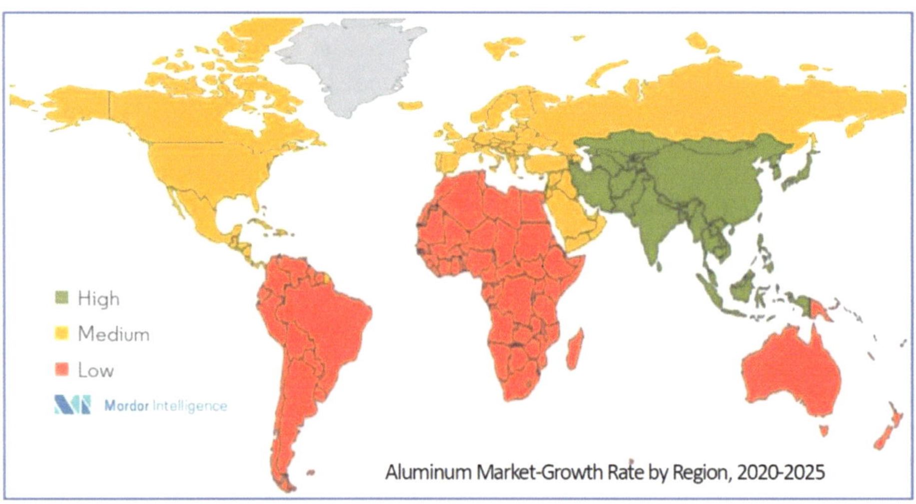

Figure 57.
Different scale of growth of aluminum markets in different geographical areas of the world in 2020–2025 due to the diversification of the development of the end-user industry.

market for civil aircraft sales by 2040 among all countries, which will have a significant impact on the demand for aluminum. For example, China is expected to purchase new aircraft from Boeing by 2036 worth USD 1.1 trillion. The growing demand for aluminum is also significantly influenced by the growing demand for cargo transportation, therefore the production of commercial aircraft in Japan will continue to increase, including the production of new generation F-35A fighters and other military aircrafts. Thus, the demand for aluminum in the Asia-Pacific region will increase rapidly in the next two decades due to the rapidly expanding end-user industry.

Aluminum is the only material that more than covers its own collection costs in the consumer disposal stream and is 100% recyclable, retaining its properties for an indefinite period of time. More than 70% of the aluminum produced from the beginning of industrial use of aluminum has been recycled and is still used today. In view of the expected growth of the global aluminum market in 2025 to USD 235.8 billion, and the related expected increase in the production capacity of bauxite mines around the world, one should take into account the probable increase in the share of secondary aluminum from new scrap, proportional to the total consumption. However, the share of aluminum produced from old scrap may decline, as in countries with high economic growth, the share of aluminum used to build new infrastructure with long-term use is significant. It could undoubtedly have a negative impact on greenhouse gas emissions, unless the stricter requirements of the climate policy force an increase in the efficiency of aluminum smelters and the related reduction of energy consumption for aluminum production. Recycling aluminum saves over 90% of energy costs compared to primary production, and thus reduces the use of fossil fuels, including oil, the energy of which would be used to produce fresh aluminum. For the recovery of post-consumer scrap, it is important to diversify the end use of aluminum. For example, statistically, an average of 113,000 aluminum cans are recycled in the world every minute, and their value per year amounts to over USD 800 million. Aluminum used in cars and vehicles can be easily recycled. Many applications in the transport sector, but also in the electronics sector, have a much shorter real life-time and therefore they are more recyclable than electrical systems and construction applications. According to the Aluminum Association, more than half a million tons of aluminum used in cars are recycled annually [419].

The challenges posed by the developing industry in the Industry 4.0 stage and the expectations of the automotive and aviation industries force constant progress in the development of new materials using aluminum. The undoubted advantages of these materials include the density almost three times lower than that of steel, which is the main premise for applications in the area of transport. Newer and newer aluminum alloys with other alloying elements are being developed and implemented. Of course, there are also alloys of other elements in which aluminum is an alloying element. Aluminum matrix composite materials ACMs have a steady share of the systematically growing group of materials using aluminum. This chapter presents a generalization of the results of numerous own studies and provides a detailed overview of the structure and properties and manufacturing technology of several new AMCs selected for presentation due to their advanced manufacturing methods and unexpected properties and the resulting wide application possibilities (**Figure 58**).

Among these new advanced aluminum matrix composites materials, there are basically three groups of such materials [66–80], including:

1. composite materials manufactured by gas-pressure infiltration with liquid aluminum alloys, respectively, eutectic AlSi12 and hypoeutectic AlSi7Mg0.3, constituting the matrix, in which the reinforcement has a skeleton structure,

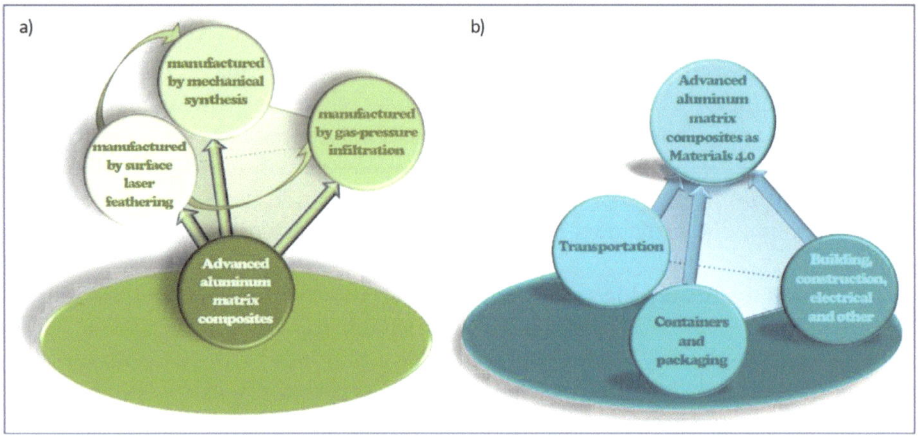

Figure 58.
The scheme (a) of three groups of the new advanced aluminum matrix composites materials presented in the chapter manufactured using advanced technologies of Industry 4.0 stage; (b) the main areas of predicted applications of advanced aluminum matrix composites.

obtained alternatively by sintering Al_2O_3 powder with the participation of carbon fibers enabling, after their removal, obtaining a porous structure, sintering mixtures of HNTs halloysite nanotubes with carbon fibers, after their removal, providing a porous mullite skeleton and selective laser sintering of titanium powder to obtain a skeleton with the porosity of at least 50% with pores with a size of 50–500 μm,

2. nanocomposite materials manufactured by mechanical synthesis using high-energy mechanical milling of a mixture of an aluminum alloy powder for plastic deformation AlMg1SiCu with alternatively halloysite HNTs nanotubes and multi-wall carbon nanotubes MWCNTs and the next two-stage press consolidation and subsequent cold pressing in the first order hot at an elevated temperature in order to produce bars of the assumed diameter,

3. composite surface layers on the substrate of aluminum alloys, alternatively AlMg3, AlMg5, AlMg9, manufactured by surface laser feathering by introducing powder, alternatively tungsten carbide WC/W_2C or silicon carbide SiC with the use of fluxes as agents supporting the operation of the laser beam and facilitating the selective remelting of the aluminum alloys, without the need for additional heat and surface treatment.

The structure, technology, and properties of these materials were presented, as well as the application perspectives, especially in transport and biomedical devices, as well as in connection with the implementation of the Industry 4.0 stage of the Industrial Revolution.

Acknowledgements

The author would like to acknowledge for the cooperation in research realisation his former PhD students Dr Marek Kremzer, Dr Magdalena Kujawa, Dr Anna Achtelik-Franczak, Dr Błażej Tomiczek, Dr Magdalena Macek-Sroka, and Dr Wojciech Pakieła.

Author details

Leszek A. Dobrzański
Medical and Dental Engineering Centre for Research, Design and Production ASKLEPIOS, Gliwice, Poland

*Address all correspondence to: leszek.dobrzanski@centrumasklepios.pl

References

[1] Dobrzański LA. Engineering materials and material design. Fundamentals of materials science and metallurgy. 2nd ed changed and supplemented. Gliwice-Warszawa: WNT;2006. 1596 p. (in Polish)

[2] Mondolfo LF. Aluminum Alloys: Structure and Properties. Changed ed. London: Butterworths/Elsevier;2013. 17 p.

[3] Dobrzański LA. Metals and their alloys. Open Access Library Annal VII, vol. 2. Gliwice: International OCSCO World Press;2017. 982 p. (in Polish)

[4] Totten GE, MacKenzie DS. Handbook of Aluminum: Volume 2: Alloy production and materials manufacturing. New York, NY: Marcel Dekker Inc.;2003.

[5] Buckingham DA, Plunkert PA, Bray EL. Aluminium statistics. U.S. Geological survey [Internet]. Available from: http://minerals.usgs.gov/minerals/pubs/historicalstatistics/ds140-alumi.xlsx [Accessed: 2021-02-10]

[6] Bray EL, Weaver SM. Aluminum in November 2020. U.S. Geological survey [Internet]. Available from: https://prd-wret.s3.us-west-2.amazonaws.com/assets/palladium/production/atoms/files/mis-202011-alumi.pdf [Accessed: 2021-02-10]

[7] Richards JW. Aluminium; its history, occurrence, properties, metallurgy and applications, including its alloys. Philadelphia: H.C. Baird & co.; 1896. 750 p.

[8] Drozdov A. Aluminium: The Thirteenth Element. RUSAL Library; 2007. 240 p.

[9] Clapham JH, Power EE. The Cambridge Economic History of Europe: From the Decline of the Roman Empire. CUP Archive; 1941. 207 p.

[10] Setton KM. The papacy and the Levant: 1204-1571. 1 The thirteenth and fourteenth centuries. American Philosophical Society;1976.

[11] Örsted HC. Overview of the Royal Danish Science Society's Proceedings and the Work of its Members, from 31 May 1824 to 31 May 1825. 15-16 pp. (in Danish)

[12] Royal Danish Academy of Sciences and Letters. The philosophical and historical dissertations of the Royal Danish Science Society. 1827. xxv–xxvi pp. (in Danish)

[13] Wöhler F. Ueber das Aluminium. Annalen der Physik und Chemie. 1827; 87:146-161. DOI:10.1002/andp.18270870912

[14] Venetski S. Silver' from clay. Metallurgist. 1969;13: 451–453. DOI: 10.1007/BF00741130

[15] Evans JW. Metal Production: Electrometallurgy. In: Buschow KHJ, Cahn RW, Flemings MC, Ilschner B, Kramer EJ, Mahajan S, Veyssière P, editors. Encyclopedia of Materials: Science and Technology. 2nd ed. Pergamon; 2003. p. 1-12. DOI: 10.1016/B0-08-043152-6/01888-X

[16] Tabereaux AT, Peterson RD. Chapter 2.5 - Aluminum Production. In: Seetharaman S, editor. Treatise on Process Metallurgy, Vol. 3. Elsevier; 2014. P. 839-917. DOI: 10.1016/B978-0-08-096988-6.00023-7

[17] Panigrahi J, Sharma SC. Chapter 12 - Simulation, Control, and Optimization of Water Systems in Industrial Plants. In: Ranade VV, Bhandari VM, editors. Industrial Wastewater Treatment, Recycling and Reuse. Butterworth-Heinemann; 2014. p. 463-487. DOI: 10.1016/B978-0-08-099968-5.00012-X

[18] Haupin W. Aluminum Production and Refining. In: Buschow KHJ, Cahn RW, Flemings MC, Ilschner B, Kramer EJ, Mahajan S, Veyssière P, editors. Encyclopedia of Materials: Science and Technology. 2nd ed. Pergamon; 2003. p. 132-141. DOI: 10.1016/B0-08-043152-6/00029-2

[19] Bray EL. Aluminum in January 2021. U.S. Geological survey [Internet]. Available from: https://pubs.usgs.gov/periodicals/mcs2021/mcs2021-aluminum.pdf [Accessed: 2021-02-10]

[20] Bray EL. Bauxite and Alumina in January 2018. U.S. Geological survey [Internet]. Available from: https://s3-us-west-2.amazonaws.com/prd-wret/assets/palladium/production/mineral-pubs/bauxite/mcs-2018-bauxi.pdf [Accessed: 2021-02-10]

[21] Bray EL. Bauxite and Alumina in January 2020. U.S. Geological survey [Internet]. Available from: https://pubs.usgs.gov/periodicals/mcs2020/mcs2020-bauxite-alumina.pdf [Accessed: 2021-02-10]

[22] Bray EL. Bauxite and Alumina – 2017 [advance release]. U.S. Geological survey [Internet]. Available from: https://prd-wret.s3.us-west-2.amazonaws.com/assets/palladium/production/atoms/files/myb1-2017-bauxi.pdf [Accessed: 2021-02-10]

[23] Bray EL. Bauxite and Alumina in January 2021. U.S. Geological survey [Internet]. Available from: https://pubs.usgs.gov/periodicals/mcs2021/mcs2021-bauxite-alumina.pdf [Accessed: 2021-02-10]

[24] Types of Aluminum [Internet]. Available from: https://www.thomasnet.com/articles/metals-metal-products/types-of-aluminum/ [Accessed: 2021-02-10]

[25] Hirsch J. 23 - Aluminium sheet fabrication and processing. In: Lumley R, editor. Woodhead Publishing Series in Metals and Surface Engineering, Fundamentals of Aluminium Metallurgy. Woodhead Publishing; 2011. p. 719-746. DOI: 10.1533/9780857090256.3.719

[26] The Aluminum Association. International Alloy Designations and Chemical Composition Limits for Wrought Aluminum and Wrought Aluminum Alloys [Internet]. 2015. Available from: https://www.aluminum.org/sites/default/files/Teal%20Sheets.pdf [Accessed: 2021-02-10]

[27] ASTM B209M – 14. Standard Specification for Aluminum and Aluminum-Alloy Sheet and Plate (Metric). [Internet]. 2014. Available from: https://www.astm.org/Standards/B209M.htm [Accessed: 2021-02-10]

[28] ISO 3522:2007. Aluminium and aluminium alloys - Castings - Chemical composition and mechanical properties [Internet]. 2007. Available from: https://www.iso.org/committee/50152/x/catalogue/ [Accessed: 2021-02-10]

[29] ISO 6361-5:2011. Wrought aluminium and aluminium alloys — Sheets, strips and plates — Part 5: Chemical composition [Internet]. 2015. Available from: https://www.iso.org/obp/ui/#iso:std:iso:6361:-5:ed-1:v1:en [Accessed: 2021-02-10]

[30] Xometry Europe. Aluminium Alloys: Cross-Reference Table of Designation Standards [Internet]. 2019. Available from: https://xometry.de/en/aluminium-alloys-cross-reference-table-of-designation-standards/ [Accessed: 2021-02-10]

[31] Blaz L, Sugamata M, Kaneko J, Sobota J, Wloch G, Bochniak W, Kula A. Structure and properties of 6061+26 mass% Si aluminium alloy produced via coupled rapid solidification and KOBO-extrusion powder. Journal of Materials Processing Technology. 2009;209:

4329-4336. DOI: 10.1016/j.jmatprotec.2008.11.012

[32] Esawi AMK, Borady MAE. Carbon nanotube-reinforced aluminium strips. Composites Science and Technology. 2008;68:486-492. DOI: 10.1016/j.compscitech.2007.06.030

[33] Fogagnolo JB, Ruiz-Navas EM, Robert MH, Torralba JM. The effects of mechanical alloying on the compressibility of aluminium matrix composite powder. Materials Science and Engineering A. 2003;355:50-55. DOI: 10.1016/S0921-5093(03)00057-1

[34] Fogagnolo JB, Velasco F, Robert MH, Torralba JM. Effect of mechanical alloying on the morphology, microstructure and properties of aluminium matrix composite powders. Materials Science and Engineering A 2003;342: 131-143. DOI: 10.1016/S0921-5093(02)00246-0

[35] Gómez de Salazar JM, Barrena MI. The influence of Si and Mg rich phases on the mechanical properties of 6061 Al-matrix composites reinforced with Al_2O_3. Journal of Materials Science. 2002;37:1497-1502. DOI: 10.1023/A:1014967324577

[36] Mao J, Imai T, Dong S, Saito N, Shigematsu I, Kojima S, Ishikawa T. Superplasticity of nitride mullite short fiber-reinforced 6061 aluminium composites. Scripta Materialia. 2003;49: 1061-1066. DOI: 10.1016/j.scriptamat.2003.08.017

[37] Ruiz-Navas EM, Fogagnolo JB, Velasco F, Ruiz-Prieto JM, Froyen L. One step production of aluminium matrix composite powders by mechanical alloying. Composites Part A. 2006;37:2114-2120. DOI: 10.1016/j.compositesa.2005.11.016

[38] Sivasankaran S, Sivaprasad K, Narayanasamy R, Iyer VK. Effect of strengthening mechanisms on cold workability and instantaneous strain hardening behavior during grain refinement of AA 6061-10 wt.% TiO_2 composite prepared by mechanical alloying. Journal of Alloys and Compounds. 2010;507:236-244. DOI: 10.1016/j.jallcom.2010.07.168

[39] Sivasankaran S, Sivaprasad K, Narayanasamy R, Iyer VK. Synthesis, structure and sinterability of 6061 AA100−x–x wt.% TiO_2 composites prepared by high-energy ball milling. Journal of Alloys and Compounds. 2010; 491:712-721. DOI: 10.1016/j.jallcom.2009.11.051

[40] Son HT, Kim TS, Suryanarayana C, Chun BS. Homogeneous dispersion of graphite in a 6061 aluminum alloy by ball milling. Materials Science and Engineering A. 2003;348:163-169. DOI: 10.1016/S0921-5093(02)00749-9

[41] Wang L, Choi H, Myoung J-M, Lee W. Mechanical alloying of multi-walled carbon nanotubes and aluminium powders for the preparation of carbon/metal composites. Carbon. 2009;47:3427-3433. DOI: 10.1016/j.carbon.2009.08.007

[42] Wang RM, Surappa MK, Tao CH, Li CZ, Yan MG. Microstructure and interface structure studies of SiCp-reinforced Al (6061) metal-matrix composites. Materials Science and Engineering A. 1998;254:219-226. DOI: 10.1016/S0921-5093(98)00686-8

[43] Wu Y, Kim G-Y, Russell AM. Effects of mechanical alloying on an Al6061-CNT composites fabricated by semi-solid powder processing. Materials Science and Engineering A. 2012;538: 164-172. DOI: 10.1016/j.msea.2012.01.025

[44] Wu Y, Kim G-Y, Russell AM. Mechanical alloying of carbon nanotube and Al6061 powder for metal matrix composites. Materials Science and

Engineering A. 2012;532:558-566. DOI: 10.1016/j.msea.2011.10.121

[45] Xia X, McQueen HJ. Deformation behaviour and microstructure of a 20% Al_2O_3 reinforced 6061 Al composite. Applied Composite Materials. 1997; 4:333-347. DOI: 10.1023/A: 1008862719958

[46] Adamiak M. The structure and selected properties of the EN AW6061 aluminium alloy base composites reinforced with Ti_3Al intermetallic particles. Composites. 2006;6:70-75. (in Polish)

[47] Adamiak M. Application of mechanical milling to composites materials production. Archives of Foundry. 2006;21:361-367. (in Polish)

[48] Arik H, Ozcatalbas Y, Turker M. Dry sliding wear behavior of in situ Al-Al_4C_3 metal matrix composite produced by mechanical alloying technique. Materials and Design. 2006;27:799-804. DOI: 10.1016/j.matdes.2005.01.024

[49] Bonollo F, Ceschini L, Garagnani GL. Mechanical and impact behaviour of $(Al_2O_3)p$/2014 and $(Al_2O_3)p$/6061 Al metal matrix composites in the 25-200°C range. Applied Composite Materials. 1997;4: 173-185. DOI: 10.1007/BF02481779

[50] Cui Y, Geng L, Yao CK. Interfacial bonding mechanisms and mechanical properties of squeeze-cast 6061 Al matrix composite reinforced with self-propagating high-temperature synthesized SiC particulates. Journal of Materials Science Letters. 1997;16: 788-790. DOI: 10.1023/A:1018514021685

[51] Dobrzański LA, Tomiczek B, Adamiak M. Manufacturing of EN AW6061 matrix composites reinforced by halloysite nanotubes. Journal of Achievements in Materials and Manufacturing Engineering. 2011;49: 82-89.

[52] Fogagnolo JB, Pallone EMJA, Martin DR, Kiminami CS, Bolfarini C, Botta WJ. Processing of Al matrix composites reinforced with Al-Ni compounds and Al_2O_3 by reactive milling and reactive sintering. Journal of Alloys and Compounds. 2009;471: 448-452. DOI: 10.1016/j. jallcom.2008.03.125

[53] Fogagnolo JB, Robert MH, Ruiz-Navas EM, Torralba JM. 6061 Al reinforced with zirconium dibromide particles processed by conventional powder metallurgy and mechanical alloying. Journal of Materials Science. 2004;39:127-132. DOI: 10.1023/B: JMSC.0000007736.03608.e5

[54] Fogagnolo JB, Robert MH, Torralba JM. Mechanically alloyed AlN particle- reinforced Al-6061 matrix composites: Powder processing, consolidation and mechanical strength and hardness of the as-extruded materials. Materials Science and Engineering A. 2006; 426:85-94. DOI: 10.1016/j. msea.2006.03.074

[55] Fogagnolo JB, Ruiz-Navas EM, Robert MH, Torralba JM. 6061 Al reinforced with silicon nitride particles processed by mechanical milling. Scripta Materialia. 2002;47: 243-248. DOI: 10.1016/S1359-6462(02) 00133-1

[56] Adamczyk J. Theoretical metallography. Vol. 2: Plastic deformation, strengthening and cracking. Gliwice: Silesian University of Technology Publishing House; 2012. (in Polish)

[57] Lin YC, Li HC, Liou SS, Shie MT. Mechanism of plastic deformation of powder metallurgy metal matrix composites of Cu-Sn/SiC and 6061/SiC under compressive stress. Materials Science and Engineering A. 2004;373: 363-369. DOI: 10.1016/j. msea.2004.02.011

[58] Parvin N, Assadifard R, Safarzadeh P, Sheibani S, Marashi P. Preparation and mechanical properties of SiC-reinforced Al6061 composite by mechanical alloying. Materials Science and Engineering A. 2008;492:134-140. DOI: 10.1016/j.msea.2008.05.004

[59] Qin S, Liu C, Chen J, Zhang G, Wang W. Characterization of thermal residual strain in SiCp/6061 Al composites by ultramicrohardness. Journal of Materials Science Letters. 1999;18:1099-1100. DOI: 10.1023/A:1006616109746

[60] Shakeri HR, Wang Z. Effect of alternative aging process on the fracture and interfacial properties of particulate Al_2O_3-reinforced Al (6061) metal matrix composite. Metallurgical and Materials Transactions A. 2006;33: 1699-1713. DOI: 10.1007/s11661-002-0179-1

[61] Tomiczek B, Dobrzański LA. Composite materials with an aluminum alloy matrix EN AW6061 reinforced with nanotubes obtained from halloysite. In: Pacyna J, editor. Works of the XXXIX School of Materials Science and Engineering. Monograph. Kraków– Krynica; 2011. p. 342-346. (in Polish)

[62] Zhao N, Nash P, Yang X. The effect of mechanical alloying on SiC distribution and the properties of 6061 aluminum composite. Journal of Materials Processing Technology. 2005; 170:586-592. DOI: 10.1016/j.jmatprotec.2005.06.037

[63] Dobrzański LA, Włodarczyk A, Adamiak M. Structure, properties and corrosion resistance of PM composite materials based on EN AW-2124 aluminum alloy reinforced with the Al_2O_3 ceramic particles. Journal of Materials Processing Technology. 2005; 162:27-32. DOI: 10.1016/j.jmatprotec.2005.02.006

[64] Lu L, Lai MO, Ng CW. Enhanced mechanical properties of an Al based metal matrix composite prepared using mechanical alloying. Materials Science and Engineering A. 1998;252:203-211. DOI: 10.1016/S0921-5093(98)00676-5

[65] Włodarczyk-Fligier A, Dobrzański LA, Kremzer M, Adamiak M. Manufacturing of aluminium matrix composite materials reinforced by Al_2O_3 particles. Journal of Achievements in Materials and Manufacturing Engineering. 2008;27: 99-102.

[66] Dobrzański LA, project manager. Research Project Report UMO-2013/08/M/ST8/00818. Investigations of structure and properties of newly created porous biomimetic materials fabricated by selective laser sintering, BIOLASIN. Silesian University of Technology; 2013-2016. (in Polish)

[67] Dobrzański LA, project manager. Research Project Report UMO-2012/07/B/ST8/04070. Determining the importance of the effect of the one-dimensional nanostructural materials on the structure and properties of newly developed functional nanocomposite and nanoporous materials, NANOCOPOR. Silesian University of Technology; 2013-2016. (in Polish)

[68] Dobrzański LA, project manager. Research Project Report UMO-2011/03/B/ST08/06076. Synergism of reinforcing with mineral halloysite nanotubes and as a result of the heat treatment of newly developer composite materials with the matrix made of aluminium alloys. Silesian University of Technology; 2012-2015. (in Polish)

[69] Dobrzański LA, contractor. Research Project Report UMO-2011/01/B/ST8/06663. Improving the functional properties of elements of light casting aluminium alloys heat treated by laser remelting and/or alloying the surface with carbides and/or ceramic particles.

Silesian University of Technology; 2011-2014. (in Polish)

[70] Dobrzański LA, project manager. Research Project Report UDA-POKL.04.01.01-00-003/09-00. Opening and development of engineering and PhD studies in the field of nanotechnology and materials science, INFONANO. Silesian University of Technology; 2009-2014. (in Polish)

[71] Dobrzański LA, project manager. Research Project Report UDA-POIG.01.01.01-00.23/08-00. Foresight of surface properties formation leading technologies of engineering materials and biomaterials, FORSURF. Silesian University of Technology; 2009-2012. (in Polish)

[72] Dobrzański LA, project manager. Research Project Report R15 007 02; development project. Improving the functional properties of elements of light casting magnesium and aluminium alloys heat treated by optimizing their chemical composition and by laser remelting and/or alloying the surface with carbides and/or ceramic particles. Silesian University of Technology; 2007-2009. (in Polish)

[73] Dobrzański LA, project manager. Research Project Report KBN Nr 3 T08A 041 30. Shaping the properties of composite Al-Al_2O_3 materials by changing the porosity, wettability and shape of the Al_2O_3 phase and infiltration conditions. Silesian University of Technology; 2006-2007. (in Polish)

[74] Dobrzański LA, project manager. Research Project Report KBN Nr 3 T08A 022 27. Development of a methodology for automated evaluation of the quality and structural defects in Al and Mg alloys with the use of artificial intelligence tools. Silesian University of Technology; 2004-2006. (in Polish)

[75] Kremzer M. Structure and properties of EN AC-AlSi12 matrix composite materials produced by pressure infiltration [PhD thesis]. Dobrzański LA, supervisory guidance. Gliwice: Silesian University of Technology; 2008. (in Polish)

[76] Tomiczek B. Structure and properties of nanostructured composite materials reinforced with halloysite [PhD thesis]. Dobrzański LA, supervisory guidance. Gliwice: Silesian University of Technology; 2013. (in Polish)

[77] Pakieła W. Influence of laser fusion of silicon and tungsten carbides on the structure and properties of the surface layer of Al-Mg alloys [PhD thesis]. Dobrzański LA, supervisory guidance. Gliwice: Silesian University of Technology; 2015. (in Polish)

[78] Kujawa M. AlSi12 infiltrated composite materials reinforced with sintered halloysite nanotubes [PhD thesis]. Dobrzański LA, supervisory guidance. Gliwice: Silesian University of Technology; 2015. (in Polish)

[79] Macek M. Structure and properties of aluminum nanocomposite materials reinforced with carbon nanotubes [PhD thesis]. Dobrzański LA, supervisory guidance. Gliwice: Silesian University of Technology; 2016. (in Polish)

[80] Achtelik-Franczak A. Engineering composite materials with the reinforcement of microporous titanium selectively sintered by laser [PhD thesis]. Dobrzański LA, supervisory guidance. Gliwice: Silesian University of Technology; 2016. (in Polish)

[81] Dobrzański LA, Kremzer M. Manufacture of composite materials by the infiltration of ceramic skeletons sintered with aluminum alloy. In: Proceedings of the IV Polish-Ukrainian Scientific and Technical Conference - Mechanics and Informatics;

Khmelnitskiy, Ukraine, 2006. p. 141-145. (in Polish)

[82] Dobrzański LA, Kremzer M, Nagel A. Application of pressure infiltration to the manufacturing of aluminium matrix composite materials with different reinforcement shape. Journal of Achievements in Materials and Manufacturing Engineering. 2007; 24:183-186.

[83] Dobrzański LA, Kremzer M, Gołombek K, Nagel A. Aluminium matrix composites fabricated by pressure infiltration process. Materials Engineering. 2007;28:520-523.

[84] Dobrzański LA, Kremzer M, Nowak AJ, Nagel A. Composite materials based on porous ceramic preforms infiltrated by aluminium alloy. Journal of Achievements in Materials and Manufacturing Engineering. 2007; 20:95-98.

[85] Dobrzański LA, Kremzer M, Drak M. Modern composite materials manufactured by pressure infiltration method. Journal of Achievements in Materials and Manufacturing Engineering. 2008;30:121-128.

[86] Dobrzański LA, Kremzer M, Gołombek K. Structure and Properties of Aluminum Matrix Composites Reinforced by Al_2O_3 Particles. Materials Science Forum. 2008;591-593:188-192. DOI: 10.4028/www.scientific.net/MSF.591-593.188

[87] Dobrzański LA, Kremzer M, Nowak AJ, Nagel A. Aluminium matrix composites fabricated by infiltration method. Archives of Material Science and Engineering. 2009;36:5-11.

[88] Dobrzański LA, Kremzer M, Nagel A. Structure and properties of ceramic preforms based on Al_2O_3 particles. Journal of Achievements in Materials and Manufacturing Engineering. 2009;35:7-13.

[89] Tomiczek B, Dobrzański LA. Structure and selected properties of composite materials reinforced with halloysite nanotubes. In: Pacyna J, editor. Works of the XL School of Materials Science and Engineering. Monograph. Cracow; 2012. p. 293-298 (in Polish)

[90] Dobrzański LA, Tomiczek B, Adamiak M, Gołombek K. Mechanically milled aluminium matrix composites reinforced with halloysite nanotubes. Journal of Achievements in Materials and Manufacturing Engineering. 2012; 55:654-660.

[91] Labisz K, Tański T, Dobrzański LA, Janicki D, Korcina K. HPDL laser alloying of Al-Si-Cu alloy with Al_2O_3 powder. Archives of Materials Science and Engineering. 2013;63:36-45.

[92] Dobrzański LA, Achtelik-Franczak A, Król M. Computer aided design in Selective Laser Sintering (SLS) – application in medicine. Journal of Achievements in Materials and Manufacturing Engineering 2013;60: 66-75.

[93] Dobrzański LA, Tomiczek B, Pawlyta M. Król M. Aluminium AlMg1SiCu matrix composite materials reinforced with halloysite particles. Archives of Metallurgy and Materials. 2014;59:335-338. DOI: 10.2478/amm-2014-0055

[94] Dobrzański LA, Tomiczek B, Matula G, Gołombek K. Role of halloysite nanoparticles and milling time on the synthesis of AA 6061 aluminium matrix composites. Advanced Materials Research. 2014;939:84-89. DOI: 10.4028/www.scientific.net/AMR.939.84

[95] Dobrzański LA, Macek M, Tomiczek B. Effect of carbon nanotubes content on morphology and properties of AlMg1SiCu matrix composite

powders. Archives of Materials Science and Engineering. 2014;69:12-18.

[96] Tomiczek B, Pawlyta M, Adamiak M, Dobrzański LA. Effect of milling time on microstructure of AA6061 composites fabricated via mechanical alloying. Archives of Metallurgy and Materials 2015; 60:789-793. DOI: 10.1515/amm-2015-0208

[97] Tomiczek B, Kujawa M, Matula G, Kremzer M, Tański T, Dobrzański LA. Aluminium AlSi12 alloy matrix composites reinforced by mullite porous preforms. Materialwissenschaft und Werkstofftechnik. 2015;46:368-376. DOI: 10.1002/mawe.201500411

[98] Dobrzański LA, Reimann Ł, Kujawa M, Tomiczek B. Corrosion behaviour of aluminium matrix composites reinforced with sintered halloysite nanotubes. Journal of Achievements in Materials and Manufacturing Engineering. 2015;73: 92-99.

[99] Tomiczek AE, Dobrzański LA, Macek M. Effect of milling time on microstructure and properties of AA6061/MWCNTS composite powders. Archives of Metallurgy and Materials. 2015;60:3029-3034. DOI: 10.1515/amm-2015-0484

[100] Dobrzański LA, Macek M, Tomiczek B, Pakieła W, Kremzer M. Influence of the milling time and MWCNT content on the wear properties of the AlMg1SiCu/MWCNT nanocomposites. Archives of Materials Science and Engineering. 2015;74:77-84.

[101] Dobrzański LA, Macek M, Tomiczek B, Nuckowski PM, Nowak AJ. The influence of the dispersion method on the microstructure and properties of MWCNTs/AA6061 composites. Archives of Metallurgy and Materials. 2016;61:1229-1234. DOI: 10.1515/amm-2016-0203

[102] Kujawa M, Dobrzański LA, Matula G, Kremzer M, Tomiczek B. Manufacturing of porous ceramic preforms based on halloysite nanotubes (HNTS). Archives of Metallurgy and Materials. 2016;61:917-922. DOI: 10.1515/amm-2016-0155

[103] Pawlyta M, Tomiczek B, Dobrzański LA, Kujawa M, Bierska-Piech B. Transmission electron microscopy observations on phase transformations during aluminium/mullite composites formation by gas pressure infiltration. Materials Characterization. 2016;114:9-17. DOI: 10.1016/j.matchar.2016.02.003

[104] Labisz K, Tański T, Janicki D, Borek W, Lukaszkowicz K, Dobrzański L.A. Effect of laser feeding on heat treated aluminium alloy surface properties. Archives of Metallurgy and Materials. 2016;61:741-746. DOI: 10.1515/amm-2016-0126

[105] Kujawa M, Suwak R, Dobrzański LA, Gerle A, Tomiczek B. Thermal characterization of halloysite materials for porous ceramic preforms. Archives of Materials Science and Engineering. 2021.

[106] Dobrzański LA, Kremzer M, Dziekońska M. Al_2O_3 preforms infiltrated by liquid aluminium alloy by deposition Ni-P coating. Archives of Materials Science and Engineering. 2012;55:14-21.

[107] Kujawa M, Kremzer M, Dobrzański LA. Manufacturing of aluminium matrix composite with gas pressure infiltration method. Proceedings of the 19th International Seminar of Ph.D. Students SEMDOK. 29-31.01.2014; Terchova, Slovakia; 2014. p. 38-41.

[108] Kujawa M, Dobrzański LA, Kremzer M. Production of porous ceramic skeletons based on mineral nanotubes obtained from halloysite.

Proceedings of the XLI School of Materials Science and Engineering. 24-27.09.2013; Cracow-Krynica; 2013. p. 319-325. (in Polish)

[109] Dobrzański LA, Tomiczek B, Pawlyta M, Nuckowski P. TEM and XRD study of nanostructured composite materials reinforced with the halloysite particles. Materials Science Forum. 2014;783-786:1591-1596. DOI: 10.4028/www.scientific.net/MSF.783-786.1591

[110] Dobrzański LA, Tomiczek B, Pawlyta M, Gołombek K. EN AW-AlMg1SiCu alloy matrix composite materials reinforced with halloysite particles manufactured by mechanical milling. Materials Engineering. 2014; 198:102-105.

[111] Tomiczek B, Dobrzański LA. Composite materials based on EN AW-AlMg1SiCu aluminium alloy reinforced with halloysite particles. Journal of Achievements in Materials and Manufacturing Engineering. 2013;61: 39-46.

[112] Dobrzański LA, Kremzer M, Adamiak M. The influence of reinforcement shape on wear behaviour of aluminium matrix composite materials. Journal of Achievements in Materials and Manufacturing Engineering. 2010;42:26-32.

[113] Bolewski A. Detailed mineralogy. Warsaw: Geological Publishing House; 1975. (in Polish)

[114] Stoch L. Clay minerals. Warsaw: Geological Publishing House; 1974. (in Polish)

[115] Szymański A. Technical Mineralogy. Warsaw: PWN; 1997. (in Polish)

[116] Das S, Kordijazi A, Akbarzadeh O, Rohatgi PK. An innovative process for dispersion of graphene nanoparticles and nickel spheres in A356 alloy using pressure infiltration technique. Engineering Reports. 2020;2:e12110. DOI: 10.1002/eng2.12110

[117] Shao P, Chen G, Ju B, Yang W, Zhang Q, Wang Z, Tan X, Pei Y, Zhong S, Hussain M, Wu G. Effect of hot extrusion temperature on graphene nanoplatelets reinforced Al6061 composite fabricated by pressure infiltration method. Carbon. 2020;162: 455-464. DOI: 10.1016/j.carbon.2020.02.080

[118] Zhang Q, Dong S, Ma S, Hou X, Yang W, Zhang Y, Wu G. Microstructure and compressive behavior of lamellar Al_2O_3p/Al composite prepared by freeze-drying and mechanical-pressure infiltration method. Science and Engineering of Composite Materials. 2020;27:1-9. DOI: 10.1515/secm-2020-0001

[119] Garcia-Cordovilla C, Louis E, Narciso J. Pressure infiltration of packed ceramic particulates by liquid metals. Acta Materialia 1999;47:4461-4479. DOI: 10.1016/S1359-6454(99)00318-3

[120] Daoud A. Microstructure and tensile properties of 2014 Al alloy reinforced with continuous carbon fibers manufactured by gas pressure infiltration. Materials Science and Engineering A. 2005;391:114-120. DOI: 10.1016/j.msea.2004.08.075

[121] Cook AJ, Werner PS. Pressure infiltration casting of metal matrix composites. Materials Science and Engineering A. 1991;144:189-206. DOI: 10.1016/0921-5093(91)90225-C

[122] Albiter A, Contreras A, Salazar M, Gonzalez-Rodriguez JG. Corrosion behaviour of aluminium metal matrix composites reinforced with TiC processed by pressureless infiltration. Journal of Applied Electrochemistry. 2006;36:303-308. DOI: 10.1007/s10800-005-9073-z

[123] Alonso A, Pamies A, Narciso J, Garcia-Cordovilla C, Louis E. Evaluation of the wettability of liquid aluminum with ceramic particulates (SiC, TiC, Al_2O_3) by means pressure infiltration. Metallurgical Transactions A. 1993;24:1423-1432. DOI: 10.1007/BF02668210

[124] Altinkok N, Demir A, Ozsert I, Processing of Al_2O_3/SiC ceramic cake preforms and their liquid metal infiltration. Composites A. 2003;34: 577-582. DOI: 10.1016/S1359-835X(03) 00125-8

[125] Canadan E, Ahlatci H, Cimenoglu H. Abrasive wear behaviour of Al-SiC composites produced by pressure infiltration technique. Wear. 2001;247:133-138. DOI: 10.1016/ S0043-1648(00)00499-3

[126] Cardinal S, R'Mili M, Merle P. Improvement of high pressure infiltration behavior of alumina platelet preforms: manufacture and characterization of hybrid performs. Composites A. 1998;29:1433-1441. DOI: 10.1016/S1359-835X(98)00026-8

[127] Carreno-Morelli E, Cutart T, Schaller R, Bonjour C. Processing and characterization of aluminium-based MMCs produced by gas pressure infiltration. Materials Science and Engineering A. 1998;251:48-57. DOI: 10.1016/S0921-5093(98)00649-2

[128] Chung WS, Lin SJ. Ni-Coated SiC_p reinforced aluminum composites processed by vacuum infiltration. Materials Research Bulletin. 1996;31: 1437-1447. DOI: 10.1016/S0025-5408 (96)00150-X

[129] Cornie JA. Advanced pressure infiltration casting technology produces near-absolute net-shape metal matrix composite components cost competitively. Materials Technology. 1995;10:43-48. DOI: 10.1080/ 10667857.1995.11752588

[130] Eardley ES, Flower HM. Infiltration and solidification of commercial purity aluminium matrix composites. Materials Science and Engineering A. 2003;359:303-312. DOI: 10.1016/S0921-5093(03)00357-5

[131] Assar A-EM. Fabrication of metal matrix composite by infiltration process – part 2: experimental study. Journal of Materials Processing Technology. 1999;86:152-158. DOI: 10.1016/S0924-0136(98)00304-5

[132] Garbellini O, Morando C, Biloni H, Papacio H. Infiltration of Saffil alumina fiber with AlCu and AlSi alloys. Scripta Materialia. 1999;41:187-193. DOI: 10.1016/S1359-6462(99)00141-4

[133] Luo ZP, Song YG, Zhang SQ. A TEM study of the microstructure of SiCp/Al composite prepared by pressureless infiltration method. Scripta Materialia. 2001;45:1183-1189. DOI: 10.1016/S1359-6462(01)01148-4

[134] McCullough C, Deve HE, Chanel TE. Mechanical response of continuous fiber-reinforced Al_2O_3-Al composites produced by pressure infiltration casting. Materials Science and Engineering A. 1994;189:147-154. DOI: 10.1016/0921-5093(94)90410-3

[135] Molina JM, Saravanan RA, Apron R, Garcia-Cordovilla C, Louis E, Narciso J. Pressure infiltration of liquid aluminium into packed SiC particulate with a bimodal size distribution. Acta Materialia. 2002;50:247-257. DOI: 10.1016/S1359-6454(01)00348-2

[136] Peng LM, Cao JW, Noda K, Han KS. Mechanical properties of ceramic-metal composites by pressure infiltration of metal into porous ceramics. Materials Science and Engineering A. 2004;374:1-9. DOI: 10.1016/j.msea.2003.12.027

[137] Srinivasa Rao B, Jayram V. Pressureless infiltration of Al-Mg based

alloys into Al_2O_3 preforms: mechanism and phenomenology. Acta Materialia. 2001;49:2373-2385. DOI: 10.1016/S1359-6454(01)00158-6

[138] Wannasin J, Flemings MC. Fabrication of metal matrix composites by high-pressure centrifugal infiltration process. Journal of Materials Processing Technology. 2005;169:143-149. DOI: 10.1016/j.jmatprotec.2005.03.004

[139] Zitnansky M, Martinovic M. Infiltration of the metal through a bundle of continual fibres. Journal of Materials Processing Technology. 2000; 106:163-166. DOI: 10.1016/S0924-0136(00)00609-9

[140] Maj J, Basista M, Węglewski W, Bochenek K, Strojny-Nędza A, Naplocha K, Panzner T, Tatarková M, Fiori F. Effect of microstructure on mechanical properties and residual stresses in interpenetrating aluminum-alumina composites fabricated by squeeze casting. Materials Science and Engineering A. 2018;715:154-162. DOI: 10.1016/j.msea.2017.12.091

[141] Dobrzański LA, Kremzer M, Nagel A, Huchler B. Fabrication of ceramic preforms based on Al_2O_3 CL 2500 powder. Journal of Achievements in Materials and Manufacturing Engineering. 2006;18:71-74.

[142] Dobrzański LA, Kremzer M, Nagel A, Huchler B. Structure and properties of porous preforms manufactured on the base of Al_2O_3 powder. Archives of Foundry. 2006;21: 149-154. (in Polish)

[143] Kremzer M, Dobrzański LA. Manufacture of porous ceramic frames based on Al_2O_3 powder. Works of the XXXIV School of Materials Science and Engineering. Kraków-Krynica;2006. p. 241-244. (in Polish)

[144] Mattern A, Huchler B, Stadenecker D, Oberacker R, Nagel A, Hoffmann MJ. Preparation of interpenetrating ceramic-metal composites. Journal of European Ceramic Society. 2004;24:3399-3408. DOI: 10.1016/j.jeurceramsoc.2003.10.030

[145] Naplocha K, Janus A, Kaczmar JW, Samsonowicz Z. Technology and mechanical properties of ceramic preforms for composite materials. Journal of Materials Processing Technology. 2000;106:119-122. DOI: 10.1016/S0924-0136(00)00601-4

[146] Dobrzański LA, Kremzer M, Adamiak M. Structure and tribological properties of aluminium matrix composites fabricated by pressure infiltration method. In: Proceedings of the 16th International Federation for Heat Treatment and Surface Engineering Congress; 2007.

[147] Dobrzański LA, Kremzer M, Gołombek K. Structure and properties of aluminum matrix composites reinforced by Al_2O_3 particles. In: Proceedings of the 6th International Latin American Conference on Powder Technology; 2007.

[148] Dobrzański LA, Kremzer M, Nagel A. Aluminium EN AC – AlSi12 alloy matrix composite materials reinforced by Al_2O_3 porous performs. Archives of Material Science and Engineering. 2007;28:593-596.

[149] Dobrzański LA, Kremzer M, Nagel A, Huchler B. Composite materials based on the porous Al_2O_3 ceramics infiltrated with the EN AC – AlSi12 alloy. In: Proceedings of the 3rd Scientific Conference MMME; Gliwice-Wisła; 2005. (CD-ROM).

[150] Dobrzański LA, Kremzer M, Nagel A, Huchler B. Composite materials based on Al_2O_3 porous ceramics infiltrated with EN AC-AlSi12 alloy, Composites. 2005;5:35-41.

[151] Dobrzański LA, Kremzer M, Nagel A. Influence of reinforcements shape on the structure and properties of Al-Al_2O_3 composite materials. In: Proceedings of the 11th International Research/Expert Conference Trends in the Development of Machinery and Associated Technology; 2007. p. 1479-1482.

[152] Dobrzański LA, Kremzer M. Structure and properties of EN AC-AlSi12 alloy reinforced by ceramic fibres and particles. In: Proceedings of PM 2006 World Congress and Exhibition; Busan, Korea; 2006. (CD-ROM)

[153] Dobrzański LA, Kremzer M. Manufacture of composite materials by the infiltration of ceramic skeletons sintered with aluminum alloy. Trailer of the Khmelnytsky National University. 2006;2:141-144. (in Polish)

[154] Dobrzański LA, Krupiński M, Konieczny J. The structure and properties of composite materials with an aluminum alloy matrix reinforced with SiC particles. In: Proceedings of the 12th International Scientific Conference AMME; Gliwice-Zakopane; 2003. p. 229-232. (in Polish)

[155] Morgiel J, Naplocha K, Pomorska M, Kaczmar J. Microstructure and tensile properties of composites with high strength aluminium alloys matrix reinforced with Saffil™ fibers. Composites. 2011;11:136-141.

[156] Zhang H, Maljkovic N, Mitchell BS. Structure and interfacial properties of nanocrystalline aluminum/mulite composites. Materials Science and Engineering A. 2002;326: 317-323. DOI: 10.1016/S0921-5093(01) 01500-3

[157] Zhiping L, Yinggang S, Shaoqing Z, Miller DJ. Interfacial microstructure in a B_4C/Al composite fabricated by pressureless infiltration. Metallurgical and Materials Transactions A. 2012;43: 281-293. DOI: 10.1007/s11661-011-0817-6

[158] Shang-Nan C, Jow-Lay H, Ding-Fwu L, Horng-Hwa L. The mechanical properties and microstructure of Al_2O_3/aluminum alloy composites fabricated by squeeze casting. Journal of Alloys and Compounds. 2007;436:124-130. DOI: 10.1016/j.jallcom.2006.07.062

[159] Loehman RE, Ewsuk KG, Fahrenholtz WG, Lakshman BB. Ceramic-metal composite formation by reactive metal penetration. Key Engineering Materials. 1996;127-131: 431-438. DOI: 10.4028/www.scientific.net/KEM.127-131.431

[160] Loehman RE, Ewsuk KG, Tomsia AP. Synthesis of Al_2O_3-Al composites by reactive metal penetration. Journal of American Society. 1996;79:27-32. DOI: 10.1111/j.1151-2916.1996.tb07876.x

[161] Naga SM, El-Maghraby A, El-Rafei AM. Properties of Ceramic - Metal Composites Formed by Reactive Metal Penetration. American Ceramic Society Bulletin. 2007;86:9301-9306.

[162] Sobczak J, Sobczak N. Squeeze Cast Aluminum and Magnesium Matrix Composites Reinforced with Short Alumina Fibers - Structure and Chemistry Characterization. Composites 2001;1:2155-159.

[163] Potoczek M, Śliwa RE. Microstructure and physical properties of AlMg/Al_2O_3 interpenetrating composites fabricated by metal infiltration into ceramic foams. Archives of Metallurgy and Materials. 2011;56: 1265-1269.

[164] Valdez S, Campillo B, Pérez R, Martínez L, García H A. Synthesis and microstructural characterization of Al-Mg alloy - SiC particle composite. Materials Letters. 2008;62:2623-2625. DOI: 10.1016/j.matlet.2008.01.002

[165] Freitas M, Pianaro SA, Nadal FN, Tebcherani SM, Berg EAT. Preparation and characterization of SiC/kaolin/Al composite materials by squeeze-casting. Cerâmica. 2009;55:271-280. DOI: 10.1590/S0366-69132009000300006

[166] Breslin MC, Ringnalda J, Xu L, Fuller M, Seeger J, Daehn GS, Otani T, Fraser HL. Processing, microstructure, and properties of co-continuous alumina-aluminum composites. Materials Science and Engineering A. 1995;195:113-119. DOI: 10.1016/0921-5093(94)06510-1

[167] Li W, Long JP, Jing S, Shen BL, Gao SJ, Tu MJ. Aging characteristics of short mulite fiber reinforced Al-4.0Cu-1.85Mg metal matrix composite. Journal of Materials Engineering and Performance. 2003;12:19-22. DOI: 10.1361/105994903770343420

[168] Dobrzański LA, Dobrzański LB, Dobrzańska-Danikiewicz AD. Overview of conventional technologies using the powders of metals, their alloys and ceramics in Industry 4.0 stage. Journal of Achievements in Materials and Manufacturing Engineering. 2020;98:56-85. DOI: 10.5604/01.3001.0014.1481

[169] Kim BM, Cho YK, Yoon SY, Stevens R, Park HC. Mullite whiskers derived from kaolin. Ceramics International. 2009;35:579-583. DOI: 10.1016/j.ceramint.2008.01.017

[170] Aksay IA, Dabbs DM, Sarikaya M. Mullite for structural, electronic, and optical applications. Journal of American Ceramic Society. 1991;74:2343-2358. DOI: 10.1111/j.1151-2916.1991.tb06768.x

[171] Kong LB, Zhang TS, Chen YZ, Ma J, Boey F, Huang H. Microstructural composite mullite derived from oxides via a high-energy ball milling process. Ceramics International. 2004;30:1313-1317. DOI: 10.1016/j.ceramint.2003.12.022

[172] Puff Z, Sałaciński R. Possibilities of obtaining mullite material with the participation of Al_2O_3 nanopowder. Glass and Ceramics. 2008;59:12-14. (in Polish)

[173] Anggono J. Mullite ceramics: its properties, structure, and synthesis. Jurnal Teknik Mesin Universitas Kristen Petra. 2005;7:1-10. DOI: 10.9744/jtm.7.1.pp. 1-10.

[174] Esharghawi A, Penot C, Nardou F. Contribution to porous mullite synthesis from clays by adding Al and Mg powders. Journal of the European Ceramic Society. 2009;29:31-38. DOI: 10.1016/j.jeurceramsoc.2008.05.036

[175] Cao GH, Shen GJ, Liu ZG, Wu SQ. Interface study of mullite short fibers reinforced Al-12Si alloy composites. Journal of Materials Science Letters. 2001;20:501-503. DOI: 10.1023/A:1010959912780

[176] Zawrah MF, Aly MH. In situ formation of Al_2O_3-SiC-mulite from Al-matrix composites. Ceramics International. 2006;32:21-28. DOI: 10.1016/j.ceramint.2004.12.005

[177] She JH, Mechnich P, Schneider H, Kanka B, Schmücker M. Infiltration behaviors of porous mullite/mullite preforms in aluminum-chloride solutions. Journal of Materials Science Letters. 2000;19:1887-1891. DOI: 10.1023/A:1006726425129

[178] Naplocha K, Granat K. The structure and properties of hybrid preforms for composite. Journal of Achievements in Materials and Manufacturing Engineering. 2007;22:35-38.

[179] Kaczmar JW, Kurzawa A. Structure and properties of porous ceramic preforms made of α-alumina particles. Archives of Foundry Engineering. 2010;10:157-162.

[180] Yamanaka T, Choi YB, Matsugi K, Yanagisawa O, Sasaki G. Influence of preform preparation condition on infiltration of molten aluminum. Journal of Materials Processing Technology. 2007;187-188:530-532. DOI: 10.1016/j.jmatprotec.2006.11.217

[181] Navacerrada MA, Fernández P, Díaz C, Pedrero A. Thermal and acoustic properties of aluminium foams manufactured by the infiltration process. Applied Acoustics. 2013;74: 496-501. DOI: 10.1016/j.apacoust.2012.10.006

[182] Zhao YY, Tao XF. Behaviour of metal matrix syntactic foams in compression. Proceedings of the Materials Science and Technology Conference; 2009. p. 1785-1794.

[183] Potoczek M, Śliwa RE, Myalski J, Śleziona J. Metal-ceramic interpenetrating composites produced by pressure infiltration of metal into ceramic foams. Ores and Non-ferrous Metals. 2009;54:688-692.

[184] Sobczak N, Jaworska L, Podsiadło M, Smuk B, Nowak R, Kurtyka P, Twardowska A. Nitride and carbide preforms for infiltration process. Archives of Materials Science and Engineering. 2007;28:653-656.

[185] Luyten J, Mullens S, Cooymans J, De Wilde AM, Thijs I, Kemps R. Different methods to synthesize ceramic foams. Journal of the European Ceramic Society. 2009;29: 829-832. DOI: 10.1016/j.jeurceramsoc.2008.07.039

[186] Hufenbach W, Gude M, Czulak A, Śleziona J, Dolata-Grosz A, Dyzia M. Development of textile-reinforced carbon fibre aluminium composites manufactured with gas pressure infiltration methods. Journal of Achievements in Materials and manufacturing Engineering. 2009;35: 177-183.

[187] Jiang J, Wang Y. Microstructure and mechanical properties of the rheoformed cylindrical part of 7075 aluminum matrix composite reinforced with nano-sized SiC particles. Materials and Design. 2015;79: 32-41. DOI: 10.1016/j.matdes.2015.04.040

[188] Liu Q, Ke L, Liu F, Huang C, Xing L. Microstructure and mechanical property of multi-walled carbon nanotubes reinforced aluminum matrix composites fabricated by friction stir processing. Materials and Design. 2013; 45:343-348. DOI: 10.1016/j.matdes.2012.08.036

[189] Su H, Gao W, Feng Z, Lu Z. Processing, microstructure and tensile properties of nano-sized Al_2O_3 particle reinforced aluminum matrix composites. Materials and Design. 2012; 36:590-596. DOI: 10.1016/j.matdes.2011.11.064

[190] Bandyopadhyay A, Espana F, Balla VK, Bose S, Ohgami Y, Davies NM. Influence of porosity on mechanical properties and in vivo response of Ti6Al4 implants. Acta Biomaterialia. 2010;6:1640-1648. DOI: 10.1016/j.actbio.2009.11.011

[191] Bibb R, Thompson D, Winder J. Computed tomography characterisation of additive manufacturing materials. Medical Engineering & Physics. 2011;33: 590-596. DOI: 10.1016/j.medengphy.2010.12.015

[192] Cader M, Trojnacki M. Analysis of the applicability of additive technology to fabrication of mobile robots construction. Measurements. Automation. Robotics. 2013;2:200-207. (in Polish)

[193] Contreras A, Lopez VH, Bedolla E. Mg/TiC composites manufactured by pressureless melt infiltration. Scripta Materialia. 2004;51:249-253. DOI: 10.1016/j.scriptamat.2004.04.007

[194] Dimitrov D, Schreve K, Beer N. Advances in three dimensional printing - state of the art and future perspectives. Rapid Prototyping Journal. 2006;12: 136-147. DOI: 10.1108/ 13552540610670717

[195] Dobrzański LA, Dobrzańska-Danikiewicz AD, Malara P, Gaweł TG, Dobrzański LB, Achtelik-Franczak A. Fabrication of scaffolds from Ti6Al4V powders using the computer aided laser method. Archives of Metallurgy and Materials. 2015;60:1065-1070. DOI: 10.1515/amm-2015-0260

[196] Dobrzański LA. Overview and general ideas of the development of constructions, materials, technologies and clinical applications of scaffolds engineering for regenerative medicine. Archives of Materials Science and Engineering. 2014;69:53-80.

[197] Esen Z, Bor S. Processing of titanium foams using magnesium spacer particles. Scripta Materialia. 2007;56: 341-344. DOI: 10.1016/j. scriptamat.2006.11.010

[198] Eshraghi S, Das S. Mechanical and microstructural properties of polycaprolactone scaffolds with one-dimensional, two-dimensional, and three-dimensional orthogonally oriented porous architectures produced by selective laser sintering, Acta Biomaterialia. 2010;6:2467-2476. DOI: 10.1016/j.actbio.2010.02.002

[199] Fukuda A, Takemoto M, Saito T, Fujibayashi S, Neo M, Pattanayak DK, Matsushita T, Sasaki K, Nishida N, Kokubo T, Nakamura T. Osteoinduction of porous Ti implants with a channel structure fabricated by selective laser melting. Acta Biomaterialia. 2011;7: 2327-2336. DOI: 10.1016/j. actbio.2011.01.037

[200] Gasser A, Backes G, Kelbassa I, Weisheit A, Wissenbach K. Laser additive manufacturing: laser metal deposition (LMD) and selective laser melting (SLM) in turbo-engine applications. Laser Material Processing. 2010;2:58-63. DOI: 10.1002/ latj.201090029

[201] Karageorgiou V, Kaplan D. Porosity of 3D biomaterial scaffolds. Biomaterials. 2005;26:5474-5491. DOI: 10.1016/j.biomaterials.2005.02.002

[202] Klimek M. The use of SLS technology in making permanent dental restorations. Prosthetics. 2012;12:47-55. (in Polish)

[203] Furumoto T, Koizumi A, Alkahari MR, Anayama R, Hosokawa A, Tanaka R, Ueda T. Permeability and strength of a porous metal structure fabricated by additive manufacturing. Journal of Materials Processing Technology. 2015;219:10-16. DOI: 10.1016/j.jmatprotec.2014.11.043

[204] Kolan KC, Doiphode ND, Leu MC. Selective laser sintering and freeze extrusion fabrication of scaffolds for bone repair using 13-93 bioactive glass: a comparison. In: Proceedings of the Solid Freeform Fabrication Symposium; Austin, Texas; 2010.

[205] Li JP, Habibovic P, van den Doel M, Wilson CE, de Wijn JR, van Blitterswijk CA, de Groot K. Bone ingrowth in porous titanium implants produced by 3D fiber deposition. Biomaterials. 2007;28: 2810-2820. DOI: 10.1016/j. biomaterials.2007.02.020

[206] Mazzoli A. Selective laser sintering in biomedical engineering. Medical and Biological Engineering and Computing. 2013;51:245-256. DOI: 10.1007/ s11517-012-1001-x

[207] Melchels FPW, Domingos MAN, Klein TJ, Malda J, Bartolo PJ, Hutmacher DW. Additive manufacturing of tissues and organs. Progress in Polymer Science. 2012;37:

1079-1104. DOI: 10.1016/j.progpolymsci.2011.11.007

[208] Nouri A, Hodgson PD, Wen C. Biomimetic Porous Titanium Scaffolds for Orthopedic and Dental Applications. In: Mukherjee A, editor. Biomimetics. Learning from Nature. IntechOpen; 2010. p. 415-450. DOI: 10.5772/8787

[209] Sachlos E, Czernuszka JT. Making tissue engineering scaffolds work. Review: the application of solid freeform fabrication technology to the production of tissue engineering scaffolds. European Cells & Materials. 2003;5:29-39. DOI: 10.22203/eCM.v005a03

[210] Salmoria GV, Fancello EA, Roesler CRM, Dabbas F. Functional graded scaffold of HDPE/HA prepared by selective laser sintering: microstructure and mechanical properties. International Journal of Advanced Manufacturing Technology. 2013;65:1529-1534. DOI: 10.1007/s00170-012-4277-y

[211] Simancik F. Metallic foams - ultra light materials for structural applications. Materials Engineering. 2001;5:823-828.

[212] Song Y, Yan Y, Zhang R, Xu D, Wang F. Manufacturing of the die of an automobile deck part based on rapid prototyping and rapid tooling technology. Journal of Materials Processing Technology. 2002;120:237-242. DOI: 10.1016/S0924-0136(01)01165-7

[213] Thomas CL, Gaffney TM, Kaza S, Lee CH. Rapid prototyping of large scale aerospace structures. In: Proceedings of Aerospace Applications Conference IEEE. 1996;4:219-230.

[214] Veiseh M, Edmondson D. Bone as an Open Cell Porous Material. ME 599K: Special Topics in Cellular Solids; 2003.

[215] Weisensel L, Travitzky N, Sieber H, Greil P. Laminated object manufacturing (LOM) of SiSiC composites. Advanced Engineering Materials. 2004;6:899-903. DOI: 10.1002/adem.200400112

[216] Xie F, He X, Cao S, Qu X. Structural and mechanical characteristics of porous 316L stainless steel fabricated by indirect selective laser sintering. Journal of Materials Processing Technology. 2013;213:838-843. DOI: 10.1016/j.jmatprotec.2012.12.014

[217] Meenashisundaram GK, Wang N, Maskomani S, Lu S, Anantharajan SK, Dheen ST, Nai SML, Fuh JYH, Wei J. Fabrication of Ti + Mg composites by three-dimensional printing of porous Ti and subsequent pressureless infiltration of biodegradable Mg. Materials Science and Engineering C. 2020;108:110478. DOI: 10.1016/j.msec.2019.110478

[218] Dobrzański LA, Dobrzańska-Danikiewicz AD, Czuba ZP, Dobrzański LB, AchtelikFranczak A, Malara P, Szindler M, Kroll L. Metallic skeletons as reinforcement of new composite materials applied in orthopaedics and dentistry. Archives of Materials Science and Engineering. 2018;92:53-85.DOI: 10.5604/01.3001.0012.6585

[219] Jurczyk M. Mechanical synthesis. Poznan: Poznan University of Technology Publishing House; 1998. (in Polish)

[220] Ozdemir I, Ahrens S, Mücklich S, Wielage B. Nanocrystalline Al-Al_2O_{3p} and SiC_p composites produced by high-energy ball milling. Journal of Materials Processing Technology. 2008;205:111-118. DOI: 10.1016/j.jmatprotec.2007.11.085

[221] Suryanarayana C. Mechanical alloying and milling. Progress in

Materials Science. 2001;46:1-184. DOI: 10.1016/S0079-6425(99)00010-9

[222] Kadi S, Lellou S, Marouf-Khelifa K, Schott J, Gener-Batonneau I, Khelifa A. Preparation, characterisation and application of thermally treated Algerian halloysite. Microporous and Mesoporous Materials. 2012;158:47-54. DOI: 10.1016/j.micromeso.2012.03.014

[223] MacKenzie KJD, Meinhold RH, Brown IWM, White GV. The formation of mullite from kaolinite under various reaction atmospheres. Journal of European Ceramic Society. 1996;16: 115-119. DOI: 10.1016/0955-2219(95) 00143-3

[224] Perera DS, Allott G. Mullite morphology in fired kaolinite/halloysite clays. Journal of Materials Science Letters. 1985;4:1270-1272. DOI: 10.1007/ BF00723478

[225] Tezuka N, Low I-M, Davies IJ, Prior M, Studer A. In situ neutron diffraction investigation on the phase transformation sequence of kaolinite and halloysite to mullite. Physica B. 2006;385-386:555-557. DOI: 10.1016/j. physb.2006.05.357

[226] Goda ES, Yoon KR, El-sayed SH, Hong SE. Halloysite nanotubes as smart flame retardant and economic reinforcing materials: A review. Thermochimica Acta. 2018;669:173-184. DOI: 10.1016/j.tca.2018.09.017

[227] Miyazaki M, Kamitani M, Takashi N, Kano J, Saito F. Amorphization of kaolinite and media motion in grinding by a double rotating cylinders mill – a comparison with a tumbling ball mill. Advanced Powder Technology. 2000;11:235-244. DOI: 10.1163/156855200750172349

[228] Schmücker M, Schneider H, MacKenzie KJD. Mechanical amorphization of mullite and thermal recrystallization. Journal of Non-Crystalline Solids. 1998;226: 99-104. DOI: 10.1016/S0022-3093(98) 00366-4

[229] Valášková M, Barabaszová K, Hundáková M, Ritz M, Plevová E. Effects of brief milling and acid treatment on two ordered and disordered kaolinite structures. Applied Clay Science. 2011;54:70-76. DOI: 10.1016/j.clay.2011.07.014

[230] Vizcayno C, de Gutiérrez RM, Castello R, Rodriguez E, Guerrero CE: Pozzolan obtained by mechanochemical and thermal treatments of kaolin. Applied Clay Science. 2010;49:405-413. DOI: 10.1016/j.clay.2009.09.008

[231] Arakawa S, Hatayama T, Matsugi K, Yanagisawa O. Effect of heterogeneous precipitation on age-hardening of Al_2O_3 particle dispersion Al-4mass%Cu composite produced by mechanical alloying. Scripta Materialia. 2000;42:755-760. DOI: 10.1016/ S1359-6462(99)00426-1

[232] Couper MJ, Parson NC. Precipitation Strengthening and Alloy Design for 6061 Al-Mg-Si Alloys In: Starke, Jr. EA, Sanders, Jr. TH, Cassada WA. Aluminium Alloys – Their Physical and Mechanical Properties 1. Wiley-VCH GmbH & Co. KGaA; 2008. p. 98-104.

[233] Maisonnette D, Suery M, Nelias D, Chaudet P, Epicier T. Effects of heat treatments on the microstructure and mechanical properties of a 6061 aluminium alloy. Materials Science and Engineering A. 2011;528:2718-2724. DOI: 10.1016/j.msea.2010.12.011

[234] Hirsch J, Skrotzki B, Gottstein G, editors. Aluminium Alloys. Their Physical and Mechanical Properties. Weinheim: Wiley-VCH Verlag GmbH & Co. KGaA; 2008.

[235] Jia DC, Li XL, Zhou Y, Han HQ. Effect of SiC particulate sizes on the

mechanical properties of Al-6Ti-6Nb matrix composites prepared by a mechanical alloying technique. Journal of Materials Science Letters. 1999;18: 2029-2032. DOI: 10.1023/A: 1006706505141

[236] Stobrawa J, Rdzawski Z, Głuchowski W, Malec W. Microstructure and properties of CuNi2Si1 alloy processed by continuous RCS method. Journal of Achievements in Materials and Manufacturing Engineering. 2009;37:466-479.

[237] NaturalNano [Internet]. 2015. Available from: www.naturalnano.com [Accessed: 2021-02-10]

[238] Wang L, Hu J, Li ZJ, Fei WD. Fracture behavior of aluminium borate whisker- reinforced aluminium alloy 6061 composite. Materials Science and Engineering A. 2008;497:358-362. DOI: 10.1016/j.msea.2008.07.018

[239] Kurzydłowski K, Lewandowska M, editors. Engineering, structural and functional nanomaterials. Warsaw: PWN; 2010. (in Polish)

[240] Hebda M, Wachal A. Tribology. Warsaw: WNT; 1980. (in Polish)

[241] Naplocha K, Kaczmar JW. Wear mechanisms of fiber reinforced composite materials based on 2024 and 7075 aluminum alloys. Journal of Achievements in Materials and Manufacturing Engineering. 2011;49: 180-187.

[242] Esawi AMK, Morsi K, Sayed A, Abdel Gawad A, Borah P. Fabrication and properties of dispersed carbon nanotube–aluminum composites. Materials Science and Engineering A. 2009;508:167-173. DOI: 10.1016/j. msea.2009.01.002

[243] Liu ZY, Xiao BL, Wang WG, Ma ZY. Tensile Strength and Electrical Conductivity of Carbon Nanotube Reinforced Aluminum Matrix Composites Fabricated by Powder Metallurgy Combined with Friction Stir Processing. Journal of Materials Science & Technology. 2014;30:649-655. DOI: 10.1016/j.jmst.2014.04.016

[244] Morsi K, Esawi A. Effect of mechanical alloying time and carbon nanotube (CNT) content on the evolution of aluminum (Al)–CNT composite powders. Journal of Materials Science. 2007;42:4954-4959. DOI: 10.1007/s10853-006-0699-y

[245] Nouri N, Ziaei-Rad S, Adibi S, Karimzadeh F. Fabrication and mechanical property prediction of carbon nanotube reinforced Aluminum nanocomposites. Materials and Design. 2012;34:1-14. DOI: 10.1016/j. matdes.2011.07.047

[246] Peng T, Chang I. Mechanical alloying of multi-walled carbon nanotubes reinforced aluminum composite powder. Powder Technology. 2014;266:7-15. DOI: 10.1016/j. powtec.2014.05.068

[247] Chen B, Li Z, Shen J, Li S, Jia L, Umeda J, Kondoh K, Li J.S. Mechanical properties and strain hardening behavior of aluminum matrix composites reinforced with few-walled carbon nanotubes. Journal of Alloys and Compounds. 2020;826: 154075. DOI: 10.1016/j.jallcom.2020. 154075

[248] Guo B, Du Y, Yan N, Song M. Mechanical properties and microstructures of Al-10Mg-4.5Si matrix composites reinforced by carbon nanotubes. Journal of Alloys and Compounds. 2019;792:860-868. DOI: 10.1016/j.jallcom.2019.04.101

[249] Turan ME. Investigation of mechanical properties of carbonaceous (MWCNT, GNPs and C60) reinforced hot-extruded aluminum matrix composites. Journal of Alloys and

Compounds. 2019;788:352-360. DOI: 10.1016/j.jallcom.2019.02.25

[250] Mokdad F, Chen DL, Liu ZY, Xiao BL, Ni DR, Ma ZY. Deformation and strengthening mechanisms of a carbon nanotube reinforced aluminum composite. Carbon. 2016;104:64-77. DOI: 10.1016/j.carbon.2016.03.038

[251] Bradbury CR, Gomon J-K, Kollo L, Kwon H, Leparoux M. Hardness of Multi Wall Carbon Nanotubes reinforced aluminium matrix composites. Journal of Alloys and Compounds. 2014;585:362-367. DOI: 10.1016/j.jallcom.2013.09.142

[252] Li H, Kang J, He C, Zhao N, Liang C, li B. Mechanical properties and interfacial analysis of aluminum matrix composites reinforced by carbon nanotubes with diverse structures. Materials Science and Engineering A. 2013;577:120-124. DOI: 10.1016/j.msea.2013.04.035

[253] Pérez-Bustamante R, Pérez-Bustamante F, Estrada-Guel I, Santillán-Rodríguez CR, Matutes-Aquino JA, Herrera-Ramírez JM, Miki-Yoshida M, Martínez-Sánchez R. Characterization of Al2024-CNTs composites produced by mechanical alloying. Powder Technology. 2011;212:390-396. DOI: 10.1016/j.powtec.2011.06.007

[254] Pérez-Bustamante R, Pérez-Bustamante F, Estrada-Guel I, Licea-Jiménez L, Miki-Yoshida M, Martínez-Sánchez R. Effect of milling time and CNT concentration on hardness of CNT/Al2024 composites produced by mechanical alloying. Materials Characterization. 2013;75:13-19. DOI: 10.1016/j.matchar.2012.09.005

[255] Simões S, Viana F, Reis MAL, Vieira MF. Improved dispersion of carbon nanotubes in aluminum nanocomposites. Composite Structures. 2014;108:992-1000. DOI: 10.1016/j.compstruct.2013.10.043

[256] Esawi AMK, Morsi K, Sayed A, Taher M, Lanka S. Effect of carbon nanotube (CNT) content on the mechanical properties of CNT-reinforced aluminium composites. Composites Science and Technology. 2010;70:2237-2241. DOI: 10.1016/j.compscitech.2010.05.004

[257] Esawi AMK, Morsi K, Sayed A, Taher M, Lanka S. The influence of carbon nanotube (CNT) morphology and diameter on the processing and properties of CNT-reinforced aluminium composites. Composites Part A. 2011;42:234-243. DOI: 10.1016/j.compositesa.2010.11.008

[258] Javadi AH, Mirdamadi Sh, Faghihisani MA, Shakhesi S, Soltani R. Fabrication of well-dispersed, multiwalled carbon nanotubes-reinforced aluminum matrix composites. New Carbon Materials. 2012;27:161-165. DOI: 10.1016/S1872-5805(12)60010-9

[259] Kwon H, Park DH, Silvain JF, Kawasaki A. Investigation of carbon nanotube reinforced aluminum matrix composite materials. Composites Science and Technology. 2010;70:546-550. DOI: 10.1016/j.compscitech.2009.11.025

[260] Morsi K, Esawi AMK, Lanka S, Sayed A, Taher M. Spark plasma extrusion (SPE) of ball-milled aluminum and carbon nanotube reinforced aluminum composite powders. Composites Part A. 2010;41:322-326. DOI: 10.1016/j.compositesa.2009.09.028

[261] Poirier D, Gauvin R, Drew RAL. Structural characterization of a mechanically milled carbon nanotube/aluminum mixture. Composites Part A. 2009;40:1482-1489. DOI: 10.1016/j.compositesa.2009.05.025

[262] Ram HRA, Koppad PG, Kashyap KT. Nanoindentation studies

on MWCNT/aluminum alloy 6061 nanocomposites. Materials Science and Engineering A. 2013;559:920-923. DOI: 10.1016/j.msea.2012.08.143.

[263] Wu Y, Kim G-Y. Carbon nanotube reinforced aluminum composite fabricated by semi-solid powder processing. Journal of Materials Processing Technology. 2011;211: 1341-1347. DOI: 10.1016/j. jmatprotec.2011.03.007

[264] Housaer F, Beclin F, Touzin M, Tingaud D, Legris A, Addad A. Interfacial characterization in carbon nanotube reinforced aluminum matrix composites. Materials Characterization. 2015;110:94-101. DOI: 10.1016/j. matchar.2015.10.014

[265] Ci L, Ryu Z, Jin-Phillipp NY, Rühle M. Investigation of the interfacial reaction between multi-walled carbon nanotubes and aluminium. Acta Materialia. 2006;54:5367-5375. DOI: 10.1016/j.actamat.2006.06.031

[266] Laha T, Kuchibhatla S, Seal S, Li W, Agarwal A. Interfacial phenomena in thermally sprayed multiwalled carbon nanotube reinforced aluminum nanocomposite. Acta Materialia. 2007; 55:1059-1066. DOI: 10.1016/j. actamat.2006.09.025

[267] Zhou W, Sasaki S, Kawasaki A. Effective control of nanodefects in multiwalled carbon nanotubes by acid treatment. Carbon. 2014;78:121-129. DOI: 10.1016/j.carbon.2014.06.055

[268] Liu ZY, Xiao BL, Wang WG, Ma ZY. Modelling of carbon nanotube dispersion and strengthening mechanisms in Al matrix composites prepared by high energy ball milling-powder metallurgy method. Composites Part A. 2017;94:189-198. DOI: 10.1016/j. compositesa.2016.11.029

[269] Wang W, Li Q, Ma R, Cui Y, Cao X, Wang X. Effects of the Si contents of an infiltrating aluminium alloy on the microstructure and strength of SiC matrix composites. Ceramics International. 2017;43:12526-12533.DOI: 10.1016/j.ceramint.2017.06.125

[270] Kwon H, Estili M, Takagi K, Miyazaki T, Kawasaki A. Combination of hot extrusion and spark plasma sintering for producing carbon nanotube reinforced aluminum matrix composites. Carbon. 2009;47:570-577. DOI: 10.1016/j.carbon.2008.10.041

[271] Yan H, Qiu H. Fabrication of carbon nanotube reinforced A356 nanocomposites. Journal of Materials Research. 2016;31:2277-2283. DOI: 10.1557/jmr.2016.258

[272] Zhou W, Yamamoto G, Fan Y, Kwon H, Hashida T, Kawasaki A. In-situ characterization of interfacial shear strength in multi-walled carbon nanotube reinforced aluminum matrix composites. Carbon. 2016;106: 37-47. DOI: 10.1016/j.carbon.2016. 05.015

[273] Zhou W, Yamaguchi T, Kikuchi K, Nomura N, Kawasaki A. Effectively enhanced load transfer by interfacial reactions in multi-walled carbon nanotube reinforced Al matrix composites. Acta Materialia. 2017;125: 369-376. DOI: 10.1016/j. actamat.2016.12.022

[274] Liu X, Liu E, Li J, He C, Zhao N. Investigation of the evolution and strengthening effect of aluminum carbide for in-situ preparation of carbon nanosheets/aluminum composites. Materials Science and Engineering A. 2019;764:138139. DOI: 10.1016/j. msea.2019.138139

[275] Raviathul Basariya M, Srivastava VC, Mukhopadhyay NK. Microstructural characteristics and mechanical properties of carbon nanotube reinforced aluminum alloy composites produced by ball milling.

Materials and Design. 2014;64:542-549. DOI: 10.1016/j.matdes.2014.08.019

[276] Choi HJ, Lee SM, Bae DH. Wear characteristic of aluminum-based composites containing multi-walled carbon nanotubes. 2010;270:12-18. DOI: 10.1016/j.wear.2010.08.024

[277] Afzal N, Rafique M, Javaid W, Ahmad R, Farooq A, Saleem M, Khaliq Z. Influence of carbon ion implantation energy on aluminum carbide precipitation and electrochemical corrosion resistance of aluminum. Nuclear Instruments and Methods in Physics Research Section B: Beam Interactions with Materials and Atoms. 2018;436:84-91. DOI: 10.1016/j.nimb.2018.09.008

[278] Yolshina LA, Muradymov RV, Korsun IV, Yakovlev GA, Smirnov S.V. Novel aluminum-graphene and aluminum-graphite metallic composite materials: Synthesis and properties. Journal of Alloys and Compounds. 2016; 663:449-459. DOI: 10.1016/j.jallcom.2015.12.084

[279] Chen B, Shen J, Ye X, Imai H, Umeda J, Takahashi M, Kondoh K. Solid-state interfacial reaction and load transfer efficiency in carbon nanotubes (CNTs)-reinforced aluminum matrix composites. Carbon. 2017;114:198-208. DOI: 10.1016/j.carbon.2016.12.013

[280] Dobrzański LA, Dobrzańska-Danikiewicz AD. Surface treatment of engineering materials. Open Access Library, vol. 5. Gliwice: International OCSCO World Press; 2011. 480 p. (in Polish)

[281] Feng W, Patel SH, Young M-Y, Zunino III JL, Xanthos M. Smart polymeric coatings—recent advances. Advances in Polymer Technology. 2007; 26:1-13. DOI: 10.1002/adv.20083

[282] Firouzi A, Del Gaudio C, lamastra FR, Montesperelli G, Bianco A. Electrospun polymeric coatings on aluminum alloy as a straightforward approach for corrosion protection. Journal of Applied Polymer Science. 2015;132:41250. DOI: 10.1002/app.41250

[283] Wankhede RG, Morey S, Khanna AS, Birbilis N. Development of water-repellent organic–inorganic hybrid sol–gel coatings on aluminum using short chain perfluoro polymer emulsion. Applied Surface Science. 2013;283:1051-1059. DOI: 10.1016/j.apsusc.2013.07.066

[284] Galvão TLP, Bouali AC, Serdechnova M, Yasakau KA, Zheludkevich ML, Tedim J. Chapter 16 - Anticorrosion thin film smart coatings for aluminum alloys. In: Makhlouf ASH, Abu-Thabit NY, editors. Advances in Smart Coatings and Thin Films for Future Industrial and Biomedical Engineering Applications. Elsevier; 2020. p. 429-454. DOI: 10.1016/B978-0-12-849870-5.00007-0

[285] Xiong G, Elam JW, Feng H, Han CY, Wang H-H, Iton LE, Curtiss LA, Pellin MJ, Kung M, Kung H, Stair PC. Effect of Atomic Layer Deposition Coatings on the Surface Structure of Anodic Aluminum Oxide Membranes. The Journal of Physical Chemistry B. 2005;109:14059-14063. DOI: 10.1021/jp0503415

[286] Sharma K. Spatial atomic layer deposition on flexible porous substrates: ZnO on anodic aluminum oxide films and Al_2O_3 on Li ion battery electrodes. Journal of Vacuum Science & Technology A. 2016;34:01A146. DOI: 10.1116/1.4937728

[287] Hennessy JJ, Balasubramanian K, Moore CS, Jewell AD, Nikzad S, France KC, Quijada MA. Performance and prospects of far ultraviolet aluminum mirrors protected by atomic layer deposition. Journal of

Astronomical Telescopes, Instruments, and Systems. 2016;2:041206. DOI: 10.1117/1.JATIS.2.4.041206

[288] Elam JW, Xiong G, Han CY, Wang H-H, Birrell JP, Welp U, Hryn JN, Pellin MJ, Baumann TF, Poco JF, Satcher Jr. JH. Atomic Layer Deposition for the Conformal Coating of Nanoporous Materials. Journal of Nanomaterials. 2006;2006:64501. DOI: 10.1155/JNM/2006/64501

[289] Kennedy E, Byrne G, Collins DN. A review of the use of high power diode lasers in surface hardening. Journal of Materials Processing Technology. 2004; 155-156:1855-1860. DOI: 10.1016/j.jmatprotec.2004.04.276

[290] Dobrzański LA, Bonek M, Hajduczek E, Klimpel A, Lisiecki A. Application of high power diode laser (HPDL) for alloying of X40CrMoV5-1 steel surface layer by tungsten carbides. Journal of Materials Processing Technology. 2004;155-156:1956-1963. DOI: 10.1016/j.jmatprotec.2004.04.058

[291] Dobrzański LA, Bonek M, Piec M, Jonda E. Diode laser modification of surface gradient layer properties of a hot-work tool steel. Materials Science Forum. 2006;532-533:657-660. DOI: 10.4028/www.scientific.net/MSF.532-533.657

[292] Chi Y, Gu G, Yu H, Chen C. Laser surface alloying on aluminum and its alloys: A review. Optics and Lasers in Engineering. 2018;100:23-37. DOI: 10.1016/j.optlaseng.2017.07.006

[293] Dobrzański LA, Dobrzańska-Danikiewicz AD. Applications of Laser Processing of Materials in Surface Engineering in the Industry 4.0 Stage of the Industrial Revolution. Materials Performance and Characterization. 2019;8:1091-1129. DOI: 10.1520/MPC20190203

[294] Dobrzański LA, Dobrzański LB, Dobrzańska-Danikiewicz AD, Kraszewska M. Manufacturing powders of metals, their alloys and ceramics and the importance of conventional and additive technologies for products manufacturing in Industry 4.0 stage. Archives of Materials Science and Engineering. 2020;102:13-41. DOI: 10.5604/01.3001.0014.1452

[295] Dobrzański LA, Dobrzański LB, Dobrzańska-Danikiewicz AD. Manufacturing technologies thick-layer coatings on various substrates and manufacturing gradient materials using powders of metals, their alloys and ceramics. Journal of Achievements in Materials and Manufacturing Engineering. 2020;99:14-41. DOI: 10.5604/01.3001.0014.1598

[296] Yue TM, Yan LJ, Chan CP, Dong CF, Man HC, Pang GKH. Excimer laser surface treatment of aluminum alloy AA7075 to improve corrosion resistance. Surface and Coatings Technology. 2004;179:158-164. DOI: 10.1016/S0257-8972(03)00850-8

[297] Zhang QY, Li XQ, Wang DY, Tian YS, Aleem BJA. Laser surface treatment of Aluminum alloys. Mater Heat Treatment. 2010;39:120-125.

[298] Ji H, Xu Y, Lu ZS, Zhou DR. Progress in strengthening of aluminium alloy surface by laser. Material Science and Technology. 2003;11:220-224.

[299] Gordani GR, ShojaRazavi R, Hashemi SH, Isfahani ARN. Laser surface alloying of an electroless Ni–P coating with Al-356 substrate. Optics and Lasers in Engineering. 2008;46: 550-557. DOI: 10.1016/j.optlaseng.2008.02.002

[300] Staia MH, Cruz M, Dahotre NB. Microstructural and tribological characterization of an A-356 aluminum alloy superficially modified by laser alloying. Thin Solid Films. 2000; 377-378:665-674. DOI: 10.1016/S0040-6090(00)01448-6

[301] Chuang YC, Le SC, Lin HC. Effect of temperature on the sliding wear behavior of laser surface alloyed Ni-base on Al–Mg–Si alloy. Applied Surface Science. 2006;253:1404-1410. DOI: 10.1016/j.apsusc.2006.02.014

[302] Vaziri SA, Shahverdi HR, Torkamany MJ, Shabestari SG. Effect of laser parameters on properties of surface-alloyed Al substrate with Ni. Optics and Lasers in Engineering. 2009; 47:971-975. DOI: 10.1016/j.optlaseng.2009.04.007

[303] Liu W, Li X, Yu XT, Mo YM. Laser surface alloyed technology on Al alloys. Science and Technology of Overseas Building Materials. 2006;4:56-58.

[304] Kadolkar P, Dahotre NB. Effect of processing parameters on the cohesive strength of laser surface engineered ceramic coatings on aluminum alloys. Materials Science and Engineering A. 2003;342:183-191. DOI: 10.1016/S0921-5093(02)00286-1

[305] Nath S, Pityana S, Majumdar JD. Laser surface alloying of aluminium with WC+Co+NiCr for improved wear resistance. Surface and Coatings Technology. 2012;206:3333-3341. DOI: 10.1016/j.surfcoat.2012.01.038

[306] Geng H, Niu Y, Tian X, Chen C, Hou X. Laser surface strengthening process of Al-Si base alloy. Journal of Materials Science and Technology. 2001;17:115-118.

[307] Chong PH, Man HC, Yue TM. Microstructure and wear properties of laser surface-cladded Mo·WC MMC on AA6061 aluminum alloy. Surface and Coatings Technology. 2001;145: 51-59. DOI: 10.1016/S0257-8972(01)01286-5

[308] Almeida A, Petrov P, Nogueira I, Vilar R. Structure and properties of Al–Nb alloys produced by laser surface alloying. Materials Science and Engineering A. 2001;303:273-280. DOI: 10.1016/S0921-5093(00)01838-4

[309] Rajamure RS, Vora HD, Srinivasan SG, Dahotre NB. Laser alloyed Al–W coatings on aluminum for enhanced corrosion resistance. Applied Surface Science. 2015;328:205-214. DOI: 10.1016/j.apsusc.2014.12.037

[310] Watkins KG, McMahon MA, Steen WM. Microstructure and corrosion properties of laser surface processed aluminium alloys: a review. Materials Science and Engineering A. 1997;231:55-61. DOI: 10.1016/S0921-5093(97)00034-8

[311] Yilbas BS, Karatas C, Karakoc H, Aleem BJA, Khan S, Al-Aqeeli N. Laser surface treatment of aluminum based composite mixed with B_4C particles. Optics and Laser Technology. 2015;66: 129-137. DOI: 10.1016/j.optlastec.2014.08.014

[312] Staia MH, Cruz M, Dahotre NB. Wear resistance of a laser alloyed A-356 aluminum/WC composite. Wear. 2001; 251:1459-1468. DOI: 10.1016/S0043-1648(01)00789-X

[313] Man HC, Kwok CT, Yue TM. Cavitation erosion and corrosion behaviour of laser surface alloyed MMC of SiC and SiN on Al alloy AA6061. Surface and Coating Technology. 2000; 132:11-20. DOI: 10.1016/S0257-8972(00)00729-5

[314] Vreeling JA, k VO, Hamstra GA, Pei YT, De Hosson JTM. In-situ microscopy investigation of failure mechanisms in Al/SiC_p metal matrix composite produced by laser embedding. Scripta Materialia. 2000;42: 589-595. DOI: 10.1016/S1359-6462(99)00404-2

[315] Man HC, Zhang S, Cheng FT, Yue TM. In situ synthesis of TiC reinforced surface MMC on Al6061 by laser surface alloying. Scripta Materialia.

2002;46:229-234. DOI: 10.1016/S1359-6462(01)01230-1

[316] Tomida S, Nakata K, Saji S, Kubo T. Formation of metal matrix composite layer on aluminum alloy with TiC-Cu powder by laser surface alloying process. Surface and Coating Technology. 2001;142:585-589. DOI: 10.1016/S0257-8972(01)01172-0

[317] Dubourg L, Ursescu D, Hlawka F, Cornet A. Laser cladding of MMC coatings on aluminium substrate: influence of composition and microstructure on mechanical properties. Wear. 2005;258:1745-1754. DOI: 10.1016/j.wear.2004.12.010

[318] Dobrzański LA. Significance of materials science for the future development of societies. Journal of Materials Processing Technology. 2006; 175:133-148. DOI: 10.1016/j.jmatprotec.2005.04.003

[319] Dobrzański LA, Dobrzański LB. Approach to the design and manufacturing of prosthetic dental restorations according to the rules of the Industry 4.0. Materials Performance and Characterization. 2020;9:394-476. DOI: 10.1520/MPC20200020

[320] Dobrzański LA, Dobrzański LB. Dentistry 4.0 Concept in the Design and Manufacturing of Prosthetic Dental Restorations. Processes. 2020;8:525. DOI: 10.3390/pr8050525

[321] Dobrzański LA. Role of materials design in maintenance engineering in the context of industry 4.0 idea. Journal of Achievements in Materials and Manufacturing Engineering. 2019;96: 12-49. DOI: 10.5604/01.3001.0013.7932

[322] Dobrzański LA, Dobrzańska-Danikiewicz AD. Why are Carbon-Based Materials Important in Civilization Progress and Especially in the Industry 4.0 Stage of the Industrial Revolution?. Materials Performance and Characterization. 2019;8:337-370. DOI: 910.1520/MPC20190145

[323] Dobrzański LA. Chapter 5 - Effect of heat and surface treatment on the structure and properties of the Mg-Al-Zn-Mn casting alloys. In: Dobrzański LA, Totten GE, Bamberger M, editors. Magnesium and Its Alloys. Boca Raton, FL: CRC Press; 2019. p. 91-202.

[324] Dobrzański LA. Fundamentals of material design methodology. Gliwice: Silesian University of Technology Publishing House; 2009. (in Polish)

[325] Dobrzańska-Danikiewicz AD. The State of the Art Analysis and Methodological Assumptions of Evaluation and Development Prediction for Materials Surface Technologies. Journal of Achievements in Materials and Manufacturing Engineering. 2011; 49:121-141.

[326] Materials Selection in Mechanical Design. Oxford, UK: Pergamon Press; 1992.

[327] Waterman NA, Ashby MF, editors. The Materials Selector. 3 vols. London: Chapman & Hall; 1997.

[328] Ashby MF. Materials Selection in Mechanical Design. 4th ed. Burlington, MA: Butterworth-Heinemann; 2011.

[329] Ashby MF, Jones DRH. Engineering Materials 1: An Introduction to Properties, Applications and Design. Waltham, MA: Butterworth-Heinemann; 2012.

[330] Ashby MF, Jones DRH. Engineering Materials 2: An Introduction to Microstructures and Processing. Waltham, MA: Butterworth-Heinemann; 2013.

[331] Oztemel E, Gursev S. Literature Review of Industry 4.0 and Related Technologies. Journal of Intelligent

Manufacturing. 2020;31: 127-182. DOI: 10.1007/s10845-018-1433-8

[332] Dobrzański LA. Structural Phenomena Accompanying the Production of Composite and Nanocomposite Materials Using Selected Technologies. Invited lecture at the 22nd Physical Metallurgy and Materials Science Conference: Advanced Materials and Technologies; 9-12 June 2019; Bukowina Tatrzańska, Poland.

[333] Dobrzański LA. Role of Materials Design in Maintenance Engineering in the Context of Industry 4.0 Idea. Invited lecture at the International Maintenance Technologies Congress and Exhibition; 26-28 September 2019; Denizli, Turkey.

[334] Dobrzański LA. The role of materials engineering in stage 4.0 of the technological revolution. Invited lecture at the 24th Seminar of the Polish Society of Materials Science; 12-15 May 2019; Jachranka, Poland. (in Polish)

[335] Dobrzański LA. The importance of materials engineering in stage 4.0 of the technological revolution. Invited lecture in the Institute of Fundamental Technological Research, Polish Academy of Sciences; 27 May 2019; Warsaw, Poland. (in Polish)

[336] Dobrzański LA. Stage 4.0 of the Technological Revolution in the Context of the Development of Engineering & Dental Materials. Invited lecture in the Lviv Polytechnic National University; 3 June 2019; Lviv, Ukraine.

[337] Jose R, Ramakrishna S. Materials 4.0: Materials Big Data Enabled Materials Discovery. Applied Materials Today. 2018;10: 127-132. DOI: 10.1016/j.apmt.2017.12.015

[338] United Nations General Assembly. Transforming Our World: The 2030 Agenda for Sustainable Development: A/RES/70/1. New York: United Nations; 2015.

[339] Dobrzański LA. Fundamentals of materials science and metallurgy. Engineering materials with the basics of material design. Warsaw, Poland: WNT; 2002. (in Polish)

[340] Dobrzański LA. Metallic engineering materials. Warsaw, Poland: WNT; 2004. (in Polish)

[341] Dobrzański LA. Fundamentals of materials science. Gliwice, Poland: Silesian University of Technology Publishing House; 2012. (in Polish)

[342] Dobrzański LA, Dobrzańska-Danikiewicz AD. Surface engineering of materials: a compendium of knowledge and an academic textbook. Gliwice: International OCSCO World Press; 2018. 1138 p. (in Polish)

[343] Dobrzański LA. Materials Design as an Important Element of Advanced Products' Engineering Design and Manufacturing. In: 17th International Congress of Mechanical Engineering. São Paulo, Brazil: Associação Brasileira de Ciências Mecânicas; 2003. (CD-ROM)

[344] Dobrzański LA. Contemporary development trends in materials science and material engineering. Materials Engineering. 2003;24:271-278. (in Polish)

[345] Dobrzański LA. Significance of Materials Science and Engineering for Advances in Design and Manufacturing Processes, Computer Integrated Manufacturing. In: Skołud B, Krenczyk D, editors. Advanced Design and Management. Warsaw, Poland: WNT; 2003. p. 128-140.

[346] Dobrzański LA. The importance of the development of materials science and materials engineering to improve

the quality of life of modern societies. Wisnyk Technologiczno University of Podilla, Khmelnytsky, Ukraine. 2003;I: 34-47. (in Polish)

[347] Dobrzański LA. Heat Treatment as Fundamental Technological Process of Formation of Structure and Properties of Metals. In: 8th Seminar of the International Federation for Heat Treatment and Surface Engineering. Winterthur, Switzerland: International Federation for Heat Treatment and Surface Engineering, 2001. p. 1-12.

[348] Dobrzański LA. Significance of Materials Science for Advances in Products Design and Manufacturing. Invited lecture at the General Assembly of the International Academy for Production Engineering; 22-28 August 2004; Cracow, Poland.

[349] Dobrzański LA. Computational Materials Science as a Method of Design of Contemporary Engineering Materials and Products. In: 10th Seminary of Research and Education Programmes in Materials Engineering. Myczkowce, Poland: Polish Society of Materials Science, 2004. p. 55-88. (in Polish)

[350] Ruehle M, Dosch H, Mittemeijer E, van de Voorde MH, editors. European White Book on Fundamental Research in Materials Science. Stuttgart, Germany: Max-Planck-Institute für Metallforschung; 2001.

[351] Dieter GE, editor. ASM Handbook Volume 20: Materials Selection and Design. Materials Park, OH: ASM International; 1997.

[352] Japan Business Federation. Society 5.0: Co-Creating the Future (Excerpt). Tokyo: Keidanren; 2018.

[353] Japan Business Federation. Toward Realization of the New Economy and Society (Outline). Tokyo: Keidanren; 2016.

[354] Japan Business Federation. Japan's Initiatives–Society 5.0. Tokyo: Keidanren; 2016.

[355] Harayama Y. Society 5.0: Aiming for a New Human-Centered Society. Hitachi Review. 2017;66:8-13.

[356] Government of Japan Cabinet Office. Society 5.0 [Internet]. 2019. Available from: http://web.archive.org/web/20190710182953/https://www8.cao.go.jp/cstp/society5_0/index.html [Accessed: 2021-02-10]

[357] Fukuyama M. Society 5.0: Aiming for a New Human-Centered Society. Japan SPOTLIGHT. 2018;July/August: 47-50.

[358] Kagermann H, Wahlster W, Helbig J. Recommendations for Implementing the Strategic Initiative INDUSTRIE 4.0: Final Report of the Industrie 4.0 Working Group. Bonn, Germany: Federal Ministry of Education and Research; 2013.

[359] Kagermann H. Chancen von Industrie 4.0 Nutzen. In: Industrie 4.0 in Produktion, Automatisierung und Logistik. Wiesbaden, Germany: Springer Fachmedien Wiesbaden; 2014. p. 603-614.

[360] Hermann M, Pentek T, Otto B. Design Principles for Industrie 4.0 Scenarios: A Literature Review. Dortmund, Germany: Technische Universität Dortmund; 2015.

[361] Rüßmann M, Lorenz M, Gerbert P, Waldner M, Justus J, Engel P, Harnisch M. Industry 4.0: The Future of Productivity and Growth in Manufacturing Industries. Boston, MA: Boston Consulting Group; 2015.

[362] European Commission. Commission Sets Out Path to Digitise European Industry. IP/16/1407 [Internet]. 2016. Available from: http://web.archive.org/web/

20200403145542/https://ec.europa.eu/commission/presscorner/detail/en/IP_16_1407 [Accessed: 2021-02-10]

[363] Danieluk K. Digital Innovation Hubs on Smart Factories in New EU Member States [Internet]. 2017. Available from: https://ec.europa.eu/futurium/en/implementing-digitising-european-industry-actions/digital-innovation-hubs-smart-factories-new-eu [Accessed: 2021-02-10]

[364] Lu BH, Bateman RJ, Cheng K. RFID Enabled Manufacturing: Fundamentals, Methodology and Applications. International Journal of Agile Systems and Management. 2006;1:73-92. DOI: 10.1504/IJASM. 2006.008860

[365] Giusto D, Iera A, Morabito G, Atzori L, editors. The Internet of Things. New York: Springer; 2010.

[366] Zhu Q, Wang R, Chen Q, Liu Y, Qin W. IOT Gateway: Bridging Wireless Sensor Networks into Internet of Things. In: 2010 IEEE/IFIP International Conference on Embedded and Ubiquitous Computing. Piscataway, NJ: The Institute of Electrical and Electronics Engineers; 2010. p. 347-352.

[367] Wan J, Yan H, Liu Q, Zhou K, Lu R, Li D. Enabling Cyber-Physical Systems with Machine-to-Machine Technologies. International Journal of Ad Hoc and Ubiquitous Computing. 2013;13:187-196. DOI: 10.1504/IJAHUC.2013.055454

[368] Gubbi J, Buyya R, Marusic S, Palaniswami M. Internet of Things (IoT): A Vision, Architectural Elements, and Future Directions. Future Generation Computer Systems. 2013;29: 1645-1660. DOI: 10.1016/ j.future.2013.01.010

[369] Zhong RY, Li Z, Pang LY, Pan Y, Qu T, Huang GQ. RFID-Enabled Real-Time Advanced Planning and Scheduling Shell for Production Decision Making. International Journal of Computer Integrated Manufacturing. 2013;26:649-662. DOI: 10.1080/0951192X.2012.749532

[370] Wu D-Z, Greer MJ, Rosen DW, Schaefer D. Cloud Manufacturing: Strategic Vision and State-of-the-Art. Journal of Manufacturing Systems. 2013;32:564-579. DOI: 10.1016/j.jmsy.2013.04.008

[371] Moreno-Vozmediano R, Montero RS, Llorente IM. Key Challenges in Cloud Computing: Enabling the Future Internet of Services. IEEE Internet Computing. 2013;17: 18-25. DOI: 10.1109/MIC.2012.69

[372] Lee J, Kao H-A, Yang S. Service Innovation and Smart Analytics for Industry 4.0 and Big Data Environment. Procedia CIRP. 2014; 16:3-8. DOI: 10.1016/j.procir.2014.02.001

[373] Bi Z, Xu LD, Wang C. Internet of Things for Enterprise Systems of Modern Manufacturing. IEEE Transactions on Industrial Informatics. 2014;10:1537-1546. DOI: 10.1109/TII.2014.2300338

[374] Brettel M, Friederichsen N, Keller M, Rosenberg M. How Virtualization, Decentralization, and Network-Building Change the Manufacturing Landscape: An Industry 4.0 Perspective. International Journal of Mechanical, Aerospace, Industrial, Mechatronic, and Manufacturing Engineering. 2014;8: 37-44.

[375] Buer S-V, Strandhagen JO, Chan FTS. The Link between Industry 4.0 and Lean Manufacturing: Mapping Current Research and Establishing a Research Agenda. International Journal of Production Research. 2018;56:2924-2940. DOI: 10.1080/00207543.2018.1442945

[376] Patel P, Cassou D. Enabling High-Level Application Development for the Internet of Things. Journal of Systems and Software. 2015;103:62-84. DOI: 10.1016/j.jss.2015.01.027

[377] Zhang Y, Zhang G, Wang J, Sun S, Si S, Yang T. Real-Time Information Capturing and Integration Framework of the Internet of Manufacturing Things. International Journal of Computer Integrated Manufacturing. 2015;28:811-822. DOI: 10.1080/0951192X.2014.900874

[378] Qiu X, Luo H, Xu G, Zhong R-Y, Huang GQ. Physical Assets and Service Sharing for IoT-Enabled Supply Hub in Industrial Park (SHIP). International Journal of Production Economics. 2015;159:4-15. DOI: 10. 1016/j.ijpe.2014.09.001

[379] Posada J, Toro C, Barandiaran I, Oyarzun D, Stricker D, de Amicis R, Pinto EB, Eisert P, Döllner J, Vallarino I. Visual Computing as a Key Enabling Technology for Industrie 4.0 and Industrial Internet. IEEE Computer Graphics and Applications. 2015;35:26-40. DOI: 10.1109/MCG.2015.45

[380] Farooq MU, Waseem M, Mazhar S, Khairi A, Kamal T. A Review on Internet of Things (IoT). International Journal of Computer Applications. 2015;113:1-7. DOI: 10.5120/19787-1571

[381] Lee J, Bagheri B, Kao H-A. A Cyber-Physical Systems Architecture for Industry 4.0-Based Manufacturing Systems. Manufacturing Letters. 2015;3:18-23. DOI: 10.1016/j.mfglet.2014.12.001

[382] Almada-Lobo F. The Industry 4.0 Revolution and the Future of Manufacturing Execution Systems (MES). Journal of Innovation Management. 2015;3:16-21. DOI: 10.24840/2183-0606_003.004_0003

[383] Wamba SF, Akter S, Edwards A, Chopin G, Gnanzou D. How 'Big Data' Can Make Big Impact: Findings from a Systematic Review and a Longitudinal Case Study. International Journal of Production Economics. 2015;165:234-246. DOI: 10.1016/j.ijpe.2014.12.031

[384] Yin YH, Nee AYC, Ong SK, Zhu JY, Gu PH, Chen LJ. Automating Design with Intelligent Human-Machine Integration. CIRP Annals. 2015;64:655-677. DOI: 10.1016/j.cirp.2015.05.008

[385] Colin M, Galindo R, Hernández O. Information and Communication Technology as a Key Strategy for Efficient Supply Chain Management in Manufacturing SMEs. Procedia Computer Science. 2015;55:833-842. DOI: 10.1016/j.procs.2015.07.152

[386] Hozdic E. Smart Factory for Industry 4.0: A Review. International Journal of Modern Manufacturing Technologies. 2015;7:28-35.

[387] Zhong RY, Huang GQ, Lan S, Dai QY, Zhang T, Xu C. A Two-Level Advanced Production Planning and Scheduling Model for RFID-Enabled Ubiquitous Manufacturing. Advanced Engineering Informatics. 2015;29:799-812. DOI: 10.1016/j.aei.2015.01.002

[388] Wang S, Wan J, Zhang D, Li D, Zhang C. Towards Smart Factory for Industry 4.0: A Self-Organized Multi-Agent System with Big Data Based Feedback and Coordination. Computer Networks. 2016;101:158-168. DOI: 10.1016/j.comnet.2015.12.017

[389] Monostori L, Kádár B, Bauernhansl T, Kondoh S, Kumara S, Reinhart G, Sauer O, Schuh G, Sihn W, Ueda K. Cyber-Physical Systems in Manufacturing. CIRP Annals. 2016;65:621-641. DOI: 10.1016/j.cirp.2016.06.005

[390] Zhong RY, Lan S, Xu C, Dai Q, Huang GQ. Visualization of RFID-Enabled Shopfloor Logistics Big Data in Cloud Manufacturing. International

Journal of Advanced Manufacturing Technology. 2016;84:5-16. DOI: 10.1007/s00170-015-7702-1

[391] Georgakopoulos D, Jayaraman PP, Fazia M, Villari M, Ranjan R. Internet of Things and Edge Cloud Computing Roadmap for Manufacturing. IEEE Cloud Computing. 2016;3:66-73. DOI: 10.1109/MCC. 2016.91

[392] Zhong RY, Newman ST, Huang GQ, Lan S. Big Data for Supply Chain Management in the Service and Manufacturing Sectors: Challenges, Opportunities, and Future Perspectives. Computers & Industrial Engineering. 2016;101:572-591. DOI: 10.1016/j.cie.2016.07.013

[393] Misra G, Kumar V, Agarwal A, Agarwal K. Internet of Things (IoT) – A Technological Analysis and Survey on Vision, Concepts, Challenges, Innovation Directions, Technologies, and Applications (An Upcoming or Future Generation Computer Communication System Technology). American Journal of Electrical and Electronic Engineering. 2016;4:23-32. DOI: 10.12691/ajeee-4-1-4

[394] Schumacher A, Erol S, Sihn W. A Maturity Model for Assessing Industry 4.0 Readiness and Maturity of Manufacturing Enterprises. Procedia CIRP. 2016;52:161-166. DOI: 10.1016/j.procir.2016.07.040

[395] Sipsas K, Alexopoulos K, Xanthakis V, Chryssolouris G. Collaborative Maintenance in Flow-Line Manufacturing Environments: An Industry 4.0 Approach. Procedia CIRP. 2016;55:236-241. DOI: 10.1016/j.procir.2016.09. 013

[396] Qin J, Liu Y, Grosvenor R. A Categorical Framework of Manufacturing for Industry 4.0 and Beyond. Procedia CIRP. 2016;52:173-178. DOI: 10.1016/j.procir.2016.08.005

[397] DIN SPEC 91345:2016-04. Referenzarchitekturmodell Industrie 4.0 (RAMI4.0). Berlin: Beuth Verlag; 2016.

[398] Dobrzański LA. Comparative Analysis of Mechanical Properties of Scaffolds Sintered from Ti and Ti6Al4V Powders. Lecture at the22nd Winter International Scientific Conference on Achievements in Mechanical and Materials Engineering; 6-9 December 2015; Zakopane, Poland.

[399] Adolphs P, Auer S, Bedenbender H, Billmann M, Hankel M, Heidel R, Hoffmeister M, Huhle H, Jochem M, Kiele-Dunsche M, Koschnick G, Koziolek H, Linke L, Pichler R, Schewe F, Schneider K, Waser B. Structure of the Administration Shell: Continuation of the Development of the Reference Model for the Industrie 4.0 Component. Berlin: Federal Ministry for Economic Affairs and Energy; 2016.

[400] Bahrin MAK, Othman MF, Azli NHN, Talib MF. Industry 4.0: A Review on Industrial Automation and Robotic. Jurnal Teknologi. 2016;78:137-143. DOI: 10.11113/jt.v78.9285

[401] Mosterman PJ, Zander J. Industry 4.0 as a Cyber-Physical System Study. Software & Systems Modeling. 2016;15:17-29. DOI: 10.1007/s10270-015-0493-x

[402] Harrison R, Vera D, Ahmad B. Engineering Methods and Tools for Cyber-Physical Automation Systems. Proceedings of the IEEE. 2016;104:973-985. DOI: 10.1109/JPROC.2015.2510665

[403] Stock T, Seliger G. Opportunities of Sustainable Manufacturing in Industry 4.0. Procedia CIRP. 2016;40:536-541. DOI: 10.1016/j.procir.2016.01.129

[404] Pfeiffer S. Robots, Industry 4.0 and Humans, or Why Assembly Work Is More than Routine Work. Societies. 2016;6:16. DOI: 10.3390/soc6020016

[405] Gorkhali A, Xu LD. Enterprise Application Integration in Industrial Integration: A Literature Review. Journal of Industrial Integration and Management. 2016;1:1650014. DOI: 10.1142/S2424862216500147

[406] Zhong RY, Xu X, Klotz E, Newman ST. Intelligent Manufacturing in the Context of Industry 4.0: A Review. Engineering. 2017;3:616-630. DOI: 10.1016/J.ENG.2017.05.015

[407] Thoben K-D, Wiesner S, Wuest T. 'Industrie 4.0' and Smart Manufacturing – A Review of Research Issues and Application Examples. International Journal of Automation Technology. 2017;11: 4-16. DOI: 10.20965/ijat.2017.p0004

[408] Xu X. Machine Tool 4.0 for the New Era of Manufacturing. International Journal of Advanced Manufacturing Technology. 2017;92: 1893-1900. DOI: 10.1007/s00170-017-0300-7

[409] Li B-H, Hou B-C, Yu W-T, Lu X-B, Yang C-W. Applications of Artificial Intelligence in Intelligent Manufacturing: A Review. Frontiers of Information Technology & Electronic Engineering. 2017;18:86-96. DOI: 10.1631/FITEE.1601885

[410] Lu Y. Industry 4.0: A Survey on Technologies, Applications and Open Research Issues. Journal of Industrial Information Integration. 2017;6:1-10. DOI: 10.1016/j.jii.2017.04.005

[411] Vaidya S, Ambad P, Bhosle S. Industry 4.0 – A Glimpse. Procedia Manufacturing. 2018;20:233-238. DOI: 10.1016/j.promfg.2018.02.034

[412] Kumar K, Zindani D, Davim JP. Industry 4.0: Developments towards the Fourth Industrial Revolution. Singapore: Springer Nature Singapore Pte Ltd.; 2019.

[413] Łobaziewicz M. Zarządzanie inteligentnym przedsiębiorstwem w dobie przemysłu 4.0. Toruń, Poland: Towarzystwo Naukowe Organizacji i Kierownictwa; 2019.

[414] Ardito L, Petruzzelli AM, Panniello U, Garavelli AC. Towards Industry 4.0: Mapping Digital Technologies for Supply Chain Management-Marketing Integration. Business Process Management Journal. 2019;25:323-346. DOI: 10.1108/BPMJ-04-2017-0088

[415] Boston Consulting Group. Embracing Industry 4.0 and Rediscovering Growth [Internet]. 2019. Available from: https://www.bcg.com/pl-pl/capabilities/operations/embracing-industry-4.0-rediscovering- growth.aspx [Accessed: 2021-02-10]

[416] Tietenberg T, Lewis L, Environmental and Natural Resource Economics. London: Pearson Education; 2003.

[417] Tay SI, Lee TC, Hamid NAA, Ahmad ANA. An Overview of Industry 4.0: Definition, Components, and Government Initiatives. Journal of Advanced Research in Dynamical and Control Systems. 2018;10:1379-1387.

[418] Dobrzański LA, Dobrzański LB, Dobrzańska-Danikiewicz AD, Dobrzańska J. The Concept of Sustainable Development of Modern Dentistry. Processes. 2020;8:1605. DOI: 10.3390/pr8121605

[419] Aluminum Market - Industry Analysis, Market Size, Share, Trends, Application Analysis, Growth And Forecast 2019-2025 [Internet]. Available

from: https://www.industryarc.com/Research/Aluminium-Market-Research-503190 [Accessed: 2021-02-10]

[420] Aluminum Market - Growth, Trends, and Forecast (2020-2025) [Internet]. 2020. Available from: https://www.researchandmarkets.com/reports/5025633/aluminum-market-growth-trends-and-forecast [Accessed: 2021-02-10]

[421] The Aluminum Association. The Economic Impact of Aluminum. Aluminum Drives Modern Manufacturing [Internet]. Available from: https://www.aluminum.org/aluminum-advantage/economic-impact-aluminum [Accessed: 2021-02-10]

Chapter 2

The Theoretical Overview of the Selected Optimization and Prediction Models Useful in the Design of Aluminum Alloys and Aluminum Matrix Composites

Halil Ibrahim Kurt, Engin Ergul, Necip Fazil Yilmaz and Murat Oduncuoglu

Abstract

The growing attention regarding aluminum alloy matrix composites within the aerospace, automotive, defense, and transportation industries make the development of new engineering materials with the improved mechanical properties. Currently, materials are selected because of their abilities to satisfy engineering demands high for strength-to-weight ratio, tensile strength, corrosion resistance, and workability. These properties make aluminum alloys and aluminum matrix composites (AMCs) an excellent option for various industrial applications. Soft computing methods such as the artificial neural network (ANN), adaptive-neuro fuzzy inference systems (ANFIS), and Taguchi with ANOVA are the most important approaches to solve the details of the mechanism and structure of materials. The optimal selection of variables has important effects on the final properties of the alloys and composites. The chapter presents original research papers from our works and taken from literature studies dealing with the theory of ANN, ANFIS, and Taguchi, and their applications in engineering design and manufacturing of aluminum alloys and AMCs. Also, the chapter identifies the strengths and limitations of the techniques. The ANFIS and ANN approaches stand out with wide properties, optimization, and prediction, and to solving the complex problems while the Taguchi experimental design technique provides the optimum results with fewer experiments.

Keywords: aluminum matrix composites, hybrid, modeling, engineering approaches, optimization, ANN, Taguchi, ANOVA, genetic algorithms, ANFIS

1. Introduction

Since the early 1920s, iron-based materials, which have an indisputable advantage in the industry, have gradually begun to leave their places to materials with high specific strength like metal matrix composites. Metal matrix composites have interesting physical and mechanical properties. In metal matrix composites, the properties of the matrix material and the properties of reinforcing materials are

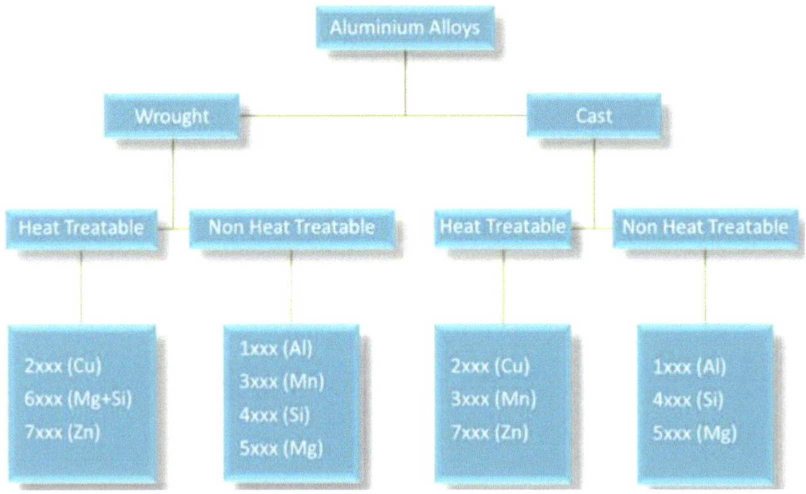

Figure 1.
Classification of aluminum alloys.

combined, resulting in higher mechanical and performance properties. Production costs are also an important factor, as well as the physical and mechanical properties of structural materials. Although high technology materials exhibit high physical and mechanical properties, high production costs restrict their use. Metal matrix composites are widely used in aerospace, automobile, military, and biomedical applications because of their high specific strength and considerably low density.

In the fabrication of MMCs, aluminum (Al) is one of the most popular matrix materials because of its low density, good corrosion resistance, and strengthening capability. Aluminum (Al) is a chemical element with the atomic number 13 and symbol Al. It is a non-magnetic and ductile metal that seems a color of silver in the boron group. Al is a metal with an atomic weight of 26.981 g/mol, melting temperature of 660°C and a density of 2.7 g/cm^3. The development of new materials will be of greater importance in future technological advances. Aluminum alloys and aluminum matrix composites can combine the beneficial properties of aluminum and other metals, ceramics and production techniques.

The mechanical, physical and chemical properties of aluminum alloys vary depending on the alloy elements and microstructure. Aluminum alloys are divided into two groups as wrought (forged) and cast alloys (**Figure 1**). This classification is as follows:

Given the composite materials, it can be said that one of the most important is the method of production. Production methods are classified according to the temperature of the metallic matrix during production. Aluminum matrix composites can be produced by many techniques such, as stir casting, compo casting, powder metallurgy, additive manufacturing, cold spray, friction stir processing, and infiltration, etc. All these methods have different advantages and disadvantages in terms of cost, appropriateness, labor, training, efficiency, time, temperature, and simplicity, etc. Therefore, the production methods of metal matrix composite materials can be divided into four groups;

1. Liquid phase production methods

    - Vortex addition technique

    - Compo-casting

- Pressure-less infiltration process
- Ultrasonic infiltration

2. Solid phase production methods

- Powder metallurgical methods
- Mechanical alloying
- Diffusion bonding
- Spark plasma sintering (SPS)

3. Gaseous state fabrication

- Chemical vapor deposition (CVD)
- Physical vapor deposition (PVD)

4. In-situ production method

- Internal oxidation process
- Unidirectional solidification process

Aluminum matrix composites can be produced with different kinds of reinforcement like MgO, SiC, Al_2O_3, B_4C, CNT and fibers are used to fabricate composites. **Figure 2** shows the reinforcement types for aluminum matrix composites. The type and ratio of matrix material and reinforcement and process parameters are some of the most variables in composite fabrication.

The most popular reinforcements for fabrication of aluminum matrix composites are carbon nanotubes (CNTs) and silicon carbide (SiC). For individual MWCNTs, they can achieve an elastic module approaching 1 TPa and tensile strength of 100 GPa. Especially, CNT are used to advance to strength of aluminum alloys. SiC is a material with low thermal conductivity; low thermal expansion coefficient, high thermal shock resistance, hard, semiconductor and greater

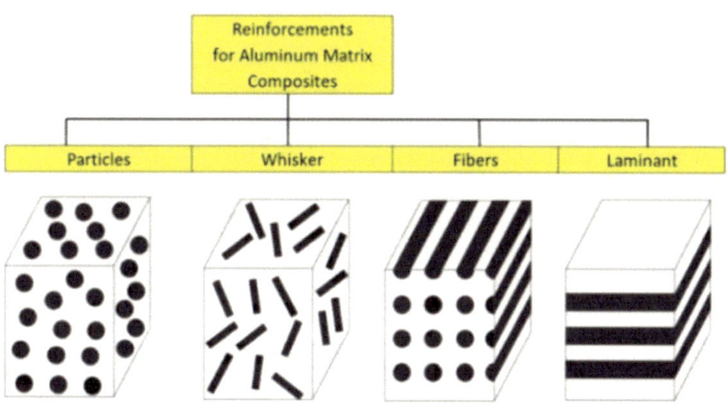

Figure 2.
Reinforcement types.

refractive index value than diamond and is one of the most suitable reinforcement materials to make highly wear-resistant composites. The role of the reinforcement in a composite material is fundamentally one of increasing the mechanical properties. The final properties of the composites reinforcement depend on the individual properties of the reinforcement selected and the matrix.

Aluminum matrix composites are utilized in a wide range of components for advanced manufacturing. Mechanical behavior of aluminum matrix composites can be examined in detail. Increased mechanical properties compared to non-reinforced aluminum enables the applications of these materials in different engineering fields to increase. The low costs of particle reinforced aluminum matrix composites in some applications compared to other composites make them very attractive materials. The fact that these materials exhibit good mechanical behavior in high-temperature applications is another important point for their commercial success.

Thus, it is important to select the optimal levels of the parameters and variables. Therefore, experimental parameter relationships can be accurately predicted and the need for materials and time can be eliminated using these approaches, ANFIS and Taguchi, are widely used on the complicated and nonlinear systems of the different engineering applications. These approaches are widely used in advanced engineering applications as they are a combination of various calculation methods. They also include process design, numerical modeling, estimation and optimization, and the control process. Modeling, estimation and optimization of features are useful and the most important part of engineering to solve complex problem. Important advantages of the calculation techniques are determined during the design and optimization of the process variables to be used during the production phase. These methods are useful for selecting the desired parameters optimally, placing them in systems, analyzing results, digitizing production, minimizing power consumption, and solving real-life problems in the process.

2. Production methods and data collection

The prediction or optimization methods require large experimental data sets that are expensive to produce. Here we describe methods to discover material parameters in the absence of experimental data. In effect, this algorithm strategy starts with the ability to "learn" and from its experience to accelerate the evolutionary process. This algorithm is tested against several problems and demonstrates that it matches and typically exceeds the efficiency and reproducibility of standard. The success of these methods in a range of problems is to accelerate materials design in the absence of a lack of experimental data.

In addition to experimental effort, modeling with artificial intelligence (AI) is one of the most important approaches to solve the details of the system and make life easier. The purpose of artificial intelligence is to obtain results with high efficiency using knowledge and make it a reality. The most commonly used artificial intelligence techniques to solve complex problems are the adaptive network-based fuzzy inference systems (ANFIS), Taguchi and artificial neural networks (ANN), and these systems are also called soft computing methods. The use of soft computing techniques is powerful modeling techniques related to the statistical approach for predicting parameters.

Over the last decades, the interest of the modeling techniques in different fields of materials science has been increased [1–7]. It is aimed to find the optimum solution with better performance by using human intuition, thinking and decision-making ability, eliminating uncertainties with simple and low-cost solutions and in solving complex and difficult to solve problems. In our previous works, new formulations are developed for the ultimate tension strength (UTS), wettability, critical angle, wear properties of aluminum alloys and composites and welding properties of produced metals using ANN, Taguchi, ANOVA and ANFIS, respectively.

3. Optimization methods

3.1 The artificial neural networks (ANNs)

Artificial intelligence is defined in the world of science as the ability of a computer or a computer-assisted machine to perform tasks related to higher logic processes such as human qualities, finding solutions, and understanding, making sense, generalizing and learning from past experiences. Learning ability underlies the logic of artificial intelligence. The greatest contribution of artificial intelligence will be to implement the most correct way they have learned very quickly. Artificial intelligence technologies consist of expert systems, fuzzy logic artificial, neural networks and machine learning and genetic algorithms [8].

ANNs are computer software where basic functions such as generating new data from the data collected by the brain by learning, remembering, and generalizing by imitating the learning path of the human brain. ANN is synthetic structure that mimics biological neural networks. ANNs; inspired by the human brain, it has emerged as a result of the mathematical modeling of the learning process.

3.1.1 ANNs structure

Since artificial neural networks are modeling of biological neural networks, first of all, it is necessary to look at the structure of the biological nervous system. The structure of neurons, the basic building block of the biological nervous system, consists of four main parts; dendrite, axon, nucleus and connections (**Figure 3a**). It has a tree-rooted structure located at the end of the dendrites nerve cell. The task of dendrites is to transmit signals from other neurons or sense organs to which it is attached to the nucleus. The nucleus collects the signals coming from the dendrite and transmits them to the axon. These collected signals are processed by the axon and sent to the connections at the other end of the neuron. Connections transmit newly produced signals to other neurons.

Biological nerve cell and artificial neural network simulations are given in **Figure 3a** and **b**. As shown in **Figure 3b**, n data is entered into a cell. The entered data is multiplied by weights and all data are collected and then bias is added, resulting in clear judgment. The net input is passed through the activation function and data output is obtained.

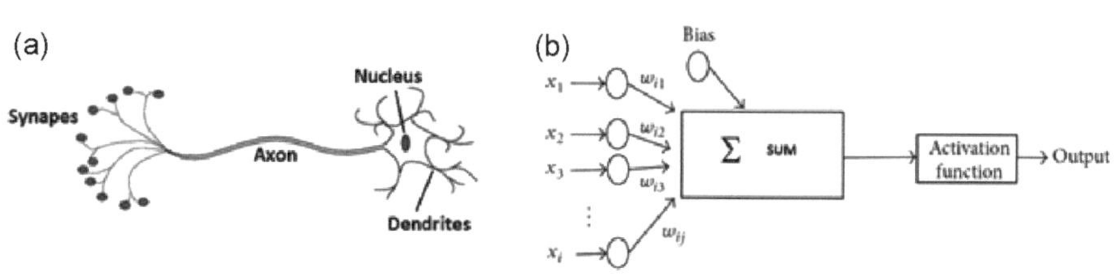

Figure 3.
(a) Biological nerve cell and (b) artificial neural network.

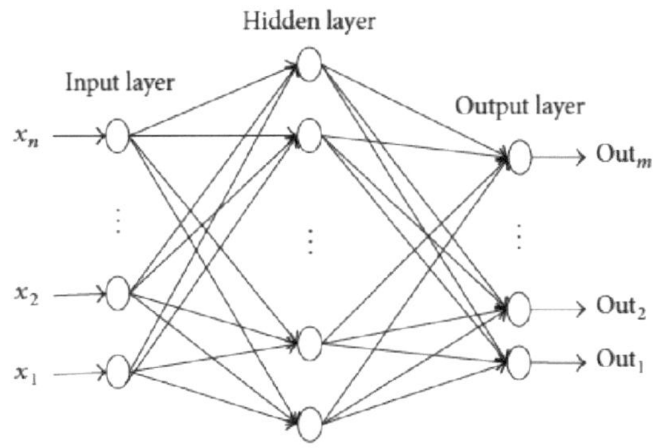

Figure 4.
Layers in ANN.

Artificial neural networks are structures formed by the binding of artificial nerve cells. Artificial neural networks are examined in three main layers; input layer, hidden layers and output layer (**Figures 3b** and **4**) [9].

Input layer: it is the layer where inputs from the outside world come to the artificial neural network. Although there are as many cells in this layer as the number of entries from the outside world, the inputs are usually transmitted to the lower layers without any processing.

Hidden layer(s): information from the input layer comes to this layer. The number of hidden layers can vary from network to network. The number of neurons in the intermediate layers is independent of the number of inputs and outputs. Although increasing the number of intermediate layers and the number of neurons in these layers increases the computational complexity and duration, the artificial neural network can also be used in the solution of more complex problems.

Output layer(s): it is the layer that produces the outputs of the network by processing the information from the intermediate layers. The outputs produced in this layer are sent to the outside world. New weight values of the network are calculated by using the output produced in this layer in feedback networks.

Artificial nerve cells are similar to biological nerve cells. Artificial neurons also form artificial neural networks by bonding between them. Just like biological neurons, artificial neurons have sections where they receive input signals, collect and process these signals, and transmit outputs. An artificial nerve cell consists of five parts;

i. **Inputs:** inputs are data coming to neurons.

ii. **Weights:** the information coming to the artificial nerve cell is transmitted to the nucleus by multiplying the weight of the connections they arrive before reaching the nucleus through the inputs. The values of these weights can be positive, negative or zero.

iii. **Transfer function (combining function):** transfer function is a function that calculates the net input of that cell by multiplying the incoming inputs by multiplying the weights of an artificial nerve cell. Some transfer functions are given in **Table 1**.

iv. **Activation function:** this function designates the response that the cell will produce in response to this input by processing the net input to the cell. The activation function is usually chosen as a nonlinear function which are a feature of ANNs, come from non-linear feature. Today, "Sigmoid function and Tangent Hyperbolic Function" are the most widely used as the activation functions in general. **Table 2** shows the activation functions.

v. **Outputs:** the value of the activation function is the output value of the cell.

$Net = \sum_{i=1}^{N} X_i * W_i$	The weight values are multiplied by the inputs and the net values are calculated by adding together the found values.
$Net = \prod_{i=1}^{N} X_i * W_i$	The weight values are multiplied by the inputs and then the net input is calculated by multiplying the values found by each other.
$Net = Max(X_i * W_i)$	After the weights of n inputs are multiplied by the inputs, the largest of them is considered as net input.
$Net = Min(X_i * W_i)$	After the weights of n inputs are multiplied by the inputs, the smallest of them is considered as net input.

Table 1.
Transfer function.

Sigmoid function	$F(Net) = \frac{1}{1+e^{-Net}}$	The sigmoid activation function is continuous and derivable function. It is the most commonly used function in ANN applications due to its non-linearity. The function produces a value between 0 and 1 for each of the input values.
Tangent hyperbolic function	$F(Net) = \frac{e^{Net}+e^{-Net}}{e^{Net}-e^{-Net}}$	Tangent hyperbolic function is a function similar to sigmoid function. In the sigmoid function, the outcome values range from 0 to 1, while the output values of the hyperbolic tangent function range from −1 to 1.

Table 2.
Activation function.

3.1.2 Classification of ANNs

ANNs have the following key features such as, non-linearity, parallel operation, learning, generalization, error tolerance and flexibility, working with missing data, using multiple variables and parameters and adaptability. Artificial neural networks applications are mostly used in prediction, classification, data association, data interpretation and data filtering processes. In ANNs, according to their structure; artificial neural networks are divided into two as forward and feedback depending on the way the neurons they contain. There is only a link from one layer to the next layers. In contrast to feed-forward (FF) neural networks, feed-out of a cell is not only input to the layer of the cell that comes after it. It can also be linked as input to any cell in its previous layer or its layer. With this structure, feedback neural networks display a nonlinear dynamic behavior. ANNs are divided into three as consulting, advisor-less and reinforced learning according to learning algorithms. Artificial neural networks are divided into two as static and dynamic learning according to learning time. According to layers, single layer networks consist only of input and output. In multilayer sensors, the structure to which many neurons, which are structurally nonlinear activation functions, are connected with certain superiority, are called multilayer sensors.

3.1.3 Training and testing of ANNs

Although the structure of an ANN and the number of nerve cells vary, there are no accepted rules for the formation of an artificial neural network. While artificial neural networks with less than the required hidden layers are insufficient in solving complicated functions, artificial neural networks with too many hidden layers encounter undesirable instability. The problem encountered after determining the number of hidden layers is in deciding how many neurons will be present in each layer. There is no problem with the input layer; this number is equal to the number of inputs in the system. Likewise, the output layer can be determined by the desired output number. The main problem is to specify the neurons number in the hidden layers. The traditional matrix algorithm says that the matrix dimensions must be either equal to the inputs number or the number of outputs. Unfortunately, there is no mathematical test about how many neurons will be found in the hidden layer in the most efficiently. The decision should be made by applying the trial and error method [10].

In the learning process of artificial neural networks, inputs are received from the external environment; a reaction output is generated by passing through the activation function. This output is again compared to the output given by experience. Errors are found with various learning algorithms and the real output is tried to be approached. In general, 80% of the samples are given to the network and the network is trained. Then the remaining 20% is given and the behavior of the network is examined. Thus, the network is tested.

It is the step of finding examples that have already occurred for the event that the network wants to learn. As the samples are collected for training the network (training set), the samples (test set) must be collected to test the network. After learning the network event, the performance of the network is measured by showing the examples in the test set. His success against the examples he has never seen reveals whether the network has learned well.

3.2 Adaptive-neuro fuzzy inference systems (ANFIS)

ANFIS is a kind of artificial neural network that is based on Takagi-Sugeno fuzzy inference system. The technique is developed in the early 1990s [1]. Various methods have been developed to increase the effectiveness of fuzzy systems and to contribute to the adaptation technique. One of them is the ANFIS technique, in which the identification process is performed with a fuzzy model, the operation of which takes place within the adaptive network structure. Neural adaptive learning techniques allow developing a model that "learns" the system by using the data set for the fuzzy modeling procedure. The fuzzy model to be used in the identification of the system has acquired the ability to update itself by using the environmental information about the system and by utilizing the input and output data related to the system thanks to its adaptive network structure. Essentially, the ANFIS structure consists of the representation of Sugeno fuzzy systems as a network structure with neural learning capabilities. This network consists of a combination of nodes, each placed in layers, to perform a certain function. 52 fuzzy inference system selection of membership functions is arbitrary, it depends on the user. The form of membership functions also depends on the parameters. However, it cannot be easily noticed how some form of membership function should be based on the data in some models [2].

3.2.1 General architecture and operation

Neural adaptive learning techniques enable to develop a model that "learns" the system by using the data set for the fuzzy modeling procedure. In other words, ANFIS creates a fuzzy inference system (FIS) by editing the membership function parameters using the input/output data set back-propagation (BP) algorithm alone or in combination with the least-squares method. This arrangement allows the system to learn the related system with the help of data modeled by our fuzzy system. In other words, it adapts to the data it will model. It is therefore adaptable. Thanks to its adaptive network structure functioning, it has acquired the ability to update itself by using environmental information about the system as well as utilizing the input and output data related to the system. It also includes advanced data analysis techniques such as ANFIS, numerical grouping and rule sets.

ANFIS consists of six layers. The first layer is called the input layer. Input signals in this layer are transferred to other layers. Layer 2 is the fuzzification layer. Each output consists of membership degrees depending on the input values and the used membership function. Layer 3 is the rule layer; each node in this layer refers to the rules and the number created according to the Sugeno fuzzy logic inference system. Layer 4 is the normalization layer, accepting all nodes and calculating the normalized level of each rule. The 5th layer is the purification layer and the weighted result values of a rule given in each node are estimated. Layer 6 is the total layer with only one node ($\sum$). In this layer, the output value of each node is added and the actual output of the system is obtained.

3.2.2 Learning algorithm

ANFIS's learning algorithm is a hybrid learning algorithm that consists of using the least-squares method and the back-propagation learning algorithm. This learning algorithm is based on error back-propagation. There are two parts to a step in the learning process; in the first part, input samples are produced and the preliminary parameters are accepted as constant and the best final parameters are determined with the least mean square method. In the second part, the input samples are reproduced and the preliminary parameters are replaced by the gradient descent method, with the final parameters considered constant. This process is repeated later [3].

3.3 Taguchi

Taguchi design is a set of methodologies that take into account the variability inherent in the material and manufacturing processes at the design stage. Taguchi has not brought theoretical innovations to experimental design. However, it has made innovations in applications in production and has enabled the method to be accepted in the manufacturing sector with successful applications.

Traditional experimental designs are difficult to use, especially when dealing with a large number of experiments and increasing the number of processing parameters. Therefore, the Taguchi experimental design method ensures that more than one factor is taken into account at the same time, but it also ensures that the most optimum result is obtained by performing fewer experiments Design of experiment (DOE) in Taguchi is used to design the experimental run layout, to study the effect of level change in the process parameters on the output performance, because any change in the input parameters affects the output functional performance. It is important to know that all factors do not effect on the performance in the same manner [4].

3.3.1 S/N (signal/noise) ratio

Taguchi employs the signal-to-noise (S/N) ratio as its preferred sort characteristic. S/N ratio is employed as a finite value in place of a standard deviation. In its simplest form, the S/N ratio is the ratio of the mean (signal) to the standard deviation (noise). S/N ratio properties can be divided into three categories as shown in **Table 3**.

Regardless of the sort characteristic category, a higher S/N ratio comes up to better sort characteristics. Hence, the optimum degree of process variables is the degree by the highest S/N ratio.

The initiative of the Taguchi technique is to fix the sort characteristic to be optimized. The sort characteristic is a variable that has an important effect on the product class of the variation. It is the output value to be sighted. The next step is to describe noise values that can harm system performance and class. Noise values are parameters that cannot be checked or are very expensive to check. The third feature is the definition of control parameters that are considered to have important effects on quality features. Control variables are conception factors that can be adjusted and continued. The levels of each test are determined at this stage. The level number of each parameter defines the test area. The matrix experiment and the analysis procedure should be identified. First, the suitable orthogonal array for noise and check variables are selected. Taguchi has provided many standard orthogonal arrays for this aim. After choosing the proper arrays, a procedure must be defined to simulate the change in class characteristic owing to noise factors. Taguchi suggest an orthogonal array-based simulation to interpret the mean and variance of a product's output owing to alterations in noise factors. The next step is to make a matrix experiment and record the outcomes. The Taguchi technique can be employed in any situation in which there is a checkable operation. The checkable operation can be a real equipment test, mathematical equation, or computer pattern that can adequately pattern the reply of many yield or operation. After the experiments are made, the configuration of the most suitable parameter in DOE should be designated. To check up on the outcomes, the S/N ratio which is a calculation of performance to select the check levels that can deal with noise and considers both average and variability is employed as a performance criterion in the Taguchi technique. As a last step, experimental validation is made using the optimum levels predicted for the examined check variables. We can say that the Taguchi method is a powerful tool that can provide simultaneous improvements in quality and cost.

3.3.2 Analysis of variance (ANOVA)

Taguchi technique cannot judge and designate influence of individual factors on all operation, while the importance level and the contribution of individual factors can be very well specified by ANOVA [6]. Analysis of variance is a statistical

The lowest is the best.	$\frac{S}{N} : -10\log\left(\frac{1}{n}\sum_{i:1}^{n} y_i^2\right)$	In such problems, the target value of Y is zero. The smallest value represents the signal to noise ratio for its best condition.
The biggest is the best.	$\frac{S}{N} : -10\log\left(\frac{1}{n}\sum_{i:1}^{n} \frac{1}{y_i^2}\right)$	In this case, the value of Y is a non-negative measurable property with an ideal goal as infinity. The greatest value refers to the signal/noise ratio for the best case.
Nominal is the best.	$\frac{S}{N} : 10\log\left(\frac{\bar{y}}{s_y^2}\right)$	In this case, the nominal value is the target when we have a characteristic with double tolerance. So if all the parts are brought to this value, the variation is zero and the best. The target value represents the signal to noise ratio for its best condition.

y = response value, y′ = mean of the response value, s = standard deviation, and n = number of trails for given experiment.

Table 3.
S/N ratios.

instrument that is used to designate the difference or similarity between two or more data groups. ANOVA formally helps to find the significance of all main variables by comparing the mean square versus a calculation of the test faults at a specific class of confidence. The goal of experimentation is to find possible methods to reduce the deviation of the required quality as much as possible. This can be reached by identifying those parameters which play a significant role in the performance characteristic [5, 6].

4. Potential application of ANN, ANFIS, and Taguchi approaches for aluminum alloys and aluminum matrix composites

The effects of temperature, time, and the additions of magnesium and copper on the wetting behavior of Al/TiC are studied theoretically [7]. The R values of training and test sets are 0.911 and 0.903, respectively. The formulation is presented in explicit form. The proposed model shows good agreement with test results and can be used to find the wetting behavior of Al/TiC. The contribution of input parameters on the output is revealed with sensitivity analysis is shown in **Figure 5**. In the input parameters, the time and temperature have a stronger effect on the wetting of TiC system.

Effects of friction stir processing (FSP) parameters and reinforcements on the wear behavior of 6061-T6 based hybrid composites are investigated [11]. The different neuron numbers are used to determining the optimal architecture of the system. The system parameters affect the learning rate and so the prediction rate. It is obtained with 17 neurons through MSE, MAE and R values. The R^2 values of training and test sets are 0.998 and 0.995 which are quite high. The sensitivity analysis for the studied AMCs is given in **Figure 6**. The change in the applied load will be affected the wear volume loss of the composites. The applied load increases the wear in the composites.

A mathematical formulation is derived and given clearly to calculate the wear volume loss of the composites. The influence of input variables on the wear volume loss of the composites is also investigated using the prepared formulation. The wear volume loss of the composites significantly enhanced with increasing sliding distance and tool traverse and rotational speeds. A minimum wear volume loss for the hybrid composites with complex reinforcements is specified at the inclusion ratio of 50% TiC +50% Al_2O_3 because of improved lubricant ability, as well as resistance to brittleness and wear. It is clear that the formulation can be used in prediction of wear loss of the composites and so, the time and production cost can be reduced.

The effects of FSP parameters and hybrid ratio on the UTS of Al matrix (5083) hybrid composites are investigated in detail [12]. The numbers of different neurons

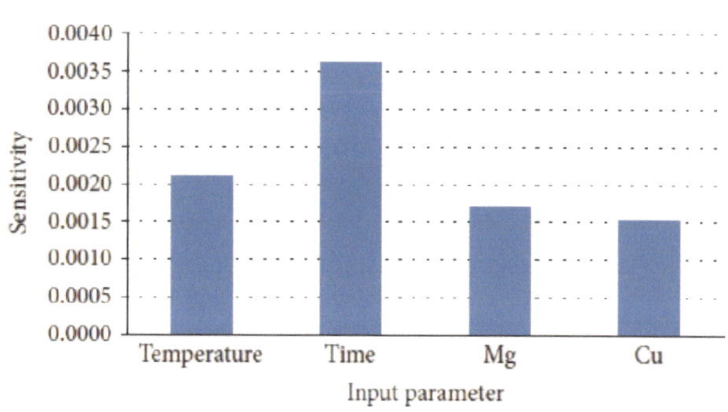

Figure 5.
Sensitivity of the input parameters for Al/TiC system.

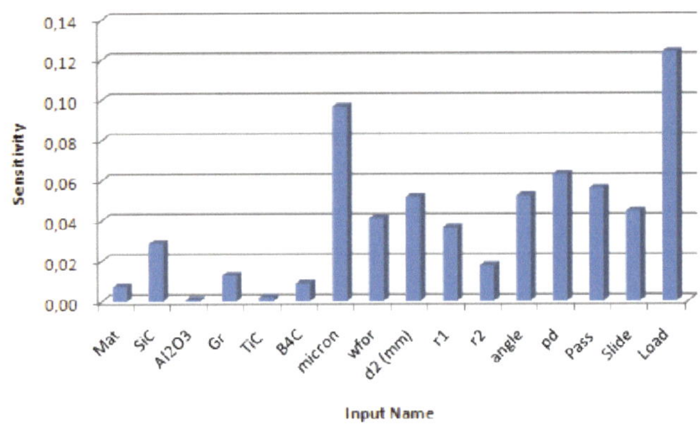

Figure 6.
Sensitivity of the input parameters for 6061 AMCs.

in one hidden layer (10–15) are used to determine the optimum model architecture. The optimal model architecture is conducted with 15 neurons. The MSE, MAE and MAPE are utilized as error-evaluation criteria, and the correlation coefficient (R) is chosen to estimate the performance of the proposed model. The maximum R and minimum error values are obtained with 15 neurons. A mathematical formulation is derived, and test results are compared with those of the model. In **Figure 7**, the error percentage of the composites is showed. The average error is 11% for training set and 4% for test set. This shows that the prediction ability of the proposed model can be accepted.

The effects of factors influencing strength, such as tool rotational and traverse speeds, and volume fractions of, carbon nanotube (CNT), aluminum oxide (Al_2O_3), graphite (Gr), silicon carbide (SiC) and zirconium oxide (ZrO_2) are also studied using the proposed formulation. Test results showed that the UTS of these composites significantly increased with increasing CNT, tool rotational and traverse speeds. In addition, the effects of complex reinforcements with different volume fractions on the 5083 AMHCs are examined. A maximum tensile strength for the hybrid composites is found at the inclusion ratio of 10% Gr + 5% ZrO_2.

The effect of different alloying elements on the ultimate tensile strength of Al-Mg_2Si composites is studied using ANN [13]. The input variables are Al, Mg, Si, copper, manganese, chromium, phosphorus, beryllium, boron, lithium, yttrium and sodium wt.% and the output is UTS in unit of MPa. Three different neuron numbers in one hidden layer (12, 13 and 14) are used. The training data set (70%) the validation data set (15%) and test data set (15%) are used. The optimal architecture is found to be 12-12-1 architecture with logistic sigmoid transfer function. The R, MSE and MAE values are used as the error criteria. The minimum MSE and MAE and the maximum R values are obtained in test set. The sensitivity of input vectors

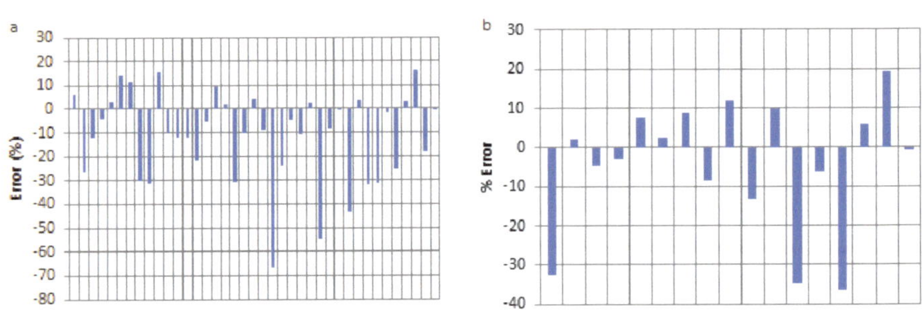

Figure 7.
Error values of the composites: (a) training set and (b) test set.

on UTS of Al-Mg_2Si composites is given in **Figure 8**. It is clear that Mg has more impact on UTS of Al-Mg_2Si composites because there are a linear relation between the size and morphology of the Mg_2Si phases and the mechanical properties of the composites containing Mg and Si elements. The results showed that all the data sets have quite high correlation and accuracy and therefore, the proposed mathematical function can be used in ANN studies.

The UTS of unrefined Al-Zn-Mg-Cu alloys and refined the alloys by Al-5Ti-1B and Al-5Zr master alloys are calculated with ANN [16]. There is no well-defined procedure to determine the optimal model structure, so the different neuron numbers in one hidden layer (5–20) are used with the trial and error approach. The optimal structure for this works is the 15-17-1 with logistic sigmoid transfer function. The R, MAE and MSE are used for the performance of datasets. The sensitivity results (**Figure 9**) display that the Mg element and heat treatment have the higher effect on the UTS of the Al-Zn-Mg-Cu Alloy. Because alloying elements interact with other metals and form intermetallic compounds, and these compounds are precipitated by heat treatment, resulting in high strength. The mathematical formula is obtained and the influences of scandium and carbon contents are researched using the formulation. The optimum additions of scandium and carbon rates are observed to be 0.5 Sc and 0.01 C wt.% to obtain the maximum UTS value. The prediction model with the obtained formulation has a high reliability rate.

The UTS, ductility, porosity, hardness and density of Al-Mg-Ti alloys are studied by ANN. The influences of input parameters are examined by the sensitivity

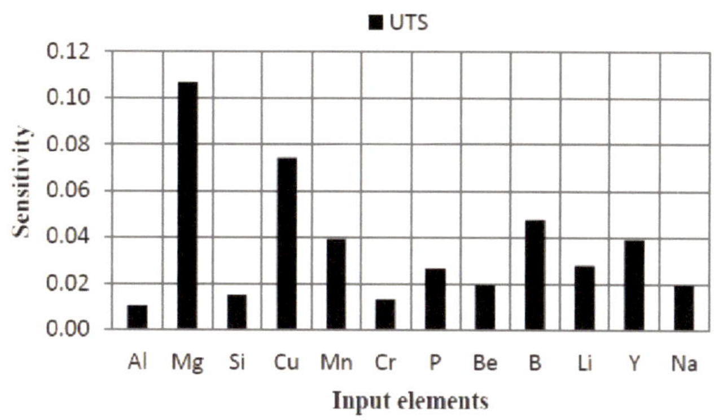

Figure 8.
Sensitivity of the input parameters for the alloy [14, 15].

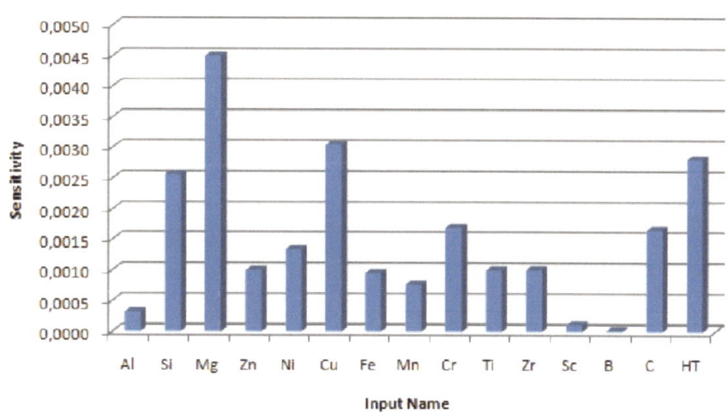

Figure 9.
Sensitivity of the input parameters for the Al-Zn-Mg-Cu alloy.

analysis. The Mg element within all input variables has the highest effect on the UTS and hardness of the alloys while the Ti element has the density and ductility. The linear correlation values for all variables are higher than 0.91 that the model accuracy is very high (**Figure 10**).

Satyanarayana et al. [17] researched the influence of reinforcement and deformation on volumetric wear of aluminum matrix composites the reinforced with red mud nano-particles using ANN and regression model. Authors used the activation function of sigmoid function, the RMSE and MAPE, four input parameters, two hidden layers with seven and six neurons, one output parameter that is volumetric wear 124 data for training set and 20 data for test set in the ANN model. The R^2 and MAPE values of regression and ANN models are 0.9775 and 0.989, and 12.96 and 7.30%, respectively, and RMSE for ANN models is 0.3177. They observed that ANN approach predicted the wear rate of the composites with excellent agreement than mathematical regression model and it could be useful to decrease time, effort and cost.

The hardness, ultimate tensile strength (UTS), and yield strength (YS) of A413/B4C composites that are produced with squeeze casting route are modeled using ANN and statistical modeling [18]. Authors used the 18 data for training, 9 data for testing, the hyperbolic tangent sigmoid function (TANSIG) and the linear transfer function (PURELIN) for the activation function, the Levenberg-Marquardt algorithm (TRAINLM) and the gradient descent with a momentum BP algorithm (TRAINGDM) for the training algorithm. There are layers of three inputs, three output and 50 + 50 neurons in two hidden layers. The data are normalized within the range (0–1) before training and testing. MSE, R and prediction percentage error are used the performance criteria of the system. They observed that the optimal architecture is 3-2-2-3 (the numbers of hidden layers and neurons are 2) with Levenberg-Marquardt algorithm and the results are in good agreement with experimental values. The R is 0.96 for hardness, 0.95 for UTS and YS, and MAPE is 1.42 for hardness, 0.62 for UTS and 0.59 for YS. It can be concluded that the cost and time could be saved with the proposed model. The full design of 3^3 (three levels and three factors) factorials are used to design connection between the input and outcome variables, and the design of experiment (DOE) with ANOVA is used to determine the significance of each factor on the responses. The pressure is 70, 105 and 140 MPa, preheating temperature of die is 150, 225 and 300°C, and B_4C rate is 4, 8 and 12 wt.%, respectively. The R^2 is 95.25% and the adjusted R^2 is 93.83% that shows a high effect of the model. The squeeze pressure with contributing of about 44–46% has a powerful effect on mechanical properties, while B_4C wt.% has about 33–43% and preheating temperature of die has about 9–16%. All P-values of the response are lower than 0.005. The optimal ratio of the pressure, B_4C rate and preheating temperature of die are seen to be 140 MPa, 12 wt.% and 225°C, respectively, to obtain the maximum mechanical properties.

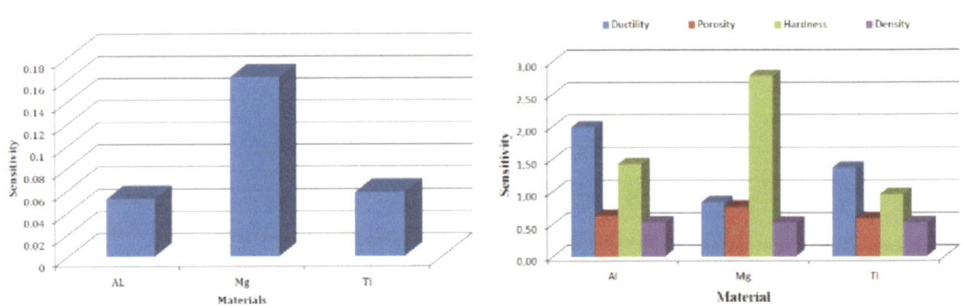

Figure 10.
Sensitivity analysis of the Al-Mg-Ti alloys.

The process parameters and experimental variables Al5059/SiC/MoS_2 composites are optimized by ANN and Taguchi [19]. The five input and the six output variables and 27 data are used in the ANN model for training and testing. The performance of output variables is determined by R^2 which is 98.12% surface roughness, 98.63% for temperature, 96.98% for radial force, 98.54% for feed force, 99.34% for material removal rate (MRR) and 98.71% for tangential force. The L27 orthogonal array (three levels and five factors) with "smaller is better" criteria for Taguchi design is performed. The optimum S/N ratios for surface roughness are found to be 5% SiC, 40 cutting speed of 1000 rpm, μm particle size, 200 mm/min feed and 0.5 mm depth of cut, respectively. For the temperature, all values are the same but cutting speed are 500 rpm. The optimum values varied for radial force, feed force, material removal rate and tangential force. The significance and influence of process variables on the quality are studied analysis of variance (ANOVA). Authors reported that the best important variable in all input parameters on the milling operation is silicon carbide addition (wt.%) in the composite followed by the feed rate, depth of cut, cutting speed and particle size of SiC.

Al2219 alloy reinforced with TiC+Al_2O_3 + Si_3N_4 is produced by squeeze method [20]. The melt temperature, die temperatures, stirring speed, feed rate and stirring time (min) are varied. To optimize the mechanical properties of hybrid composites, a statistical investigation is performed by ANFIS-gray wolf optimizer (ANFIS-GWO) and ANFIS-K-nearest neighbor (ANFIS-KNN) algorithms. They observed that the optimization results predicted the most suitable process parameters to obtain aluminum hybrid composites with high mechanical properties, and experimental and optimization results showed an optimal combination of hybrid and process parameters.

The flow stress values of 6061 Al-15% SiC metal matrix composites are predicted by an ANFIS [21]. The hot compression tests are applied to composites at various strain rates and temperatures. In the used ANFIS model, there are 17 rules with 17 membership functions (MF) which is Gaussian type for input MF, and linear parameters are 68 and non-linear parameters are 102. The samples of training and checking data are 88 and 12, respectively. The percentage mean error (PME) and RMSE are used to performance criteria. The predicted PME value of the flow stress by the ANFIS is less than 1.4%. Author declared that ANFIS with a hybrid learning algorithm can accurately be estimated the flow stress for the composites. In order to find the number of hidden nodes, there is not any way to obtain a highly system performance. The flow stress of 6061 Al-15% SiCp for plastic deformation can.

The impact resistance (IR) of Al-epoxy laminated composites by 2, 5 and 10-layers and has notch tip configuration with crack divider and crack arrester is predicted using ANFIS [22]. The experimental results of 126 are conducted for ANFIS model. The structure of ANFIS is obtained by seven input variables and the triangular and Gaussian membership functions (MFs). In this ANFIS model, the system is trained using 103 data and tested using 23 data. The best R^2 value of the Gaussian MFs model is 97.73% for training set, and the minimum R^2 value is 91.95% in the test dataset with the triangular MFs model. They concluded that both models had high R^2 values and strong potential, and IR of Al-epoxy laminated composites could be predicted by highly accurately with the proposed model under the given condition.

Kandpal et al. [23] optimized the EDM process parameters of AA6061/10% Al_2O_3 aluminum matrix composites casted by stir casting method using Taguchi route with ANOVA. The three factors with three levels in design of experiment section are set thus orthogonal array of L9 "larger is better" criteria is selected chosen. The input parameters are current and time of pulse and duty factor and the output factor is material removal rate (MRR). Authors declared that the current is

Method	Advantages	Disadvantages
ANN	Function approximation Data classification Data processing System control The solution of nonlinear problems Prediction Explicit formulation Storing information throughout the network Ability to work with missing knowledge Having error tolerance Having distributed memory Gradual deterioration Machine learning Parallel processing ability Detecting all possible interactions between variables	Training for each problem Large volumes data requirement Hardware requirement Failure to explain the behavior of the network Determining the appropriate network structure (topology) Difficulty displaying the problem to the network Not knowing the training time of the network Problem must be numerical Extremely addictive to data and applications Selecting of appropriate input variables
ANFIS	Numerical grouping Rule setting Prediction Better learning ability A much smaller convergence error Fewer adjustable parameters Parallel computation A well-structured knowledge representation A better integration Linguistic expressiveness of inaccurate inputs and system outputs Adaptability The ability to process information simultaneously Calculation to be efficient Working with linear techniques To be successful with optimization and adaptive techniques Continuity guarantee of the output Suitable for mathematical analysis There is no vagueness Reaching to the target faster The solution of nonlinear problems Function approximation Data classification Data processing System control	Training for each problem Hardware requirement Large volumes data requirement Selecting of appropriate input variables Low accuracy when there are not enough training data Cannot handle multiple output systems Long run time when the number of membership functions is large
Taguchi	Experimental design (system, tolerance, parameter) The possibility of calculating the inter-factor interactions not possible in the experimental design by changing one factor at a time Showing which factor is important Finding all available compositions	High fault tolerance Difficulty of calculating the effects of unexpected changes in experimental conditions Randomization, repetition and blocking of experiments Exponentially increasing the size of the experiments with the number and levels of factors Difficulty in explaining the high level of interactions

Table 4.
Advantages and disadvantages of the model.

the most affecting parameter on material removal rate and the optimal variables are 14 A for pulse current, 200 μs for the pulse on time and 50% for duty factor to obtain the maximum MRR.

The wear behavior (SWR: specific wear rate) of LM25/fly ash composite materials produced with stir casting route is optimized using Taguchi design of experiment with ANOVA [24]. The experimental design is carried out by the orthogonal array of L27 (three levels and four factors) "smaller is better" criteria in which sliding speed, load, reinforcement and sliding distance are input factors. The change in load compared to the other input variables would more be affected the SWR. Authors observed that the used optimization model reduced the specific wear rate and confirmed the increasing of wear resistance of the composites by the proposed optimum parameters. They also said that the Taguchi method is useful in optimizing the specific wear rate.

Table 4 shows the advantages and disadvantages of the model used. It is clear that each method has different advantages and disadvantages. ANN, ANFIS and FL techniques use the experimental results but Taguchi is used for the experiment design which provides the maximum output with the minimum experiment.

Table 5 shows the decision matrices for ANN, ANFIS and Taguchi approaches. In the decision matrices, weighting factor (WF) is designated using the information in **Table 4**. The score values where ANFIS has the highest score value are close each other. For example, ANN runs with higher data volume compared to ANFIS. This shows why ANN has the score 3 while ANFIS has the score 4. The Technique for Order Preference by Similarity to Ideal Solution (TOPSIS) method is used for the more detailed analysis in shown in **Table 6**.

The main purpose of TOPSIS, which is a multi-criteria decision making method, is to enable the organization, interpretation and analysis of information in decision making. TOPSIS method reveals the distances to positive and negative ideal solutions, revealing ideal and non-ideal solutions. In the TOPSIS, the parameters are normalized (Eq. (1)), calculated weighted normalized matrix (Eq. (2)), ideal the best and worst values, Euclidean distance from ideal best and worst (Eqs. (3) and (4)), and performance (Eq. (5)), respectively, using following equations;

$$\overline{X}_{ij} = \frac{X_{ij}}{\sqrt{\sum_{j-1}^{n} X_{ij}^2}} \tag{1}$$

$$V_{ij} = \overline{X}_{ij}\, x\, W_j \tag{2}$$

$$S_i^+ = \left[\sum_{j-1}^{m} \left(V_{ij} - V_j^+\right)^2\right]^{0.5} \tag{3}$$

$$S_i^- = \left[\sum_{j-1}^{m} \left(V_{ij} - V_j^-\right)^2\right]^{0.5} \tag{4}$$

$$C_i = \frac{S_i^-}{S_i^+ + S_i^-} \tag{5}$$

WF	1	5	4	2	3	Score
Method	Time	Features	Predicting	Volume	Design	
ANN	3	5	5	3	1	57
ANFIS	3	5	5	4	1	59
Taguchi	4	3	3	5	5	56

Table 5.
Decision (score) matrices.

WF	0.05	0.4	0.2	0.15	0.2	Volume and design (min)			Volume and design (max)		
Method	**Time**	**Features**	**Predicting**	**Volume**	**Design**	**Si^+**	**Si^-**	**Ci**	**Si^+**	**Si^-**	**Ci**
ANN	3	5	5	3	1	0.014	0.043	0.754	0.038	0.024	0.391
ANFIS	3	5	5	4	1	0.007	0.044	0.861	0.039	0.021	0.354
Taguchi	4	3	3	5	5	0.043	0.014	0.245	0.024	0.038	0.608

Table 6.
TOPSIS matrices.

The ideal best and worst values are the determining of the minimum or maximum values in the given ranges. The Ci values show the ranks of ANN, ANFIS and Taguchi methods. The maximum and minimum Ci values for the minimum data volume and design and the maximum data volume and design are observed at ANFIS and Taguchi approaches, respectively. This explains that if the analysis of broad features such as the solution of nonlinear problems, data classification, data processing, system control, or prediction will be performed, ANFIS or ANN techniques should be applied. If an experiment design with different variables will be executed, the Taguchi method should be used. So, experiment variables and its effects with less experimental study can reveal easily.

5. Conclusion

The key parameters are determined by executing the artificial neural network (ANN), adaptive-neuro fuzzy inference systems (ANFIS) and Taguchi with ANOVA. The nonlinear problems, the function approximation, data classification, data processing and system control etc., in engineering applications of AMCs can be easily carried out by soft computing approaches. Although many different methods are used for this purpose, we can say that the most popular and the most widely used methods are Taguchi, ANN and ANFIS approaches due to factors such as the minimum error, maximum accuracy, fast, cost, and time in forecasting, decision analysis, optimization, modeling and solution of complex problems and etc. One of the most important tasks in ANN and ANFIS is the determination of the number of layers, neuron, hidden layer, learning algorithm, and transfer function because there is no well-defined procedure to find the optimal parameter settings and network architecture. These variables affect the learning and forecasting abilities of the system with high accuracy. In the ANN, datasets must be normalized. ANFIS has the advantage to combine both ANN and Fuzzy knowledge. So ANFIS is more precise in term of efficiency even though ANN may outperform ANFIS model. ANFIS algorithm has a hybrid learning approaches in its structure. This helps the algorithm to be faster and more precise in term of efficiency than most of ANN algorithm.

The models have an important advantage coming from their ability to generate mathematical equations that can be easily programmed and used in applications in the production process. The theoretical analysis of material parameters is quite complicated due to various factors. The explicit formulations are proposed using these methods for estimating the parameters of composites. The training requirement for ANN includes large amounts of data, but this does not apply to ANFIS. The calculations can be made with the mathematical formulation obtained from the ANN which is an important advantage of ANN compared to the other methods. ANFIS and Neural Networks must be trained for each problem. The compatibility of experimental and theoretical results is researched by MSE, MAE, MAPE, RMSE, R and R^2 criteria because the minimum error criteria and maximum correlation coefficient are expected. The sensitivity of input parameters on the output in the stated studies, and the decision and TOPSIS matrices for three approaches are derived and discussed. Also, the advantages and disadvantages of the methods are tabulated. It can be concluded that ANFIS and ANN approaches can be used to solve the many complex problems with minimum error, control system, detect the interactions between variables, reach the target faster, predict and optimize the results with the maximum accuracy. The Taguchi is an experimental design method and provides the optimum results with fewer experiments using multiple results at the same time. The system, parameter and tolerance design is the special interest of the Taguchi.

Author details

Halil Ibrahim Kurt[1*], Engin Ergul[2], Necip Fazil Yilmaz[3] and Murat Oduncuoglu[4]

1 Department of Metallurgical and Materials Engineering, Engineering Faculty, Gaziantep University, Gaziantep, Turkey

2 İzmir Vocational High School, Dokuz Eylul University, İzmir, Turkey

3 Department of Mechanical Engineering, Engineering Faculty, Gaziantep University, Gaziantep, Turkey

4 Computer Programing Department, Technical Sciences, Gaziantep University, Gaziantep, Turkey

*Address all correspondence to: hiakurt@gmail.com

References

[1] Jang J-SR. Fuzzy modeling using generalized neural networks and kalman filter algorithm. In: AAAI. 1991

[2] Xinqing L, Tsoukalas LH, Uhrig R. A neurofuzzy approach for the anticipatory control of complex systems. In: Proceedings of IEEE 5th International Fuzzy Systems. New Orleans, LA, USA: IEEE; 1996

[3] Demuth M, Beale M. Fuzzy Logic Toolbox user's Guide for Use with MATLAB. Natick, MA, USA: The Math Works Inc; 2004

[4] Krishnaiah K, Shahabudeen P. Applied Design of Experiments and Taguchi Methods. Connaught Circus, New Delhi, India: PHI Learning Pvt. Ltd; 2012

[5] Thakur A et al. Application of Taguchi method for resistance spot welding of galvanized steel. ARPN Journal of Engineering and Applied Sciences. 2010;**5**(11):22-26

[6] Rao SR, Padmanabhan G. Application of Taguchi methods and ANOVA in optimization of process parameters for metal removal rate in electrochemical machining of Al/5% SiC composites. International Journal of Engineering Research and Applications (IJERA). 2012;**2**(3):192-197

[7] Kurt HI, Oduncuoglu M. Effects of temperature, time, magnesium, and copper on the wettability of Al/TiC system. Mathematical Problems in Engineering. 2015;**2015**:6

[8] Rabunal JR, Dorado J, Sierra AP. Encyclopedia of Artificial Intelligence. Information Science Reference; 2009

[9] Kurt HI, Oduncuoglu M. Application of a neural network model for prediction of wear properties of ultrahigh molecular weight polyethylene composites. International Journal of Polymer Science. 2015;**2015**:11. Article ID 315710. DOI: 10.1155/2015/315710

[10] Detienne KB, Detienne DH, Joshi SA. Neural networks as statistical tools for business researchers. Organizational Research Methods. 2003;**6**(2):236-265

[11] Kurt HI, Oduncuoglu M, Asmatulu R. Wear behavior of aluminum matrix hybrid composites fabricated through friction stir welding process. Journal of Iron and Steel Research International. 2016;**23**(10):1119-1126

[12] Kurt HI. Influence of hybrid ratio and friction stir processing parameters on ultimate tensile strength of 5083 aluminum matrix hybrid composites. Composites Part B: Engineering. 2016; **93**:26-34

[13] Kurt HI, Oduncuoglu M. Formulation of the effect of different alloying elements on the tensile strength of the in situ Al-Mg_2Si composites. Metals. 2015;**5**(1):371-382

[14] Kurt Hİ. Optimization of tensile strength of Al alloys with Mg and Ti. El-Cezeri Journal of Science and Engineering. 2017;**4**(1):25-31

[15] Kurt Hİ. Theoretical analysis of the ductility, porosity, hardness and density in aluminum–magnesium alloys with titanium. Anadolu Üniversitesi Bilim Ve Teknoloji Dergisi A-Uygulamalı Bilimler ve Mühendislik. 2016;**17**(1):151-158

[16] Kurt H, Oduncuoglu M, Kurt M. A mathematical formulation to estimate the effect of grain refiners on the ultimate tensile strength of Al-Zn-Mg-Cu alloys. Metals. 2015;**5**(2):836

[17] Satyanarayana G, Naidu GS, Babu NH. Artificial neural network and regression modelling to study the effect

of reinforcement and deformation on volumetric wear of red mud nano particle reinforced aluminium matrix composites synthesized by stir casting. Boletin de la Sociedad Espanola de Ceramica y Vidrio. 2018;**57**(3):91-100

[18] Soundararajan R et al. Modeling and analysis of mechanical properties of aluminium alloy (A413) reinforced with boron carbide (B_4C) processed through squeeze casting process using artificial neural network model and statistical technique. Materials Today: Proceedings. 2017;**4**(2):2008-2030

[19] Daniel SAA et al. Multi objective prediction and optimization of control parameters in the milling of aluminium hybrid metal matrix composites using ANN and Taguchi-grey relational analysis. Defence Technology. 2019;**15** (4):545-556

[20] Adithiyaa T, Chandramohan D, Sathish T. Optimal prediction of process parameters by GWO-KNN in stirring-squeeze casting of AA2219 reinforced metal matrix composites. Materials Today: Proceedings. 2020;**21**:1000-1007

[21] Kalaichelvi V et al. Prediction of the flow stress of 6061 Al–15% SiC–MMC composites using adaptive network based fuzzy inference system. Materials & Design. 2009;**30**(4):1362-1370

[22] Nazari A, Didehvar N. Analytical modeling impact resistance of aluminum–epoxy laminated composites. Composites Part B: Engineering. 2011; **42**(7):1912-1919

[23] Kandpal BC, Kumar J, Singh H. Optimisation of process parameters of electrical discharge machining of fabricated AA 6061/10% Al_2O_3 aluminium based metal matrix composite. Materials Today: Proceedings. 2018;**5**(2):4413-4420

[24] Sadhana AD et al. Wear behaviour of aluminium matrix composites (LM25/Fly ash)–A Taguchi approach. Materials Today: Proceedings. 2020. DOI: 10.1016/j.matpr.2020.03.684

Section 2

Aluminium Alloys

Chapter 3

Effect of Zr Addition and Aging Treatment on the Tensile Properties of Al-Si-Cu-Mg Cast Alloys

Jacobo Hernandez-Sandoval, Mohamed H. Abdelaziz, Agnes M. Samuel, Herbert W. Doty and Fawzy H. Samuel

Abstract

The present study focused on the tensile properties at ambient and high temperatures of alloy 354 without and with the addition of zirconium. Tensile tests were performed on alloy samples submitted to various aging treatments, with the aim of understanding the effects of the addition made on the tensile properties of the alloy. Zirconium reacts only with Ti, Si, and Al in the alloys examined to form the phases $(Al,Si)_2(Zr,Ti)$ and $(Al,Si)_3(Zr,Ti)$. Testing at 25°C reveals that the minimum and maximum quality index values, 259 and 459 MPa, are observed for the as-cast and solution heat-treated conditions, respectively. The yield strength shows a maximum of 345 MPa and a minimum of 80 MPa within the whole range of aging treatments applied. The ultimate tensile and yield strength values obtained at room temperature for T5-treated samples stabilized at 250°C for 200 h are comparable to those of T6-treated samples stabilized under the same conditions, and higher in the case of elevated-temperature (250°C) tensile testing. Coarsening of the strengthening precipitates following such prolonged exposure at 250°C led to noticeable reduction in the strength values, particularly the yield strength, and a remarkable increase in the ductility values.

Keywords: aluminum alloys, aging, thermal exposure, tensile testing, precipitation, fractography

1. Introduction

The 354 alloy belongs to the Al-Si-Cu-Mg system similar to B319 alloy that is widely used for automotive engine blocks [1]. The high silicon content in the 354 alloy improves the alloy castability whereas the presence of Cu and Mg noticeably enhances the yield strength (YS) and the ultimate tensile strength (UTS) of the 354 alloy due to the formation of intermetallic phases, mainly Al_2Cu or eutectic Al + Al_2Cu, and Mg_2Si precipitates [2, 3]. However, segregation behavior of Cu may lead to incipient melting during solution treatment which will apparently reduce the alloy strength [4]. Addition of Mg has a strong affinity to react with Sr, leading to the formation of a complex $Mg_2SrAl_4Si_3$ intermetallic phase, and hence reducing the effectiveness of Sr as a Si modifying agent [5]. In the absence of Cu, high Fe and Mg contents lead to the formation of π-$FeMg_3Si_6Al_8$ phase which is difficult to dissolve during the solution treatment process [6, 7]. In the quaternary Al-Si-Cu-Mg alloy system, *Q*-phase ($Al_4Mg_8Cu_2Si_6$) can coexist with the Al_2Cu, Mg_2Si, and Si phases depending on the levels of Cu, Mg, and Si [8–11]. The different factors that may influence the mechanical behavior of cast aluminum alloys are schematically represented in **Figure 1** [12].

Zirconium may be added to Al alloys in order to refine the grain structure due to the presence of fine coherent dispersoids (mainly Al_3Zr) which obstruct dislocation motion and in turn, enhance the elevated temperature mechanical properties of aluminum alloys [13]. In order to increase the volume fraction of Al_3Zr precipitates and based on the phase diagram of Al-Zr, the concentration of Zr in the alloys investigated in this study was kept at around 0.3 wt.% [14].

The main purpose of solution heat treatment is to obtain a supersaturated solid solution at high temperatures (below the eutectic temperature). As a result, a homogeneous supersaturated solid solution (SSSS) will form through dissolving the precipitated phases during the solidification process, such as β-Mg_2Si, θ-Al_2Cu, Q-$Al_5Cu_2Mg_8Si_6$, π-$Al_9FeMg_3Si_5$ and β-Al_5FeSi phases. The β-Mg_2Si and θ-Al_2Cu phases can be easily dissolved when the optimum solution heat treatment temperature and time are employed. The solution treatment temperature is determined according to the alloy composition and solid solubility limit; however, it must be lower than the melting point of the phases that exist in the as-cast structure to avoid incipient melting of these phases [15, 16].

Following the empirically developed concept of quality index proposed by Drouzy et al. [17, 18] Cáceres proposed a mathematical model emphasizing the significance of the quality index as follows [17, 19, 20]:

$$Qc = \left[(qn)^n \, exp^{-qn} + 0.4 log\,(100qn)\right] \tag{1}$$

where quality index *Q* can be calculated using the relative quality index (*q*), strain-hardening exponent (*n*), and the strength coefficient (K).

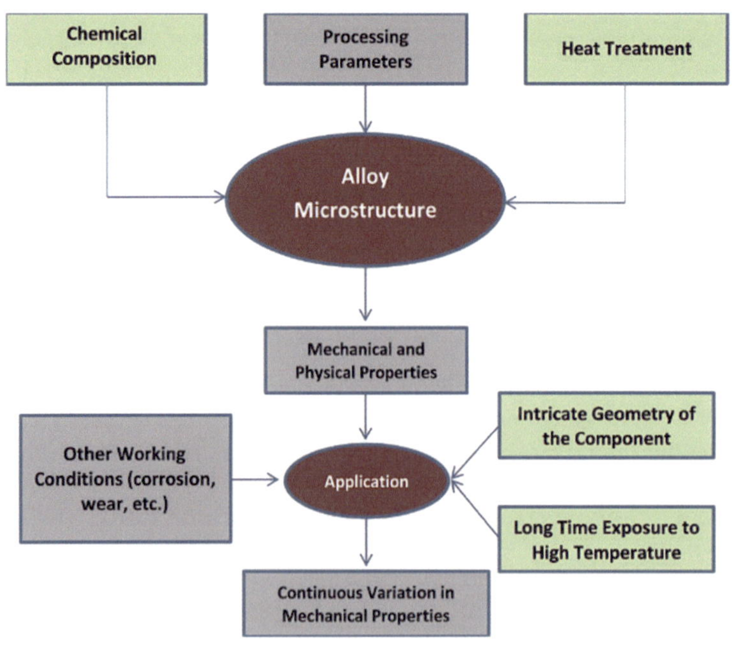

Figure 1.
Schematic representation of factors affecting alloy performance [12].

The present study was undertaken to explore the effect of Zr addition and aging conditions of the as cast tensile bars on:

1. Characterizing the microstructural features of the investigated alloys,

2. Exploring the tensile properties at both ambient and elevated temperatures, and

3. Correlating the tensile properties to the microstructural features to establish the strengthening or softening mechanisms responsible for the observed properties.

It should be noted here that the term "temperature" applies to aging temperatures as well as testing temperature.

2. Experimental procedure

Alloy 354 modified with 200 ppm of strontium (using Al-10% Sr master alloy) and grain refined using 0.20 wt.%Ti (Al-5%Ti-1%B) was used as the base alloy (alloy A). To this alloy, 0.3%Zr in the form of Al-25wt.%Zr master alloy was added (alloy B). The chemical compositions of both alloys are listed in **Table 1**. **Figure 2** shows the microstructure of the as-received base alloy ingots. Melting and casting procedures were carried out as described elsewhere.

Alloy code	Element (wt.%)								
	Si	Fe	Cu	Mn	Mg	Zr	Ti	Sr	Al
A	9.1	0.12	1.8	0.0085	0.6	—	0.18	0.02	87.6
B	9.1	0.12	1.8	0.0085	0.6	0.3	0.18	0.02	87.6

Table 1.
Chemical composition of the 354 alloys used in this study.

Figure 2.
Microstructure (200×) of the base alloy 354 used in this work.

To prepare test bars for the tensile tests, three samples for chemical analysis were also taken at the time of the casting; this was done at the beginning, in the middle, and at the end of the casting process to ascertain the exact chemical composition of each alloy. The experimental work was divided into two stages: Stage I in which the 354 alloy (alloy A) was used, and Stage II where the 354 alloy with 0.3%Zr (alloy B) was used. In Stage I, the melt temperature was kept around 750°C, whereas in Stage II, the melt temperature was superheated to 800°C, to ensure the complete decomposition of the Al-25%Zr master alloy used.

2.1 Stage I-alloy A

Tensile bars were solution heat treated at 495°C for 8 h, followed by quenching in warm water at 60°C, after which artificial aging was applied according to the plan listed in **Table 2**. After aging, the test bars were allowed to cool naturally at room temperature (25°C). All of the samples, whether as-cast, solution heat-treated, or aged, were tested to the point of fracture using an MTS servo-hydraulic mechanical testing machine at a strain rate of $4 \times 10^{-4}\ s^{-1}$.

The yield strength (YS) was calculated according to the standard 0.2% offset strain, and the fracture elongation was calculated as the percent elongation (%El) over 50 mm gauge length, as recorded by the extensometer. The ultimate tensile strength (UTS) was also obtained from the data acquisition system of the MTS machine. The average %El, YS, or UTS values obtained from the five samples tested per condition were considered to be the values representing that specific condition. An extensometer, or strain gage was used in the tests to measure the extent of deformation in the samples.

Samples for metallography were sectioned from the tensile-tested bars of all the alloys studied, about 10 mm below the fracture surface. The percentage porosity and eutectic Si-particle characteristics were measured and quantified using an optical microscope linked to a Clemex image analysis system. The microstructures of the polished sample surfaces were examined using an Olympus PMG3 optical microscope. Phase identification was carried out using electron probe microanalysis (EPMA) in conjunction with wavelength dispersive spectroscopic (WDS) analysis, using a JEOL*JXA-889001WD/ED combined microanalyzer operating at 20 kV and 30 nA, where the electron beam size was ~2 μm.

Mapping of certain specific areas of the polished sample surfaces was also carried out where required, so as to show the distribution of different elements within the phases. The fracture surfaces of tensile-tested samples were also examined using the same SEM, employing the backscattered electron (BSE) detector and EDS system. The fracture behavior was analyzed using the backscattered electron (BSE) images

Temperature (°C)	**Aging time (h) and aging condition codes**												
	2	**4**	**6**	**8**	**10**	**12**	**16**	**20**	**24**	**36**	**48**	**72**	**100**
155	1	2	3	4	5	6	7	8	9	10	11	12	13
170	14	15	16	17	18	19	20	21	22	23	24	25	26
190	27	28	29	30	31	32	33	34	35	36	37	38	39
240	40	41	42	43	44	45	46	47	48	49	50	51	52
300	53	54	55	56	57	58	59	60	61	62	63	64	65
350	66	67	68	69	70	71	72	73	74	75	76	77	78

Table 2.
Artificial aging conditions used for room temperature tension tests.

obtained, and analysis of the EDS spectra of phases observed on the fracture surface. Differential scanning calorimetry (DSC) was used to characterize the sequence of reactions occurring during the heating and/or cooling cycles of an alloy sample during a DSC scan which continuously changes with the increasing or decreasing temperature cycle to produce peaks according to the two expected reactions:

- Phase formation → heat emission → exothermic peak
- Phase dissolution → heat absorption → endothermic peak

2.2 Stage II-alloy B

For the high temperature tensile tests, samples from selected conditions were tested to fracture using an Instron Universal mechanical testing machine at a strain rate of $4 \times 10^{-4}\ s^{-1}$. The heating furnace installed on the testing machine is an electrical resistance, forced-air box type, having the dimensions 30 × 43 × 30 cm. The yield strength (YS) was calculated according to the standard 0.2% offset strain, and the fracture elongation was calculated as the percent elongation (%El) over the 25.4 mm gauge length as recorded by the extensometer. The ultimate tensile strength (UTS) was obtained from the data acquisition system of the universal machine. In order to reach and stabilize the intended test temperature during the tests, at the time that the samples were mounted in the tensile machine, the furnace was already pre-set at the required temperature; also, these samples were kept mounted in the furnace of the tensile testing machine for 30 min before the start of every test.

3. Results and discussion

3.1 Stage I-alloy A

Figure 3 shows the macrostructure revealing the grain size for alloy A, about 200 μm. A complete modification of the silicon particles in the microstructure of alloy A in the as-cast condition can be seen in **Figure 4(a)**. From **Figure 4(a)** and **(b)**, solution heat treatment has changed the morphology of the silicon particles from faceted to globular. As a consequence of solution heat treatment, there may also be observed a reduction in the number of silicon particles and a reduction in the density of the silicon phase, due to the diffusion of silicon into the aluminum matrix. The white arrows in **Figure 4(a)** show the rounded shape of the dendrites with grain refining [21], whereas **Figure 4(b)** reveals the dissolution of the Al_2Cu phase observed in **Figure 4(a)**—circled.

Zhu and Liu [22] proposed a model of the granulation of unmodified eutectic Si composed of three major stages during heat treatment: (i) the mass transport of solute, (ii) a discontinuous phase fragmentation, and lastly (iii) spheroidization. During heat treatment, the silicon atoms in the matrix at the Si particle tips diffuse to locations on the curved surfaces of the particles, leading to the dissolution of eutectic silicon at the tips. This transport of silicon atoms ultimately causes the fragmentation and spheroidization of eutectic silicon which is important from strength point of view compared to Si particles with sharp edges which act as sites for stress concentration.

The values of secondary dendrite arm spacing (SDAS), porosity, modification level, and grain size for both the as-cast (AC) and solution heat-treated (SHT) condition are listed in **Tables 3** and **4**. As can be seen, SHT resulted in (i) no noticeable change in both the SDAS and grain size, (ii) a significant decrease in the particle density due to coarsening of the eutectic Si particles, and (iii) almost complete solubility of Al_2Cu in the aluminum matrix. Since the solutionizing temperature was well below the incipient melting temperature, tensile test bars revealed negligible change in the amount of porosity, i.e., no incipient melting.

Figure 3.
Macrograph showing grain size of the tensile bars in the as-cast condition.

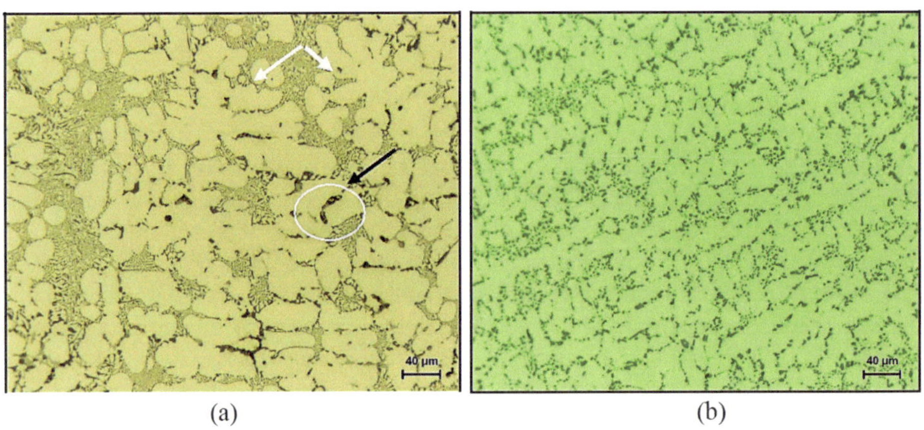

Figure 4.
Optical microstructure: (a) before, and (b) after solution heat treatment.

Alloy code condition	SDAS (μm)	Grain size (μm)	Porosity (%)		Volume fraction of intermetallics (%)	
					EPMA	
	Av*		Av	SD**	Av	SD
A-AC	19.3	201	0.14	0.06	3.08	0.32
A-SHT	23.1	192	0.12	0.05	1.27	0.10

*Average.
**Standard deviation.

Table 3.
SDAS, porosity%, grain size, level of modification, and volume fraction of intermetallics for alloy A.

Alloy code condition	Area (%)	Particle length (μm)		Roundness ratio (%)		Aspect ratio		Density (particles/ mm^2)
	Av	Av	SD	Av	SD	Av	SD	
A-AC	14.58	3.522	3.94	0.4302	0.181	2.033	0.8609	39110
A-SHT	10.868	4.286	3.144	0.554	0.1547	1.641	0.5429	12080

Table 4.
Silicon particle characterization for alloy A.

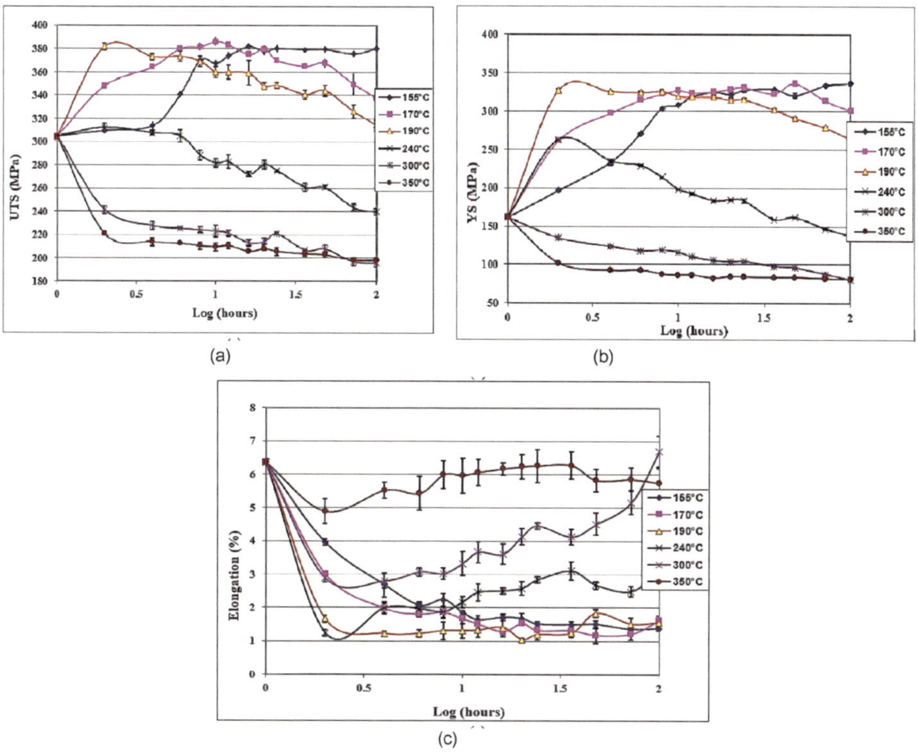

Figure 5.
Variation in alloy tensile parameters as a function of aging temperature and time: (a) UTS, (b) YS, and (c) %El.

Figure 5 illustrates the effect of aging treatment on the alloy strength parameters. The main observations inferred from this figure can be summarized as follows:

1. Solution heat treatment and artificial aging at 190°C for 2 h or at 155°C for 100 h resulted in an increase in the alloy strength by ~64% over its as-cast strength.

2. Aging at 155 or 170°C for a long period of time offered maximum resistance to softening.

3. The greatest decrease in tensile strength occurred at 240°C (312 MPa at 2 h to 240 MPa at 100 h). Similarly, a significant decrease in strength took place upon aging at 190°C for a lengthy time (from 382 MPa at 2 h to 314 MPa at 100 h) indicating the end of peak-aging or the commencement of over aging.

4. The greatest reduction in the alloy UTS and YS levels resulted when the tensile bars were aged at a temperature as high as 350°C even for a short period of 2 h.

5. In comparison to the ascendant and steady strength curves corresponding to aging temperatures of 155°C and 170°C, fluctuations in the strength curves were observed at aging temperatures of 190°C and over, similar to that reported by Reif [23] where a similar alloy was used and an ascendant monotonic strength curve was observed at an aging temperature of 155°C.

6. Although the highest ductility values were obtained after 2 h aging at 350°C (~5%), none of the aging conditions reached the higher ductility values exhibited by the solution heat-treated condition (~6.5%). This observation may be considered evidence that the mechanical behavior displayed by alloy A is common to that of the Al-Si-Cu-Mg alloys whose strength is obtained at the expense of ductility [24, 25].

In order to analyze the alloy quality by means of the Quality Index charts, the as-cast and the solution heat treated conditions plus aging conditions at 155°C, 190°C, and 350°C for aging times in the range of 2–100 h were used. From a previous study [5], K was calculated as 500 MPa.

The plastic strain and the quality index (*Q*) both exhibit a great improvement following solution heat treatment. The fact that plastic deformation (*q*) was about 0.31 in the solution-treated condition means that the alloy reached 31% of its maximum quality index value (*Q*). The importance of *q* is that it shows how much a sample is away from its maximum possible ductility $q = 1$ and indicates that it would be possible to control the microstructure, for example by reducing the SDAS, or the porosity, or intermetallic level to enhance the alloy ductility and hence, the quality index, *Q*. When the ductility increases sharply from the as-cast to the solution heat treated condition, such changes can be related to the spheroidization of silicon particles and to the uniformity of the microstructure in the solution heat-treated condition, as shown in **Figure 6(a)**.

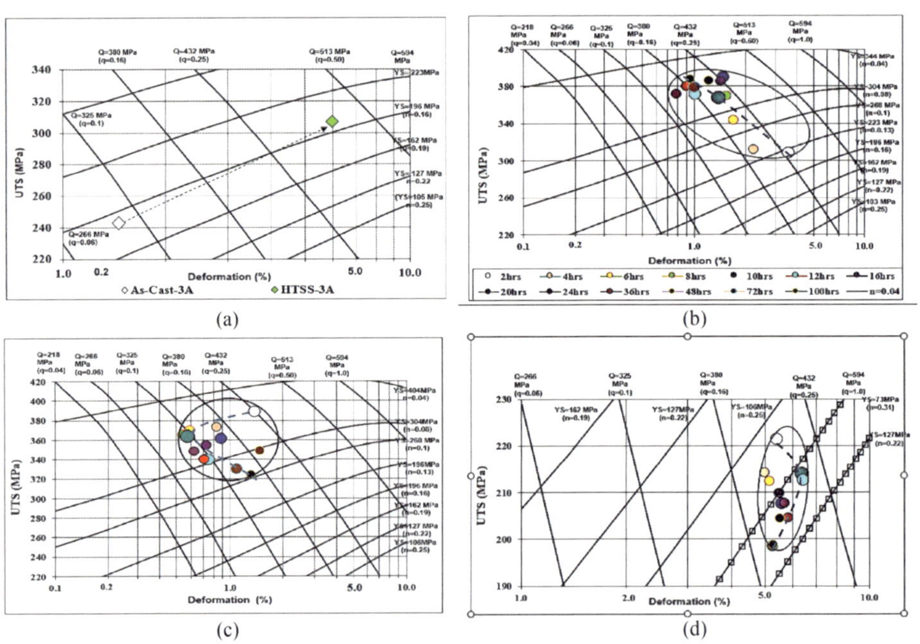

Figure 6.
Q-charts following: (a) SHT, (b) aging at 155°C, (c) aging at 190°C, and (d) aging at 350°C. Legends in (b) apply for other charts. The curved lines indicate the passage from 2 to 10 to 100 h.

From the data presented in **Figure 6(b)–(d)**, it is evident that the change in crystallographic structure of Al_2Cu phase from G-P zones (155°C) to a metastable phase (190°C) to a stable phase (350°C) is the main parameter controlling the alloy performance quality. As can be seen, at each aging temperature, all points fall within a narrow circle due the progress in the formation of the precipitated phase. The broken lines in these figures show the change in the Q-level as a function of aging temperature. The width of the circle deceased from 175 MPa (155°C) to 75 MPa (190°C) to 25 MPa (350°C), representing the hardening and softening behavior of the alloy as a function of the aging temperature and time [26]. Using aging times of 2 and 100 h as reference points, the Q, UTS and %El values are presented in **Table 5**. As can be seen, the Q values after 2 h are more-or-less same over such a large range of aging temperatures, due to the variation in both UTS and %El. However, aging for 100 h revealed highest value at 190°C compared to 155°C (under aging) and 350°C (over aging). The Q values for test bars aged at 350°C for 100 h is the same due to the balance between UTS and %El.

3.2 Stage II-alloy B

The heat treatment procedures followed for the alloy B are listed in **Table 6**. The same treatments were applied for both 25°C and 250°C tensile testing.

Figure 7 [27] shows the DSC heating curves of the alloys in the as-cast and SHT conditions, where three explicit peaks could be detected and coded 1, 2, 3. Considering the main parameter is the precipitation of Al_2Cu phase particles, thus the height of peak number 1 following SHT compared to that in the as-cast condition plays a crucial role in controlling the alloy performance after aging. In addition, it is an indication of the effectiveness of the SHT process in dissolving the initial Al_2Cu phase. In **Figure 7**, peak # 1 after solutionizing is more

Aging temperature (°C)	Q (MPa)	UTS (MPa)	%El	Aging time (h)
155	432	300	3.5	2
190	432	388	1.5	2
350	406	221	5.3	2
155	340	387	0.9	100
190	470	324	1.5	100
350	406	198	5.3	100

Table 5.
Q, UTS and %El values for alloy A after 2 and 100 h aging at different temperatures.

Heat treatment procedures and parameters			
Heat treatment	**Solution treatment**	**Quenching**	**Aging**
SHT*	495°C for 5 h	Warm water (60°C)	NA
T5 temper	N/A	N/A	180°C for 8 h
T6 temper	495°C for 5 h	Warm water (60°C)	180°C for 8 h

SHT, solution heat treatment.

Table 6.
Heat treatment procedures and parameters applied to alloys investigated in stage II.

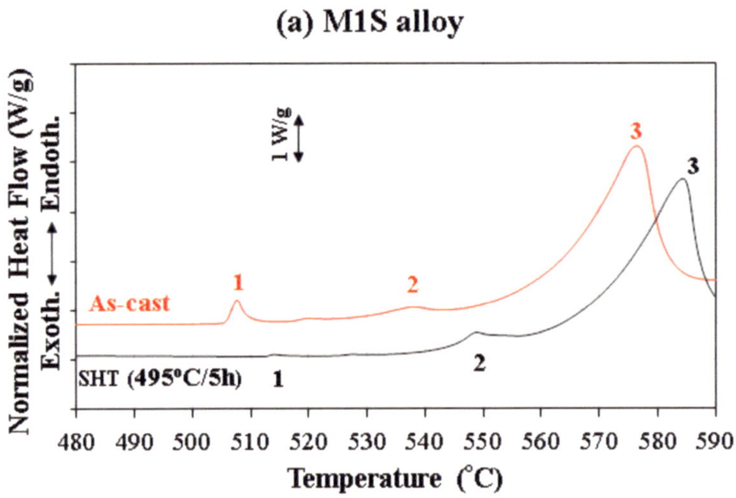

Figure 7.
Portion of the DSC heating curves of as-cast and as-quenched alloy B samples [27].

or less negligible due to dissolution of most of the Al_2Cu phase, as shown in **Figure 4(b)**.

The principle phases seen in alloy B are demonstrated in the optical as well as backscattered (BSE) images displayed in **Figure 8(a)** and **(b)** [27], respectively. **Figure 8(a)** exhibits α-Al dendrites separated by eutectic silicon colonies. The phases observed in **Figure 8(b)** were identified using EDS analysis and reference to the results of Hernandez-Sandoval [28] and Garza-Elizondo [29]. Selective EDS spectra identifying these phases are displayed in **Figure 8(c)** through **Figure 8(d)**. The existence of Al_2Cu phase in the block-like form may be attributed to the presence of Sr. in the alloy which leads to segregation of copper to localized areas [30]. The platelets of the Fe-rich β-Al_5FeSi phase are easily recognized in the BSE image, surrounded by the blocky Al_2Cu particles. The Mg-rich *Q*-phase ($Al_5Cu_2Mg_8Si_6$) is found growing out of the Al_2Cu phase as seen in the BSE image. The absence of coarse Al_3Zr precipitates [31] may be related to superheating that led to considerable dissolution of the Al_3Zr phase from the master alloy during the melting process. As a result, the coarse Zr-containing phases are rarely detected since Al_3Zr particles act as nucleation spots for these coarse phases. According to Garza-Elizondo [29], coarse Zr-rich particles may be nucleated on the undissolved Al_3Zr particles provided by the master alloy, i.e., Al-15 wt.%Zr. In the present study, superheating the melt to 800°C would significantly reduce the numbers of Al_3Zr particles in the matrix. The predicted fine zirconium trialuminide (Al_3Zr) dispersoids that may be present on a nanoscale would require a high magnification BSE image to be detected.

Figure 9(a) shows a bright-field (BF) TEM image obtained in a T6-treated sample of alloy B with the electron beam parallel to the [001] zone axis. This figure shows a high density of uniformly distributed needle-like precipitates which are oriented along <110> family of directions and aligned along the [100] planes. The length of these precipitates ranges from 50 to 150 nm, close to the reported size of θ′-Al_2Cu plates (50–100 nm long) [32, 33]. **Figure 9(b)** displays the associated selected area electron diffraction (SAED) pattern obtained from **Figure 9(a)**. The observable discrete diffraction maxima for the precipitates in SAED pattern indicate the presence of θ′-Al_2Cu, where the streaks most probably result from the presence of fine S′-Al_2CuMg particles. Computer simulation studies [34–37] on the S′-phase reflections show that they are hidden within the streaks of θ′.

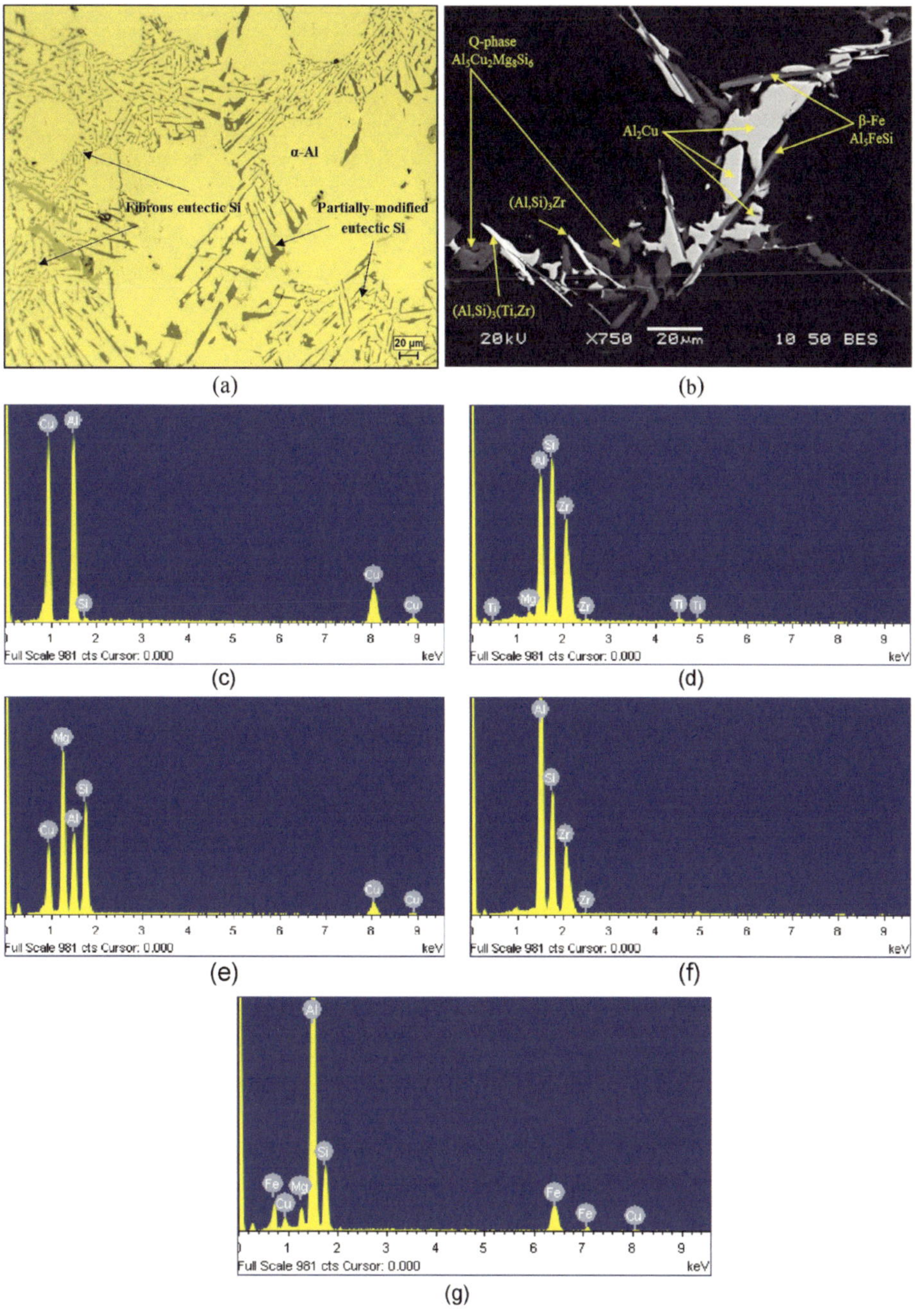

Figure 8.
(a) Optical micrograph at 200× magnification, and (b) backscattered electron image of alloy B (354 + 0.3 wt.%Zr), obtained at a low cooling rate of 0.35°C/s, showing the different phases present in the alloy; (c–g) EDS spectra corresponding to Al_2Cu, $(Al,Si)_3(Ti,Zr)$, Q-$Al_5Mg_8Cu_2Si_6$, $(Al,Si)_3Zr$, and β-Al_5FeSi phases observed in (b) [27].

In the present work, **Figure 10(a)**, the addition of ~0.3 wt.%Zr to the 354-type Al-Si-Cu-Mg cast alloy in the as-cast condition improves the ambient-temperature (25°C) strength values of the Zr-free 354 alloy (alloy A), by ~26 MPa (UTS) and 40 MPa (YS), respectively. Following SHT, the UTS and ductility values remained almost constant at ~300 MPa and ~ 6.3%, respectively, while the yield strength increased by ~33 MPa compared to alloy A. It is believed that the improved strength

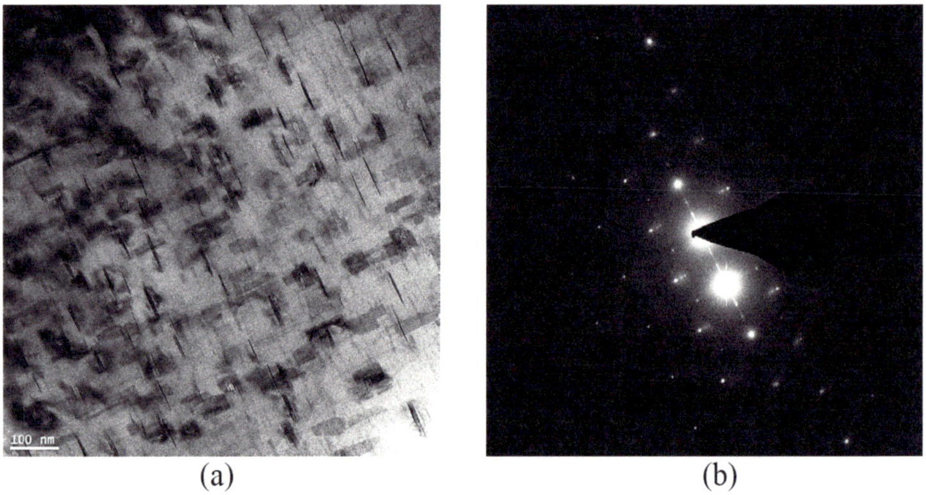

Figure 9.
(a) Bright-field TEM image of alloy B in T6-treated condition, and (b) the selected area electron diffraction (SAED) pattern.

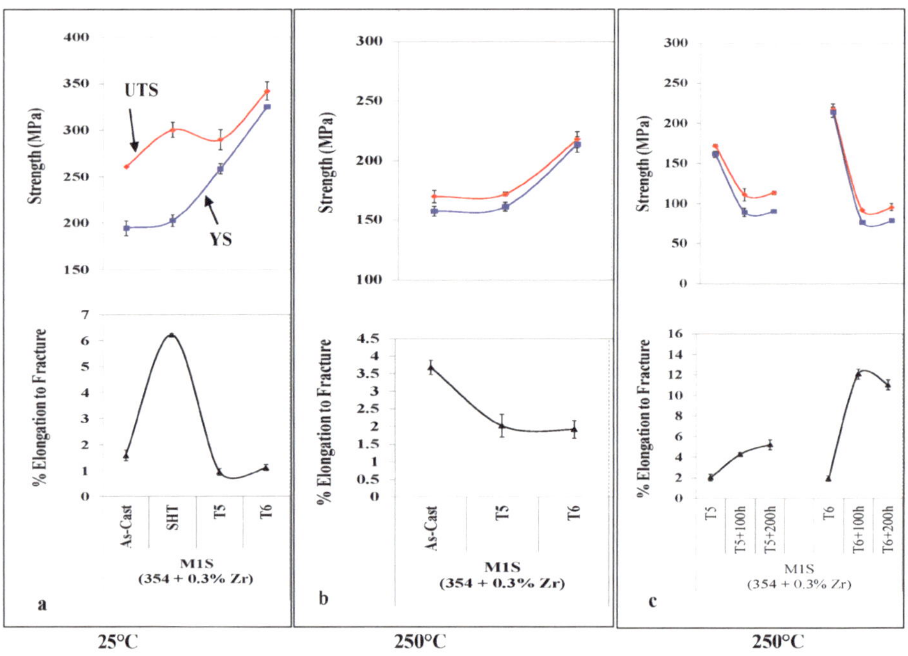

Figure 10.
(a) Ambient, and (b, c) high temperature tensile properties of alloy B.

values of alloy B emphasizes the role of Zr addition in enhancing the ambient-temperature tensile properties through the formation of fine secondary strengthening precipitates (Al_3Zr) as reported by many authors [14, 38–40]. The fact that UTS and YS in the T5 and T6 conditions are very close may be attributed to the strengthening effect of the fine dispersoids, which precipitate during the artificial aging stage of the T5, and T6 treatments as reported in **Figure 9**.

Tensile testing at 250°C, endured a significant softening due to the possible coarsening of the strengthening precipitates (Al_2Cu) that existed during tensile testing at room temperature (**Figure 11**). In addition, the T5 heat treatment did not improve the elevated-temperature strength values of the as-cast alloys but reduced the alloy ductility by ~50%. However, application of the T6 heat treatment noticeably enhanced the strength values of the as-cast condition from about 175 to 225 MPa. Another parameter to consider is the effect of thermal stability. In the present work, some tensile samples were stabilized at 250°C (following the T5 and T6 aging treatment) for a lengthy period of time, i.e., 100 and 200 h. As can be seen from **Figure 10(c)**, the stabilized T5-treated alloy B samples exhibit better strength values (UTS and YS) than those obtained in the stabilized T6-treated condition. However, the ductility values obtained after stabilization of the T5-treated samples are dramatically lower in comparison. **Figure 11(a)** shows Al_2Cu particle size and distribution in a T6 sample stabilized at 250°C for 200 h, whereas **Figure 11(b)** is the corresponding EDS spectrum.

A detailed investigation of the fracture surfaces of tensile bars of alloy B were examined in the T6-treated condition, before and after stabilization for 200 h at 250°C. The T6-temper treatment was selected due to its wide use in the automotive industry. The BSE image shown in **Figure 12(a)** [41] shows the fracture surface of the tensile-tested alloy in the T6-treated condition. The fracture surface

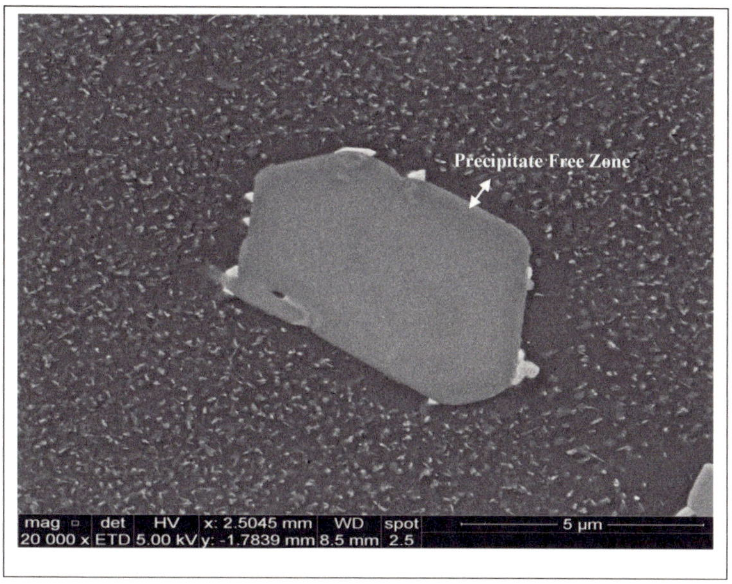

(a)

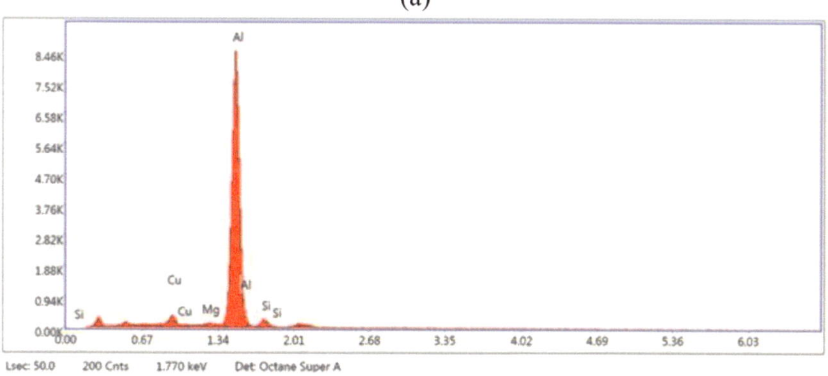

(b)

Figure 11.
(a) Backscattered electron images showing the size and distribution of precipitates in the T6-treated B alloy after stabilization at 250°C for 200 h; (b) EDS spectrum corresponding to the rod-like particles in (a).

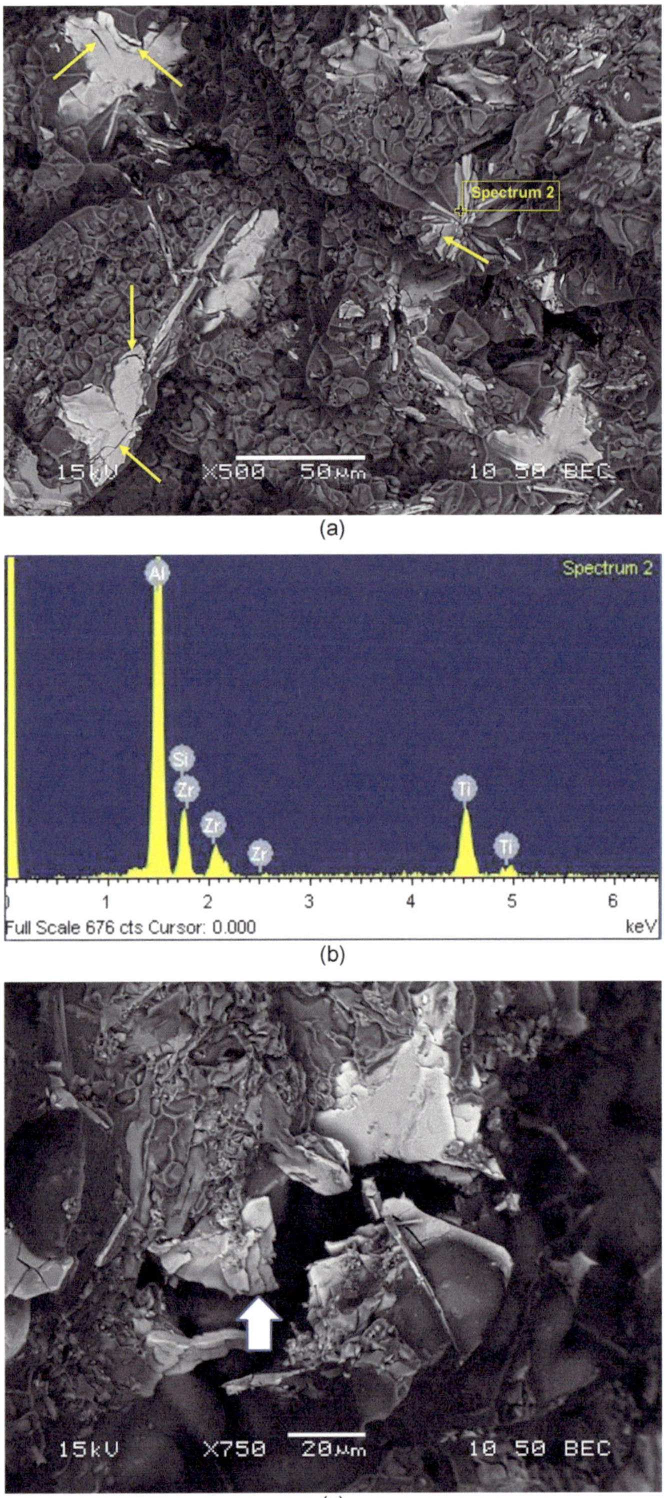

(a) (b) (c)

Figure 12.
Fracture surface of T6-treated alloy B: (a) BSE image showing a uniform dimple structure and cracked particles (arrowed), (b) EDS spectrum corresponding to the point of interest in (a), and (c) high magnification BSE image showing a cracked Al-Si-Ti-Zr particle (arrowed) [41].

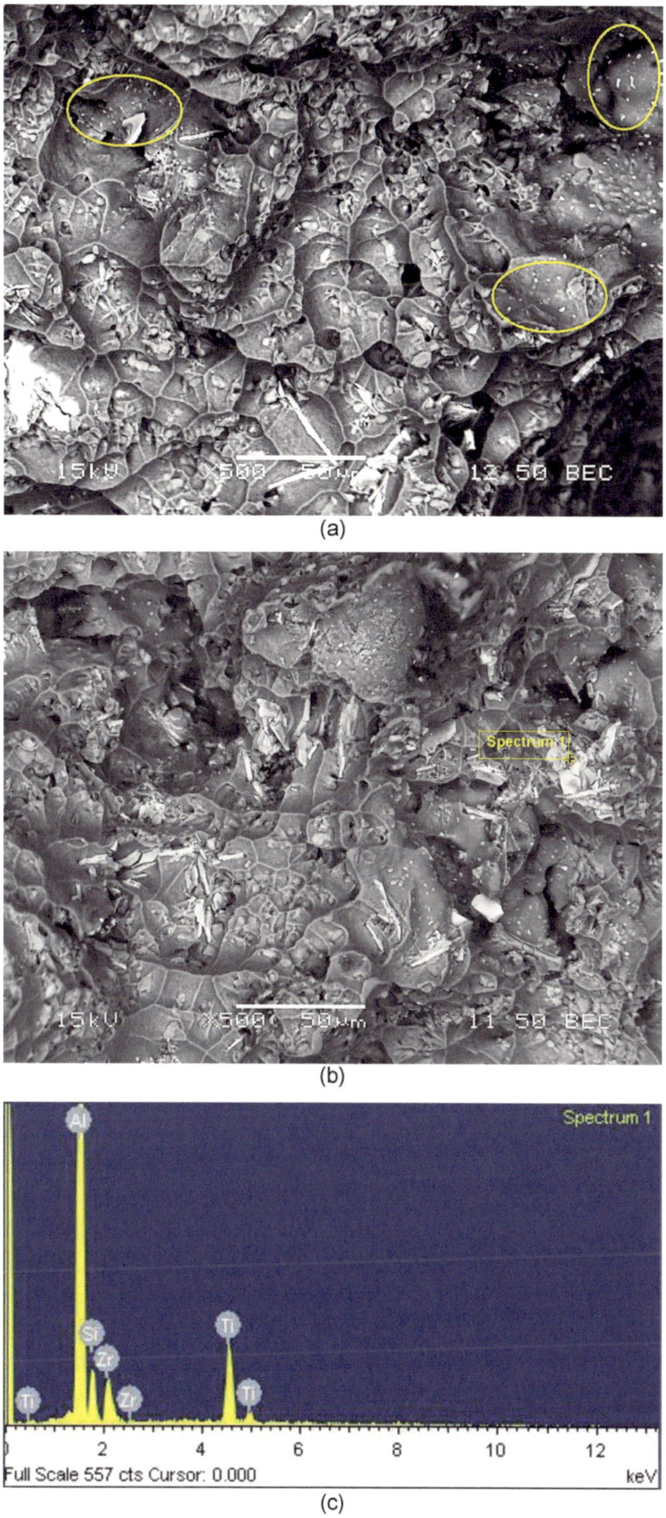

Figure 13.
Fracture surface of alloy B: (a, b) BSE images of T6-treated alloy after stabilization at 250°C for 200 h showing a coarse dimpled structure, coarsened precipitates and $Al_x(Zr,Ti)Si$ *particles involved in the crack initiation process, and (c) corresponding EDS spectrum of the phase of interest as shown in (b) [41].*

has a dimpled-structure throughout, which indicates the ductile nature of the fracture mode. In addition, the BSE image exhibits the precipitation of $Al_x(Zr,Ti)Si$ compound, in the form of star-like shape, as confirmed by the associated EDS spectrum in **Figure 12(b)**. Also, cracks can be spotted in various particles of this compound, as indicated by the arrows. The higher magnification BSE image shown in **Figure 11(c)** reveals a cracked $Al_x(Zr,Ti)Si$ phase particle.

Figure 13(a) [41] shows the fracture surface of the T6-treated B alloy tested at 250°C after stabilization for 200 h at the testing temperature. The dimple structure is coarser compared to that before stabilization at 250°C. This observation would explain the improved ductility of the alloy due to the softening behavior associated with the prolonged elevated-temperature exposure at 250°C. Coarsened precipitates appear in the interiors of the dimples, as indicated by the oval contours in **Figure 13(a)**. The BSE image and the EDS spectrum shown in **Figure 13(b)** and **(c)**, respectively, confirm the presence of $Al_x(Zr,Ti)Si$ phase particles.

4. Conclusions

Based on an analysis of the results presented in this article, the following conclusions may be made:

1. For the base 354 alloy A, solution heat treatment and artificial aging at 190°C for 2 h or at 155°C for 100 h resulted in an increase in the alloy strength by ~64% over its as-cast strength. Aging at 155 or 170°C for a long period of time offered maximum resistance to softening.

2. The Zr-rich intermetallic phases appear in two different forms, namely $(Al,Si)_2(Zr,Ti)$ in block-like form, and containing high level of silicon, and $(Al,Si)_3(Zr,Ti)$ in needle-like form, containing high level of aluminum.

3. Quality index (*Q*) charts constructed for alloy 354 characterize the tensile properties in terms of the heat treatment conditions applied. Minimum and maximum *Q* values, i.e., 259 and 459 MPa, are observed for the as-cast and solution heat-treated conditions, respectively; the yield strength shows a maximum of 345 MPa and a minimum of 80 MPa within the range of aging treatments applied.

4. DSC runs carried out on alloy B (354 alloy + 0.3%Zr) revealed peak patterns which included differences in peak heights—which reflected the amount of the precipitated phase, and shifts in the transformation temperature.

5. Melt superheating at 800°C is beneficial in terms of reducing the amount of coarse Zr-rich phases in the alloy structure, as it provides efficient dissolution of the Al_3Zr phase from the master alloy during the melting process. Coarse Zr-containing phases are rarely observed due to the limited number of Al_3Zr particles available to act as nucleation sites for these coarse phases.

6. TEM investigations confirm that the investigated alloys are strengthened primarily by θ-Al_2Cu and S-Al_2CuMg precipitates and their precursors, in addition to a secondary strengthening effect by precipitates in the form of $Al_x(Zr,Ti)Si$ which form following the addition of Zr.

7. Prolonged exposure at 250°C, resulting in coarsening of the strengthening precipitates, causes noticeable reduction in strength values, particularly the yield strength (cf. 160 and 325 MPa), and a remarkable increase in the ductility values (cf. 6.3 and 1.1%).

8. The strength values (UTS and YS) obtained at room temperature for the stabilized T5-treated alloy samples are comparable to those of the stabilized T6-treated condition, and higher in the case of elevated-temperature tensile testing.

9. The fracture surface of the T6-treated alloy B after stabilization for 1 h at 250°C reveals a dimpled-structure throughout, indicating the ductile nature of the fracture mode.

The $Al_x(Zr,Ti)_{1-x}Si$ complex compound is observed with star-like and blocky morphologies, with cracks appearing in various particles of this compound. By increasing the stabilization time up to 200 h, coarser and deeper dimples are formed, highlighting the improved ductility of the alloy due to the softening behavior associated with the prolonged exposure at 250°C.

Author details

Jacobo Hernandez-Sandoval[1,2], Mohamed H. Abdelaziz[1,3], Agnes M. Samuel[1], Herbert W. Doty[4] and Fawzy H. Samuel[1]*

1 Département des Sciences appliquées, Université du Québec à Chicoutimi, Canada

2 Facultad de Ingeniería Mecánica y Eléctrica, Universidad Autónoma de Nuevo Leon, Mexico

3 Département PEC, Université Française d'Égypte, Le Caire, Egypt

4 General Motors Global Technology Center, Warren, MI, USA

*Address all correspondence to: fhsamuel@uqac.ca

References

[1] Ibrahim MF, Samuel AM, Doty HW, Samuel FH. Effect of aging conditions on precipitation hardening in Al-Si-Mg and Al-Si-Cu-Mg alloys. International Journal of Metalcasting. 2017;**11**(2):274-286

[2] Hatch JE. Aluminum: Properties and Physical Metallurgy. USA: ASM International; 1984

[3] Samuel FH, Samuel A. Effect of heat treatment on the microstructure, tensile properties, and fracture behavior of permanent mold Al-10 Wt Pct Si-0.6 Wt Pct Mg/sic/10 P composite castings. Metallurgical and Materials Transactions A. 1994;**25**(10):2247-2263

[4] Ibrahim MF, Abdelaziz MH, Doty HW, Valtierra S, Samuel FH. Effect of microalloying elements on the heat treatment response and tensile properties of Al-Si-Mg alloys. In: Ares AE, editor. Solidification. Rijeka: IntechOpen; 2018. Ch. 01

[5] Hernandez-Sandoval J, Garza-Elizondo GH, Samuel AM, Valtierra S, Samuel FH. The ambient and high temperature deformation behavior of Al-Si-Cu-Mg alloy with minor Ti, Zr, Ni additions. Materials & Design. 2014;**58**:89-101

[6] Meyers C, Hinton K, Chou J-S. Towards the optimization of heat-treatment in aluminium alloys. Materials Science Forum. 1992;**72**:102-104

[7] Mohamed AMA, Samuel FH. A review on the heat treatment of Al-Si-Cu/Mg casting alloys. In: Czerwinsk F, editor. Heat Treatment; Conventional and Novel Applications. Rijeka: IntechOpen; 2012. Chapter 4

[8] Cayron C, Buffat P. Transmission electron microscopy study of the B′ phase (Al–Mg–Si alloys) and Qc phase (Al–Cu–Mg–Si alloys): Ordering mechanism and crystallographic structure. Acta Materialia. 2000;**48**(10):2639-2653

[9] Matsuda K, Teguri D, Sato T, Ikeno S. EFTEM observation of Q'phase in Al-Mg-Si-Cu alloy. Materials Science Forum. 2002;**396**(4):947-951

[10] Hwang J, Banerjee R, Doty H, Kaufman M. The effect of Mg on the structure and properties of type 319 aluminum casting alloys. Acta Materialia. 2009;**57**(4):1308-1317

[11] Wang G, Sun Q, Feng L, Hui L, Jing C. Influence of Cu content on ageing behavior of AlSiMgCu cast alloys. Materials & Design. 2007;**28**(3):1001-1005

[12] Abdelaziz MH. Microstructural and mechanical characterization of transition elements-containing Al-Si-Cu-Mg alloys for elevated-temperature applications [PhD thesis]. Canada: UQAC; 2018

[13] Srinivasan D, Chattopadhyay K. Metastable phase evolution and hardness of nanocrystalline Al–Si–Zr alloys. Materials Science and Engineering A. 2001;**304**:534-539

[14] Knipling KE. Development of a nanoscale precipitation-strengthened creep-resistant aluminum alloy containing trialuminide precipitates [PhD thesis]. USA: Northwestern University; 2006

[15] Ragab K. The use of fluidized sand bed as an innovative technique for heat treating aluminum based castings [PhD thesis]. Canada: UQAC; 2012

[16] Mohamed A, Samuel A, Samuel F, Doty H. Influence of additives on the microstructure and tensile properties of near-eutectic Al–10.8% Si

cast alloy. Materials & Design. 2009;**30**(10):3943-3957

[17] Drouzy M, Jacob S, Richard M. Interpretation of tensile results by means of quality index and probable yield strength-application to Al-Si7 Mg foundry alloys-France. International Journal of Cast Metals. 1980;**5**(2):43-50

[18] Jacob S. Quality index in prediction of properties of aluminum castings—A review. In: Transactions of the American Foundry Society and the One Hundred Fourth Annual Castings Congress. Vol. 208. Pittsburgh, Pennsylvania; 2000. pp. 811-818

[19] Cáceres C. A rationale for the quality index of Al-Si-Mg casting alloys. International Journal of Cast Metals Research. 2000;**12**(6):385-391

[20] Abdelaziz MH, Doty HW, Valtierra S, Samuel FH. Static versus dynamic thermal exposure of transition elements-containing Al-Si-Cu-Mg cast alloy. Materials Science and Engineering A. 2019;**739**:499-512

[21] Tahiri H, Mohamed SS, Doty HW, Valtierra S, Samuel FH. Effect of Sr–grain refining–Si interactions on the microstructural characteristics of Al–Si hypoeutectic alloys. International Journal of Metalcasting. 2018;**12**(2):343-361

[22] Zhu PY, Liu QY. Kinetics of granulation of discontinuous phase in eutectic structures. Materials Science and Technology. 1986;**2**:500-507

[23] Reif W, Dutkiewiewicz J, Ciach R. Effect of precipitates in Al-Si-Cu-Mg alloys. Materials Science and Engineering A. 1997;**234**:165-168

[24] Ammar HR, Samuel AM, Samuel FH. Porosity and the fatigue behavior of hvpoeutectic and hypereutectic aluminum-silicon casting alloys. International Journal of Fatigue. 2008;**30**(6):1024-1035

[25] Câceres CH, Taylor JA. Enhanced ductility in Al-Si-Cu-Mg casting alloys with high Si content. In: Tiryakiouglu M, Crepeau P, editors. Shape Casting: The John Campbell Symposium. California: TMS; 2005. pp. 245-254

[26] Toschi S. Optimization of A354 Al-Si-Cu-Mg alloy heat treatment: Effect on microstructure, hardness, and tensile properties of peak aged and overaged alloy. Metals. 2018;**8**:961. DOI: 10.3390/met8110961

[27] Abdelaziz MH, Elgallad EM, Doty HW, Valtierra S, Samuel FH. Melting and solidification characteristics of Zr-, Ni-, and Mn-containing 354-type Al-Si-Cu-Mg cast alloys. Philosophical Magazine. 2019;**99**(13):1633-1655

[28] Hernandez Sandoval J. Improving the Performance of 354 Type Alloy [PhD thesis]. Canada: Université du Québec à Chicoutimi; 2010

[29] Garza-Elizondo GH. Effect of Ni, Mn, Zr and Sc Additions on the Performance of Al-Si-Cu-Mg Alloys [PhD thesis]. Canada: UQAC; 2016

[30] Ibrahim MF, Samuel E, Samuel AM, Al-Ahmari AMA, Samuel FH. Metallurgical parameters controlling the microstructure and hardness of Al-Si-Cu-Mg base alloys. Materials & Design. 2011;**32**(4):2130-2142

[31] Priya P, Matthew JM, Krane JM, Johnson DR. Precipitation of Al_3Zr dispersoids during homogenization of Al-Zn-Cu-Mg-Zr alloys. In: Light Metals. 2016. pp. 213-218

[32] Tavitas-Medrano F, Gruzleski J, Samuel F, Valtierra S, Doty H. Effect of Mg and Sr-modification on the mechanical properties of 319-type aluminum cast alloys subjected to artificial aging. Materials Science and Engineering A. 2008;**480**(1):356-364

[33] Reif W, Dutkiewicz J, Ciach R, Yu S, Krol J. Effect of ageing on the evolution of precipitates in AlSiCuMg alloys. Materials Science and Engineering A. 1997;**234**:165-168

[34] Purtee M. Aging effects of an aluminum-based 319 alloy. Foundry Management & Technology. 1998;**126**(10):42-44

[35] Takeda M, Ohkubo F, Shirai T, Fukui K. Precipitation behaviour of Ai-Mg-Si ternary alloys. In: Materials Science Forum, Vol. 217-222. Stafa-Zurich, Switzerland: Trans Tech Publications, Ltd.; 1996. pp. 815-820

[36] Elwazri A, Varano R, Siciliano F, Bai D, Yue S. Characterisation of precipitation of niobium carbide using carbon extraction replicas and thin foils by FESEM. Materials Science and Technology. 2006;**22**(5):537-541

[37] Kang H, Kida M, Miyahara H, Ogi K. Age-hardening behavior of Al-Si-Cu base cast alloys. Journal of Japan Foundry Engineering Society. 1997;**69**(10):828-834

[38] Shaha SK. Development and characterization of cast modified Al-Si-Cu-Mg alloys for heat resistant power train applications [PhD thesis]. Canada: Ryerson University; 2015

[39] Knipling KE, Dunand DC, Seidman DN. Nucleation and precipitation strengthening in dilute Al-Ti and Al-Zr alloys. Metallurgical and Materials Transactions A. 2007;**38**(10):2552-2563

[40] Knipling KE, Dunand DC, Seidman DN. Precipitation evolution in Al–Zr and Al–Zr–Ti alloys during isothermal aging at 375-425 C. Acta Materialia. 2008;**56**(1):114-127

[41] Abdelaziz MH, Elgallad EM, Samuel AM, Doty HW, Samuel FH. High-temperature tensile fractography of Zr-, Ni-, and Mn-containing Al-Si-Cu-Mg cast alloys. Advances in Materials Science and Engineering. 2020

Chapter 4

The High-Speed 6xxx Aluminum Alloys in Shape Extrusion Industry

Rafał Hubicki and Maria Richert

Abstract

This chapter describes and analyzes the 6xxx aluminum alloys used in the shape extrusion sector dedicated to automotive and construction industry. The division and application of 6xxx aluminum alloys are performed. The precipitation hardening of **6xxx** (Al-Mg-Si) alloys is presented as these alloys easily undergo deformation and present the potential for new kinds of alloys for high-speed extrusion. The mechanisms of strengthening are shown with the evolution of precipitation sequences. Also some examples of industry applications of 6xxx aluminum alloys are presented.

Keywords: aluminum alloys, extrusion, aging, microstructure

1. Introduction

1.1 Characteristics of aluminum alloys

Historically, aluminum was first produced in 1825 by reducing aluminum chloride with potassium amalgam. In 1886, Héroult and Hall discovered the possibility of producing aluminum by electrolysis. In 1895, aluminum was first used as a material for the church roof. With the increase in the production of aluminum, which occurred especially after the Second World War, the scope of its use expanded. Today, aluminum alloys are widely used in transport, mechanical engineering, electrical and energy industries, food industry, chemical industry, sports, aviation, transport, yacht and shipbuilding, and many other fields. Below, in **Table 1**, the grades of aluminum alloys used in industry are presented.

Group	Series	Alloy
Al	1xxx	1050A, 1070A
Al-Cu	2xxx	2014, 2017, 2024
Al-Mn	3xxx	3103, 3003
Al-Mg-Si	6xxx	6060, 6063, 6061, 6082
Al-Si	4xxx	4032
Al-Mg	5xxx	5019, 5083, 5754
Al-Zn	7xxx	7003, 7020, 7022, 7075

Table 1.
Industrial aluminum alloy groups.

Aluminum alloys are highly ductile, so it is easy to make the desired structural elements, machine parts, and others from them.

By changing the content of alloying elements in aluminum alloys, the strength properties can be adjusted. A very effective factor influencing the strengthening of aluminum alloys is heat treatment—supersaturation and aging. This treatment is possible for Al-Mg-Si 6xxx series alloys, which show a variable solubility in the solid state. Similar possibilities are also available for the 2xxx, Al-Cu-Mg, and 7xxx, Al-Zn-Mg-Cu, alloys (**Figures 1** and **2**). However, the most popular alloy for profile extrusion is 6xxx series.

During the extrusion process, aluminum alloys are placed in the extrusion press container (cylinder) and pressed by a pressing ram (or stem—via a dummy block or pressure plate). The metal flows out through the hole in the die, which gives the shape to the extruded profile (**Figure 3**).

The state of stress in most of the plasticized area is a three-axial nonuniform compression. It is therefore possible to make large plastic deformations without affecting the consistency of the material (maximum elongation coefficients are about 300, average—about 50). This is the main advantage of extrusion processes. Large deformations require high forces. The main limitation to the scale of deformation that can be obtained in one extrusion operation is not the phenomenon of material decohesion (as in many other processes) but the strength of the tools.

When the extruded section leaves the tool, it is cooled with water or air and then drawn, still in the malleable state. This removes the stresses accumulated in the aluminum alloy and at the same time allows to achieve the expected and correct profile dimensions. The profiles are then cut and obtain ultimate strength by hot or cold hardening.

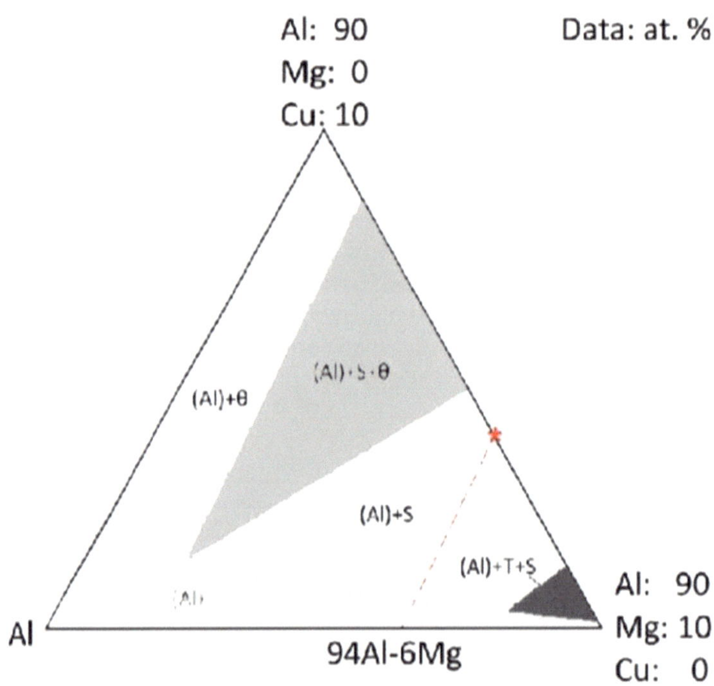

Figure 1.
Isothermal section of ternary Al-Cu-Mg phase diagram at 400°C, h ¼ CuAl 2, S¼ MgCuAl 2, and T ¼ (Cu 1-x Al x) 49 Mg 32. Along the dashed red line, the atomic fraction is constant, 94Al-6 Mg. And at the red star, the composition point is 90Al-4Cu-6 Mg, namely, the average composition of the eutectic region, shown with the dashed white box.

DOI: http://dx.doi.org/10.5772/intechopen.93239

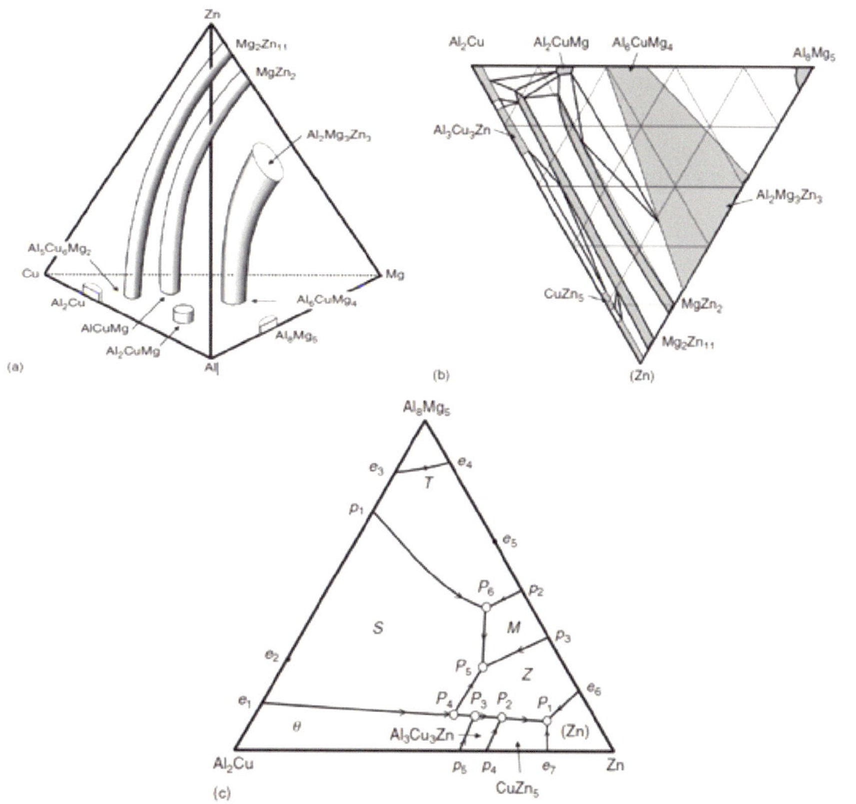

Figure 2.
Phase diagram Al-Cu-Mg-Zn: (a) polythermal diagram, (b) distribution of phase fields in the solid state in the aluminum corner, and (c) single-phase domains [1].

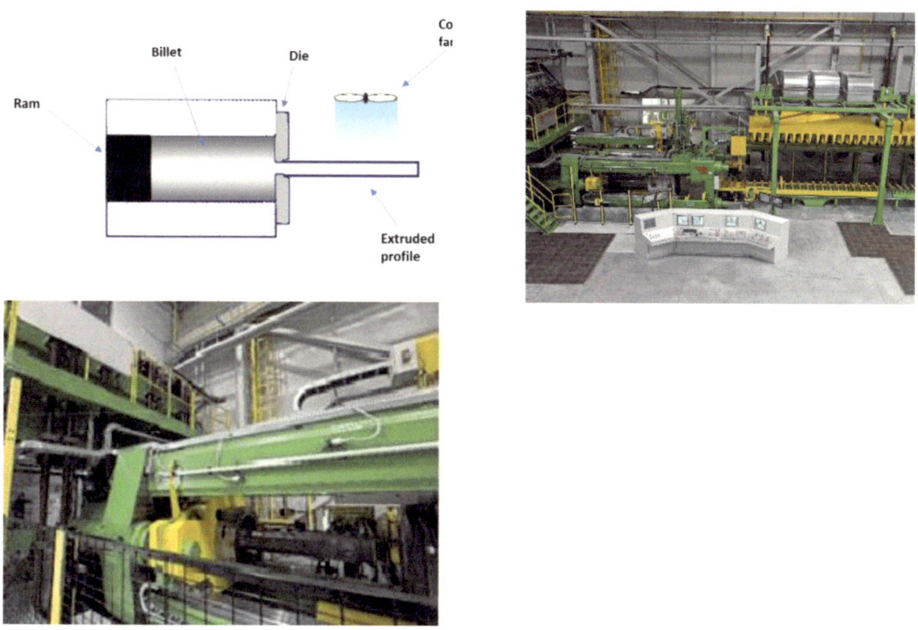

Figure 3.
Extrusion diagram and extrusion press for aluminum alloys.

In the case of large strains, hot extrusion is used, because during cold extrusion the forces are so high that the tools do not withstand the loads. Cold extrusion for large deformations can only be performed for soft materials (e.g., pure aluminum).

2. 6xxx series aluminum alloys for extrusion of shapes

The 6xxx series alloys are the most commonly used and their global consumption is the largest. The worldwide demand for aluminum is around 29 million tons per year. About 22 million tons is new aluminum and 7 million tons is recycled aluminum scrap. The use of recycled aluminum is economically and environmentally compelling. It takes 14,000 kWh to produce 1 ton of new aluminum. Conversely, it takes only 5% of this energy to remelt and recycle 1 ton of aluminum. There is no difference in the quality between virgin and recycled aluminum alloys [2].

Alloys of the 6xxx series are heat-treated as they show a variable solubility in the solid state. **Figure 4** shows a pseudo-binary system Al-Mg_2Si. The variable solid solubility curve allows for heat treatment of 6xxx series alloys. After plastic working (e.g., extrusion), the supersaturation and artificial aging of these alloys are applied. After supersaturation, which consists in rapid cooling of the alloy, it is then heated to an appropriate temperature in order to precipitate hardening phases.

The phase sequence in this alloy is as follows:

$$\begin{aligned}&\text{super-saturated solid solution (SSSS)}\\&-\text{atomic clusters}-\text{GP zones}-\beta''-\beta'/B'U-\beta/S\end{aligned} \quad (1)$$

5

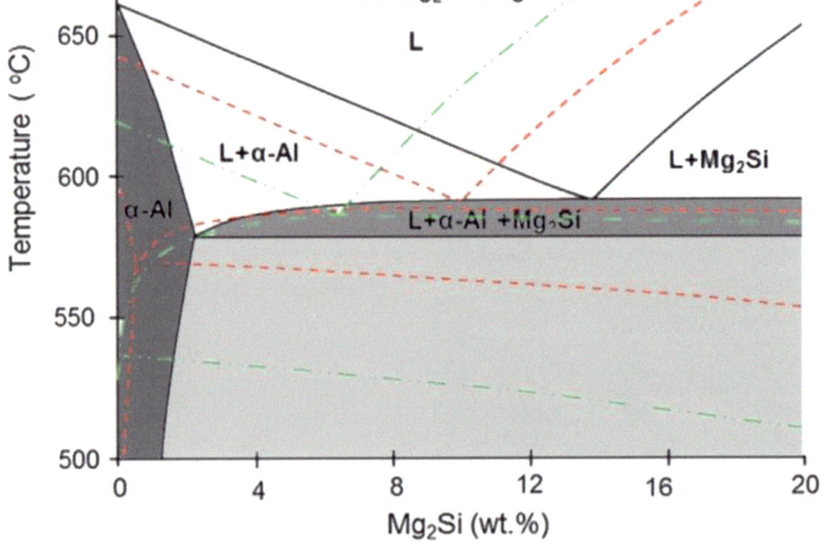

Figure 4.
Calculated equilibrium phase diagram of Al-Mg2Si pseudo-binary alloys with excess Mg.

Phase	Phase form	Formula	Spatial group	Network parameters
GP zones [3, 4]	Needles	Unknown	C2/m	a = 1.48, b = 0.405, c = 0.648, β = 105.3 deg
β″ [5, 6]	Needles	Mg_5Si_6	C2/m	a = 1.516, b = 0.405, c = 0.674, β = 105.3 deg
β′ [7, 8]	Needles	$Mg_{1,8}Si$	$P6_3$	a = b = 0.715, c = 0.405, γ = 120 deg
U1 [7, 9–11]	Needles	$MgAl_2Si_2$	P_{3m1}	a = b = 0.405, c = 0.674, γ = 120 deg
U [7–11]	Needles	MgAlSi	P_{nma}	a = 0.675, b = 0.405, c = 0.794
B [9]	Lath	Unknown	Hexagonal	a = 1.04, c = 0.405, γ = 120 deg
β	Plates	Mg_2Si	F_{m3m}	a = 0.6354
Si	Plates	Si	F_{d3m}	a = 0.5431

Table 2.
Precipitations in 6xxx series alloys.

Table 2 describes the characteristics of the precipitating phases [12].

Si plates appear in silicon-rich alloys. All phases that show the shape of needles are consistent with the direction <100> of the aluminum matrix.

The extrusion temperature of Al-Mg-Si alloys ranges between 400°C and 500°C, which causes that all phases are dissolved in the alloy to form a solid solution. Therefore, in order to prevent the precipitation of an incoherent phase of Mg_2Si, a rapid quenching of the alloy is required after extrusion.

Rapid quenching maintains the supersaturation of the alloy's solid solution caused by Mg and Si and a high concentration of vacancies, also resulting from rapid cooling. The supersaturation is followed by the aging process (**Figure 5**). It is carried out at temperatures from 165°C to 185°C until maximum hardness is reached—state T6.

Hardening achieved as a result of aging depends on the size and density of the precipitations and the volume of metastable phases obtained.

Isothermal heating at 175°C induces the precipitation of β-phase hardening the alloy. The β″-phase is the most effective hardening phase in Al-Mg-Si alloys and is

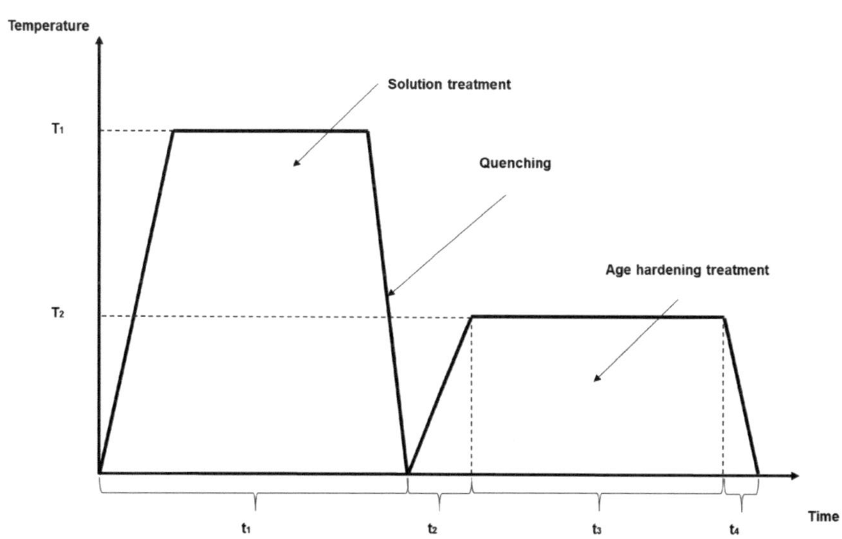

Figure 5.
Sketch of precipitation hardening phases.

formed at temperatures between 125°C and 200°C. The β″-phase composition is Mg_5Si_6; therefore, an appropriate Mg/Si ratio of 5/6 should be maintained during the casting process. The β-phase (Mg_2Si) is not formed below 200°C, so it cannot be present in alloys aged below this temperature.

Marioara et al. [12] studied phase β″ precipitation at different Mg/Si ratios. The research showed that a higher Si content in the alloy promotes the formation of a large number of fine GP zones in a shorter time than alloys with a lower Si content. After annealing for 3 h at 175°C, there was a sharp peak of hardness associated with the appearance of GP zones in these alloys, and after 17 h, a wider peak of hardness associated with the occurrence of phase β″ precipitation appeared. Alloys richer in Mg had thicker elements of microstructure than alloys with an increased Si content. They also contained less U2 precipitations but were richer in β′ precipitations. The studies carried out revealed a strong influence of Si, the content of which controls the process of phase precipitation through the formation of precipitation clusters in the initial stages of annealing.

The great interest in Al-Mg-Si alloys is related to their application in the automotive industry. Chakrabarti et al. [13] conducted a research proving that the strengthening phase involved in the Al-Mg-Si ternary alloys is the metastable β″ phase [14, 15].

Kuroda et al. studied the effect of small amounts of Ni, Co, and V on hardness after age hardening of the AlMgSi base alloy. It was found that the additives increase the hardness, with the highest value of hardness occurring after the introduction of Ni and two-stage aging [16].

The influence of Cu addition on the precipitation process in AlMgSi alloy was studied by Zandbergen et al. [17]. The occurrence of Cu in all investigated precipitations of the hardening phases of the alloy was determined.

On the other hand, the influence of Zn and Ag additives on the precipitation process in AlMgSi alloys was analyzed in the work of Saito et al. [18]. The researchers found a weak influence of Zn on the precipitation sequence of hardening phases. They found that Zn is built into the structure of precipitations replacing some elements in the network of precipitated phases.

A new alloy based on AlMgSi, with the addition of Mn that did not require aging after extrusion, was presented by Lee et al. [19]. Mn creates precipitations of 0.05–0.5 μm. The alloy containing 1 wt.% Mn had high strength properties and good ductility in comparison with the commercial 6N01 alloy. Moreover, it showed higher fatigue resistance.

Chen et al. studied the process of phase precipitation in Al-Mg-Si alloys [20]. They showed that the formation of Mg_5Si_6, hardening AlMgSi alloy, is preceded by the precipitation of $Mg_2Si_{2,6}Al_{6,4}$ and more precisely $Mg_2Si_2Al_7$. The precipitation of this phase is dependent on the temperature of the aging of the alloy.

Mechanisms of hardening with β″ particles, characterized by coherence with aluminum network, were studied by Ringdalen et al. [21]. The study concerned the alloy 6060 and included the interaction of the dislocation lattice with the precipitations.

The AlMgSi alloy, due to the possibility of heat treatment, has a high potential for modeling its structures and properties. Analyzing the process of supersaturation of this alloy and subsequent aging, it is possible to obtain the assumed properties. Hardening phases in the AlMgSi alloy nucleate heterogeneously at the grain boundaries and iron precipitations, which are always present in AlMgSi alloys

Figure 6.
Hardening phases in AlMgSi alloy.

(**Figure 6**). Precipitations of β″, the hardening phase of AlMgSi alloy, can be found at **Figure 7**. Their growth to larger sizes is associated with the formation of Mg_2Si phase. Large Mg_2Si particles are not an obstacle to displaced dislocations and do not increase the strength properties of the alloy.

An example of research on the heat treatment of AlMgSi alloys is the article by Richert et al. [22] concerning the new, multistage aging of the 6xxx series alloys. The studies of Ryen et al. [23] showed the presence in alloys AA6063 and AA6068, after aging precipitations as B′, β″, U2, and U1, β′ and some unidentifiable crystal structures. Detailed studies of the precipitations in aluminum alloys are also presented in the work by Andersen et al. [24]. From this work, data were taken from the composition of the precipitations formed during aging in AlMgSi alloys:

β-Mg_2Si,
U1-$MgAl_2Si_2$ 'A'
β″-$Mg_4(Al_xMg_{1-X})Si_4$
U2-$Mg_4Al_4Si_4$
β′-Mg_9Si_5
B′ ~ $Mg_{48}Al_{16}Si_{36}$
GP-$Mg_4(Al_xMg_{1-x})Si_4$

The author found that the early stages in Al-Mg-Si and Al-Mg-Cu systems have isostructural GP zones being 1D strings along <100>Al, identical to the eye-like units of the β″-phase, where Cu can replace Si. Calculations show GP zones can take different compositions, the most stable being Mg_4Si_4Mg and Mg_4Cu_4Al, with Mg and Al as central interstitial columns. Solute clusters for the GP zones are likely short defect-free needles using a vacancy to produce a central interstitial column. The needle-shaped β″-phase is the most important hardening precipitate in the AA6xxx system.

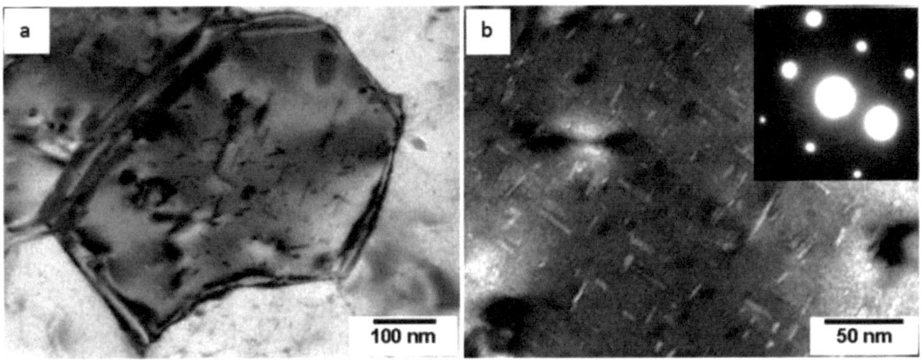

Figure 7.
Nanometric precipitations in AlMgSi alloy [25].

3. High-speed aluminum alloy

3.1 Studies on the production of a high-speed extrusion alloy

3.1.1 Research methodology

In this work, tests were carried out on 6060 alloy with three different chemical compositions, falling within the definition of the standard EN-AW 755-2 (**Table 3**).

Alloy 1 with the lowest Si and Mg contents, Alloy 2 with medium Si and Mg contents, and Alloy 3 with the highest Si and Mg contents were prepared as follows (**Figure 8**):

1. The billets were continuously cast from 6060 alloy, with a diameter of 228 mm, from three melts, differing in the content of alloying components:

    - Alloy 1 (low alloying content)

    - Alloy 2 (with medium content of alloying elements—standard composition)

    - Alloy 3 (high alloying content)

2. The billets were homogenized in accordance with the technology used in Grupa Kęty S. A.

3. The extrusion of profiles (four-hole die) was performed on a 35 MN press as follows:

    - From each alloy, three billets were extruded with the standard speed of the extrusion ram for this shape, and three billets with the extrusion ram speed increased by 20%.

    - The temperatures of the extrusion billets and the press container were the same for all billets.

    - All extruded sections were saturated on the press runout table and then artificially aged in the furnace to T66.

    - Three samples of each extruded section were taken for each hole to check the strength properties and material hardness of the sections.

	Si	Fe	Cu	Mn	Mg	Cr	Ni	Zn	Ti
Alloy 1	0.38	0.20	0.01	0.04	0.36	0.01	0.01	0.02	0.02
Alloy 2	0.45	0.22	0.01	0.04	0.40	0.01	0.01	0.02	0.01
Alloy 3	0.49	0.21	0.01	0.05	0.45	0.01	0.00	0.01	0.02

Table 3.
Chemical composition of billets tested.

Figure 8.
Aluminum alloy billets before loading onto the loading ramp of the press [26].

The tests of mechanical properties of extruded sections were carried out on the Zwick Z300 fatigue tester.

1. In the second stage, only Alloy 1 was tested as promising the highest ductility and thus the possibility to use the highest extrusion rate for the manufactured sections.

The following tests were carried out:

- Alloy 1 was extruded with the following set speed rates: standard (7 billets); standard + 20% (7 billets); standard + 30% (8 billets); standard + 40% (7 billets); standard + 50% (9 billets); and standard + 60% (9 billets),
- The temperatures of extrusion billets and the container were the same for all variants of extrusion.
- From each extruded shape according to each variant of the ram speed, 12 samples with 1000 mm length (72 samples in total) were taken in order to check the strength properties and for hardness measurement.
- All extruded sections were saturated on the runout of the press and then subjected to different variations of artificial aging to T66.
- After extrusion of the sections and their heat treatment, the yield point and tensile strength of the samples were determined on the Zwick Z300 fatigue tester, and hardness tests were carried out on the Zwick ZHU250 testing machine.

- On the basis of the results presented in **Table 4** and the results of properties tests, the aging variant to achieve the properties of the alloy, according to PN-EN 755-2,2016-05 standard, was selected.

- After extrusion and aging, hardness tests were performed on the Zwick ZHU250, and the yield point and tensile strength were determined on the Zwick Z300 fatigue tester machine.

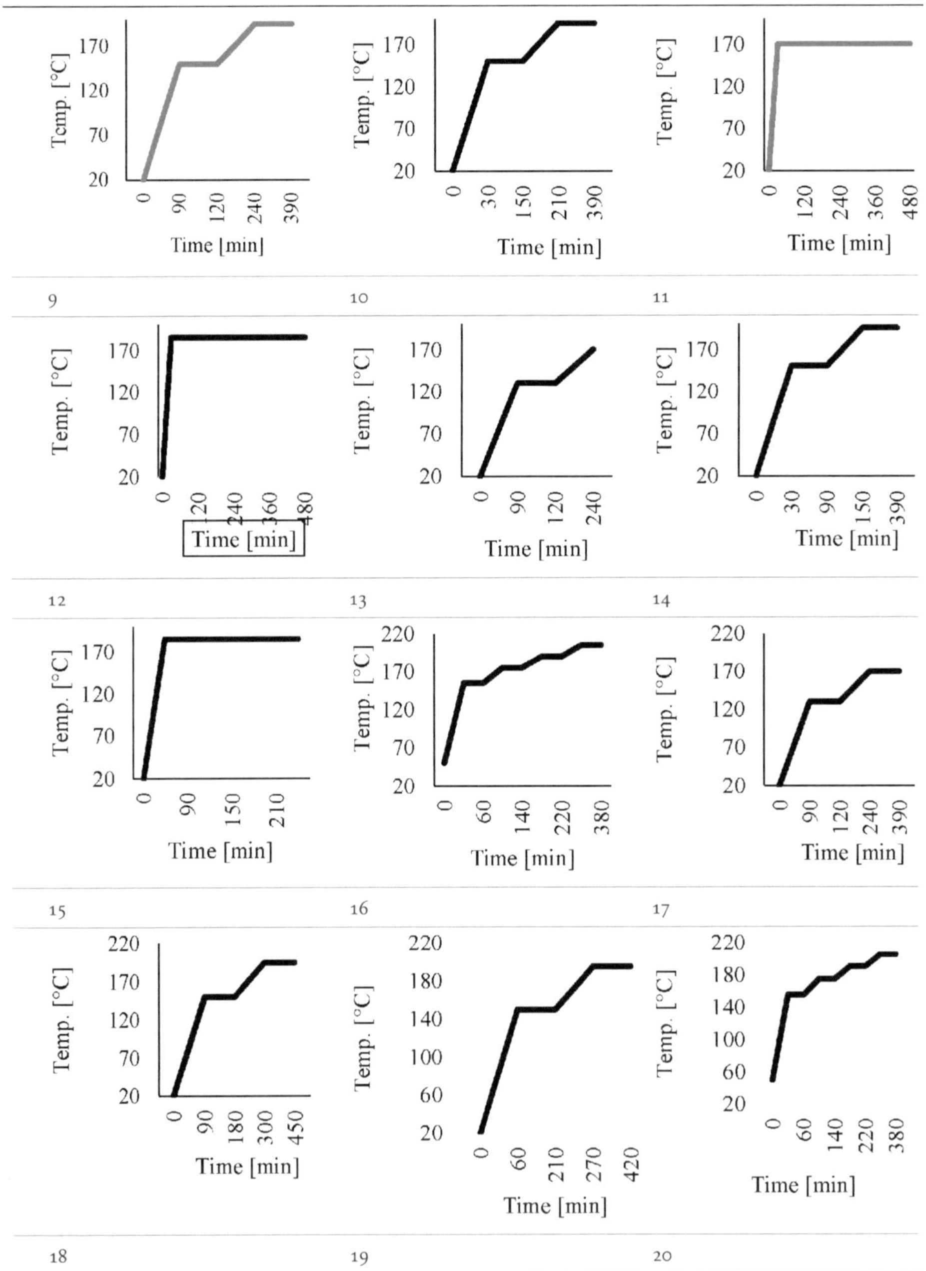

Table 4.
Aging curves of Al-Mg-Si alloy with the lowest composition of alloying elements, second stage of research.

- Observations of the sample structure on the Olympus GX53 and Olympus GX71 and GX51 optical microscopes with digital image capture were performed.

- Samples for microscopic examination and microhardness testing were mounted in the EpoFix epoxy resin. The samples prepared in this way were subjected to grinding (on abrasive papers with the following grades: 220, 500, 800, 1200, 2000, and 4000) and two-stage mechanical polishing according to Struers' procedure using a diamond-paste suspension DiaDuo with a grain size of 3 μm and a colloidal silicon oxide suspension OPS for finish polishing with a grain size of ¼ μm.

Two methods of etching were used:

1. Etching in 0.5% HF acid solution—this method highlights the particles present in the microstructure.

2. Anodizing in Barker's reagent composed of 1.8 ml HBF_4 + 100 ml H_2O—color contrast was obtained from individual grains during polarized light observation.

- Microhardness measurements were performed using the Vickers method with the SHIMADZ HMV-G hardness tester. Microhardness tests were carried out on metallographic specimens at a load of 100 g (HV0.1), and the time to withstand the load was 10 s.

- Observations of the microstructure and morphology of the precipitations with the scanning electron microscope JEOL JSM-6460LV were also performed, using the EDS microanalyzer.

- The share of Mg_2Si phase in the tested samples was determined using the X-ray method. X-ray examinations were performed on PANalytical Empyrean polycrystalline diffractometer with Cu $K\alpha1$ = 1.5417 Å, at U = 40 kV and I = 30 mA, equipped with PANalytical High Score Plus program integrated with ICDD PDF4+ 2016 crystallographic database. PIXcel detector was used in the tests. The test range was 10–120° of 2θ angle.

- In the third stage of experimental testing, Alloy 1 was subjected to different aging variants in order to answer the question on which aging variant allows to achieve the assumed alloy properties, compliant with the standard, despite the lowest values of Mg and Si components. The evaluation of the properties of the samples after aging was performed by hardness and microhardness measurements.

3.1.2 Comparative results of tested alloys

The research revealed that after casting and homogenization, Alloy 1 has a correct structure with equiaxed grains and broken lattices of iron precipitations (**Figure 9**).

Figure 10 shows the yield point ($R_{0.2}$) and tensile strength (Rm) of extruded profiles made of Alloy 1 with the lowest alloying content and other alloy compositions with higher content of alloying elements, i.e., Alloy 2 and Alloy 3. According to the requirements of the standard, the extruded section should have minimum $R_{0.2}$ = 160 MPa and Rm = 215 MPa. The shape made of Alloy 2 with the average content of alloying elements and Alloy 3 with the highest content of alloying elements obtained the property level required by the standard EN755-2 (for products made of alloy 6060, in the T66 state). The properties of Alloy 3 are too high. And the shape with the lowest level of alloying elements failed to obtain such a level.

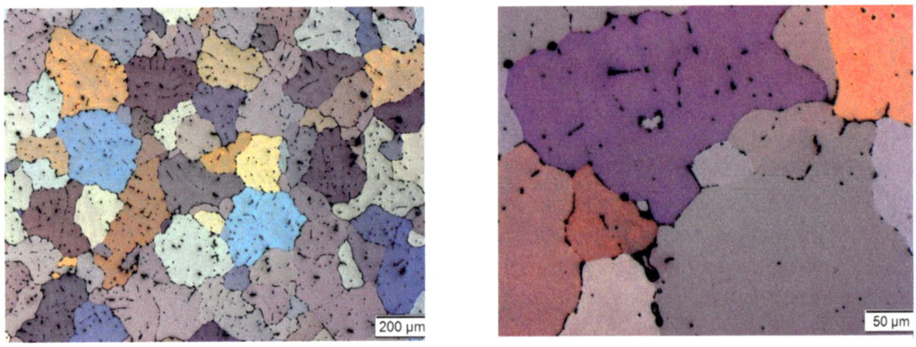

Figure 9.
The structure of Alloy 1 billet with the lowest content of Mg and Si alloying elements.

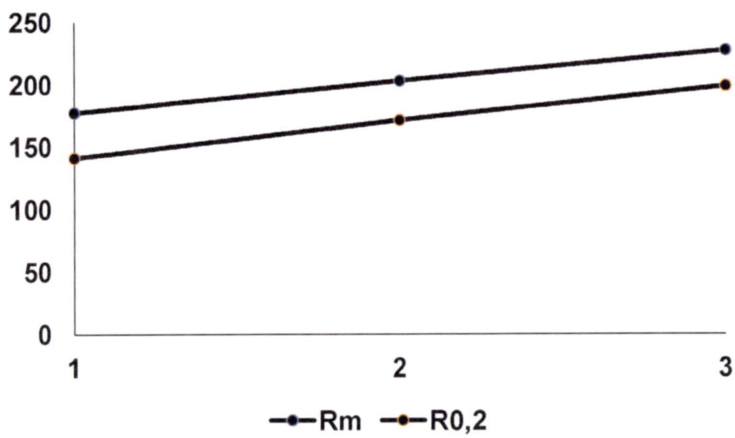

Figure 10.
Yield point and tensile strength of extruded sections from billets with three different contents of the main alloying elements.

Figure 11 illustrates the plasticity of the alloys tested. The highest value of elongation was achieved by the section with the lowest content of alloying elements. This result indicates that the alloy potentially has a plasticity level allowing to use higher extrusion rates in relation to the other alloying components.

The delta (**Figure 11**), i.e., the difference between $R_{0.2}$ and Rm, shown in the diagram (**Figure 11**) illustrates the "yield strength reserve" and confirms the favorable deformation conditions for the alloy with the lowest level of alloying elements with respect to the potential of utilizing its higher plasticity.

The yield point and tensile strength level of Alloy 1 must be achieved by means of a new, finely developed heat treatment.

The next figure (**Figure 12**) shows the elongation of the extruded sections and their hardness. The obtained result confirms that the alloy with the lowest content of alloying elements (Alloy 1) has the highest yield strength reserve.

3.1.3 Studies on the effects of the extrusion rate on the properties, structures, and phase compositions of the alloy with the lowest content of alloying elements

The potential possibility of using the cheapest Alloy 1 with the lowest content of alloying components while optimizing the extrusion rate for the most efficient

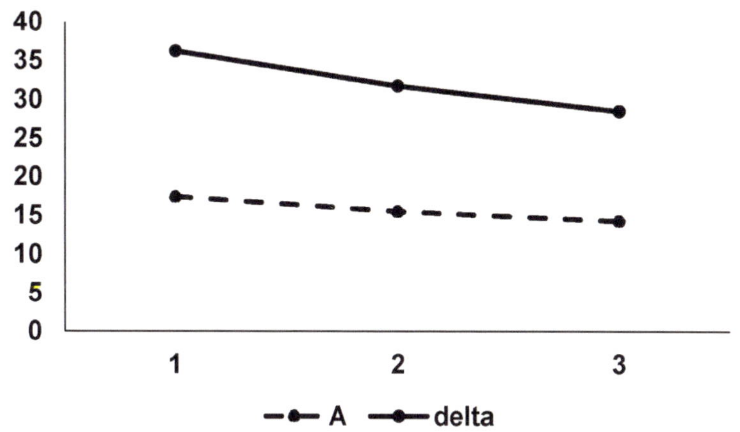

Figure 11.
Elongation and yield strength reserve of extruded sections from billets with three different contents of the main alloying elements.

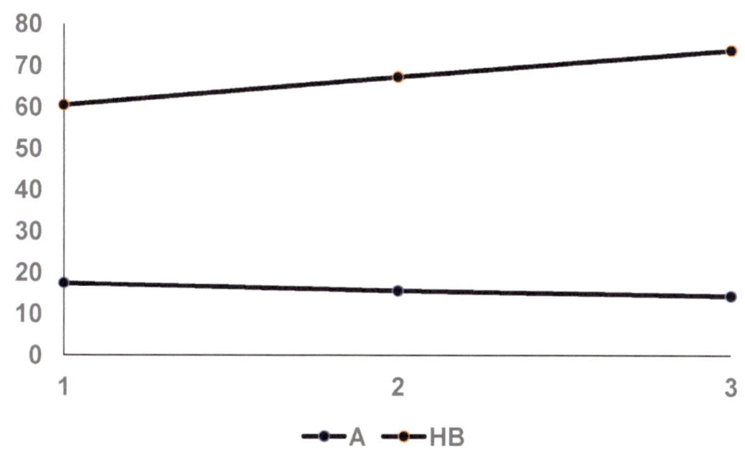

Figure 12.
Elongation and hardness of extruded sections from billets with three different contents of the main alloying elements.

process created grounds for research into heat treatment optimization to achieve the properties required by the standard. Alloy 1 has been tested for its structures, properties, and phase compositions.

Table 5 shows the macrostructure and microstructure of extruded Alloy 1 shapes with respect to the increasing extrusion rate. It is characterized by the equiaxed grains of medium size (medium chord parameter) in the range from 46 μm to 54 μm. Homogenization of the structure and grain size with a maximum grain size difference of 15% results from the hot working, activating the processes of structure recovery during extrusion.

Results of phase composition tests of extruded shapes are presented in **Figure 13**.

There is a final slight decrease in the content of the precipitated phases at the highest extrusion rate applied standard value +60%. At the standard rate +20%, the highest phase content of Mg_2Si and Mg_5Si_6 was recorded. The obtained result sets the direction for actions aimed at obtaining sections with assumed properties at the highest efficiency of the extrusion process. An increase in the extrusion rate above 20%, without changing the subsequent heat treatment applied after extrusion to the

Deformation rate	Section macrostructure	Section microstructure	Grain size, average grain diameter µm
Standard value—X		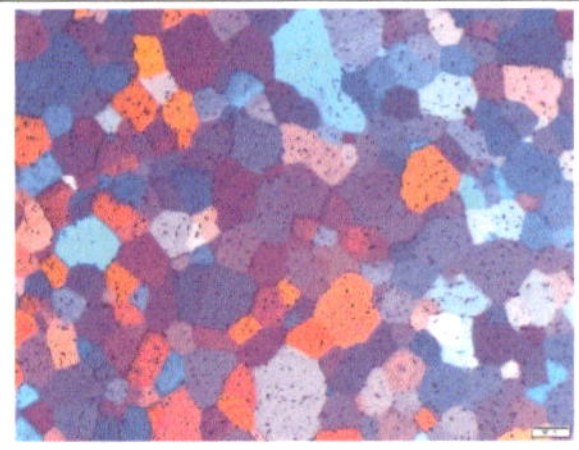	53
X+20%			46
X+30%	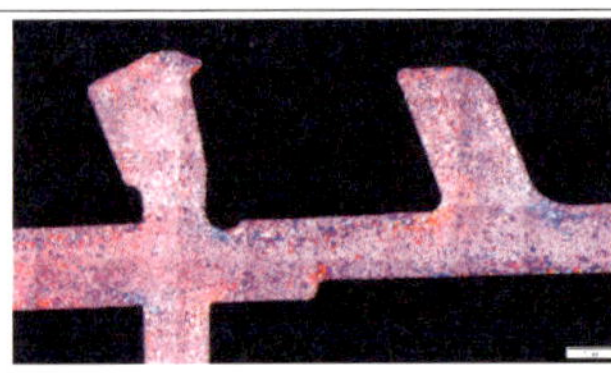	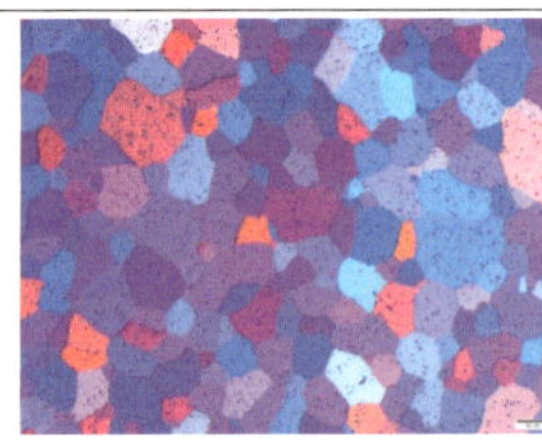	54
X+40%	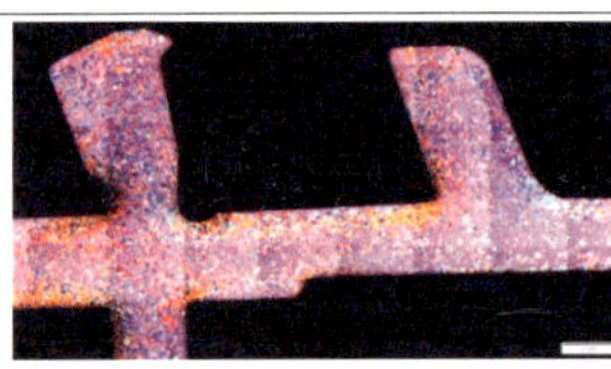	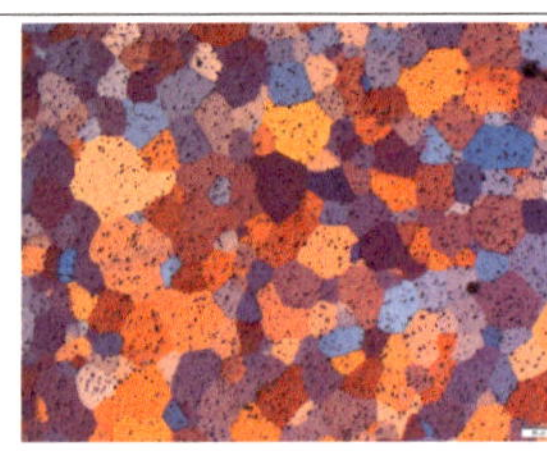	52
X+60%		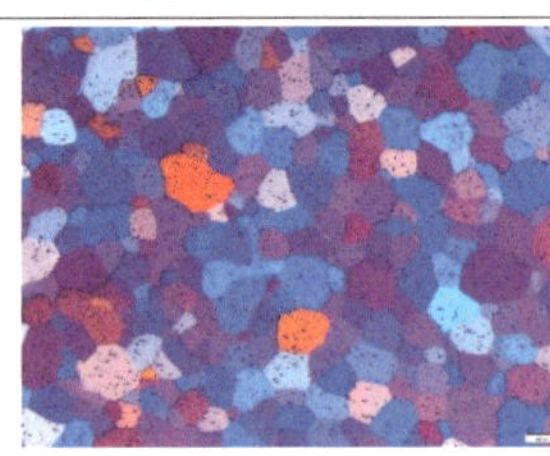	48

Table 5.
Structure of Alloy 1 sections depending on the extrusion rate.

T6 state, is an optimal variant ensuring the highest efficiency of the process. On the other hand, if the applied heat treatment allowed for a significant increase in the hardening phase, then higher extrusion rates could be applied.

3.1.4 Effect of heat treatment on the properties of extruded sections made of Alloy 1

The alloy meets the criteria of a high-speed extrusion alloy as the tests resulted in a significant increase in the extrusion rate of sections in relation to the rate currently used. However, due to the low content of Mg and Si alloying elements, the

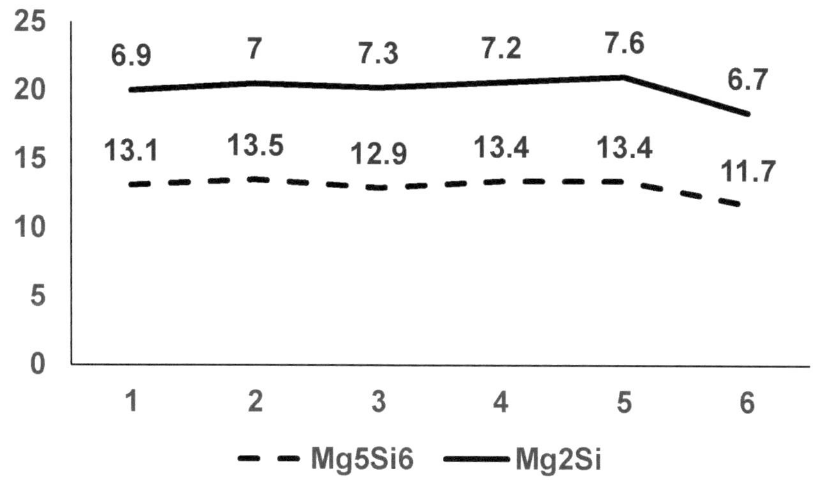

Figure 13.
Phase changes in extruded sections.

Aging curve	Press ram/stem speed	Rm [MPa]	$R_{0.2}$ [MPa]
Aging option 19	Standard value—X	212.0	188.5
	X+20%	218.8	195.3
	X+30%	201.0	176.0
	X+40%	211.0	186.8
	X+50%	200.5	176.3
	X+60%	216.0	183.8

Table 6.
Final results of strength properties after heat treatment for Alloy 1: stage 2.

classical heat treatment did not allow to obtain the recommended level of properties according to the standard. Therefore, it was necessary to carry out a number of studies to develop a new heat treatment that would ensure that the yield point, tensile strength, and hardness value of the standard were achieved. **Table 6** contains data on the parameters of the applied heat treatment variants.

The analysis of the obtained data showed that in variant 19 of heat treatment, the yield point reached the level assumed in the standard. The obtained result is a great technological achievement and promises a better economic result.

The research has shown that it is possible to produce a cheaper alloy (low content of base alloying elements), which allows to use a 60% higher extrusion rate and, using an appropriate heat treatment, achieve a comparable level of properties to alloys containing higher levels of Mg and Si elements.

The higher ductility of the new high-speed extrusion alloy enables the production of a greater tonnage of products per unit of time than the currently used alloys, which translates into increased production efficiency.

4. Summary: 6xxx aluminum alloys for a high-speed extrusion

At the beginning of the twenty-first century, commercial concerns involved in the production and sale of extrusion sections made of aluminum alloys started to show interest in increasing the productivity of the extrusion process through a radical change in the technology. The subgrades of alloy 6060 with a reduced content of alloying elements, called lean alloys or high-speed alloys, began to form. They were characterized by an increased plasticity in relation to classical alloy 6060, as a result of which the speed of the process increased by 20% in comparison with alloys with classical chemical composition. In order to obtain certain strength properties, it was necessary to use unconventional aging cycles of the material. Because the alloys required an increase in extrusion rate, lean alloys were used primarily for extrusion of sections with relatively uncomplicated cross sections. These types of sections were mainly used in the construction industry, so they were sometimes called building alloys. By 2007, thanks, among other things, to the exchange of workers between the companies, the production technology of building alloys became widely known, and the hydro concern started to produce billets from these alloys for sale to the general public, designing and producing new generations of alloys for fast extrusion at the same time.

Designing alloys for high-speed extrusion requires a deep knowledge of the hardening mechanisms and plastic deformation processes. During hot deformation, which takes place in extrusion, it is important to know the effect of temperature on the extrusion processes. The processes of structure recovery during hot deformation are dynamic recrystallization and dynamic recovery. In aluminum and its alloys, which exhibit high stacking-fault energy, dynamic recovery processes take place.

All effects in the extrusion process are temperature and speed controlled. It is shown that the mechanism of deformation is either dynamic recovery or dynamic recrystallization, according to whether the alloy is of high or low stacking-fault energy. This leads to extrudate structures which contain either subgrains, ideally with no static recrystallization occurring, or grains formed by successive dynamic, metadynamic, and static recrystallization. It is shown that either type of structure is related to, and can be controlled by, the prevailing Zener-Hollomon (Z-H) parameter *Z*. There is thus also a relationship between properties and *Z* [27].

The Zener-Hollomon parameter is given by equation:

$$Z = \dot{e} \backslash \exp(Q/RT) \tag{2}$$

where $\dot{e}$ is the strain rate, Q is the activation energy, R is the gas constant, and T is the temperature.

The Zener-Hollomon parameter decreases with a decreasing strain rate and an increasing temperature. In the actual industrial production, to avoid the occurrence of cracks, low Z conditions (lower strain rates and higher temperatures) are usually preferred; however, when the deformation rate is too low, the temperature drop is serious, which is harmful for deformation [28].

In case of high-speed extrusion alloys Z-H parameter should increase, the strain rate will increase and temperature should be lowered. The decrease of subgrain/grain size should be expected.

Deformation temperature and strain rate are important factors controlling the hot deformation flow stress. The experimental results show that the dynamic softening is accelerated with the increase of deformation temperature and decrease of strain rate. Strain effects on flow stress and hence on extrusion pressure are predominant for hot extrusion (due to strain rate sensitivity). Therefore, it is rather difficult to predict the extrusion force in hot extrusion. We can estimate the strain rate at any location x in the billet from the geometrical considerations. Let a cylindrical billet have an initial radius of R_0 and extruded radius of Rf. be semi-cone angle of the die.

We can formulate the strain rate at any location x from the entry of die as

$$\dot{\varepsilon} = \frac{d\varepsilon}{dt} = -2\frac{v_0 R_0^2}{\left(R_0 - x\tan\alpha\right)^3}\tan\alpha \tag{3}$$

The average strain rate undergone by a billet is given by

$$\varepsilon = \frac{6v_0 D_0^2 \tan\alpha}{\left(D_0^3 \text{-} D_f^3\right)} - \ln R \tag{4}$$

where v_0 is the velocity of ram.

It is generally believed that the average grain size (D_A) decreases with increasing strain rate and decreasing temperature and is independent of initial grain size and accumulated strain.

The effects of temperature and strain rate can be expressed in terms of the second-order function of the Zener-Hollomon parameter, Z, in an exponent-type function of temperature and strain rate, as follows [29]:

$$D_A = B_o + B_1 \ln Z + B_2 \left(\ln Z\right)^2 \tag{5}$$

where B_1 and B_2 are polynomial coefficients. Generally, the grain size of as-extruded material decreases with the increase of the Zener-Hollomon parameter. The fact that the Z-H parameter determines the microstructural evolution reveals that the foundation of an ultrafine-grained structure is formed dynamically and the process is closely related to the thermal activation process during and after the deformation.

In general, dynamic recrystallization does not seem to occur in aluminum alloys with Mg below 4%. The high recovery processes develop in such aluminum alloys—with subgrain lattice formation.

The size of stable subgrains that result from dynamic recovery in aluminum alloys depends upon hot working conditions. McQeen et al. [30] expressed this by equation:

$$d_s = a + b\log Z \tag{6}$$

where d_s is the subgrain size, Z is the Zener-Hollomon parameter, and a and b materials are constant.

The subgrain size usually reaches a limiting value of a few microns in aluminum alloys. With continued deformation, there is usually no further decrease in the subgrain size.

For hot extrusion, the extrusion pressure p is directly proportional to stain rate. As strain rate increases, the extrusion pressure also increases, almost linearly. As ram speed increases, the extrusion pressure also increases, due to increasing strain rate. However, the extrusion pressure is reduced with increased working temperature in hot extrusion (**Figure 14**). The extrusion speed has some limit depending on billet temperature. It means that the technological application of high-speed alloys requires some experimentally developed extrusion parameters.

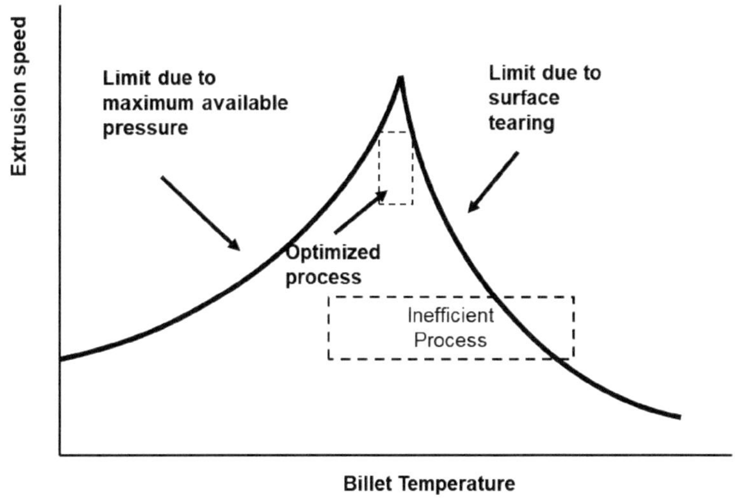

Figure 14.
Temperature dependence to extrusion speed.

In the analysis of hardening of polycrystalline materials, the component (σ_y), described by the model of the Hall-Petch (H-P) equation, acquires a particular importance as it allows to determine the value of lower yield point as a function of grain size:

$$\acute{o}_y = \acute{o}_o + k_y d^{-1/2} \quad (7)$$

where σ_y: normal stress corresponding to the yield point; σ_0: stress of internal friction of the mobile dislocations; k_y: slope factor, characterizing the resistance of grain boundaries to dislocation motion; d: diameter of grain or subgrain.

For this reason, the selection of conditions of extrusion rate and temperature determining the grain size is of great importance in the technology of extrusion of aluminum alloys.

At present, high-speed extrusion alloys are a desired alternative to conventional alloys due to the prospect of higher tonnage throughput per unit of time. The expected effect of this action is an increase in the company's profit. In the international literature, it is difficult to find data on the chemical composition of alloys for a high-speed extrusion because they are confidential information, which results from the well-understood interest of the company. There is also a noticeable lack of wider interest of scientists in this subject, who usually strive to disseminate the results, as it is contrary to companies' confidentiality requirements. Therefore, this type of research is usually conducted in a narrow circle of professionals closely cooperating with the industry. Among the literature items and reports that appear on the websites, there is information about progress in the production of new aluminum alloys for specific purposes, e.g., for the automotive, construction, high-speed rail, and other industries. The purpose of the alloys determines their production and properties.

The research conducted on fast extrusion alloys is a promising direction for the development of section production and provides prospects for improving extrusion technology. Great possibilities of increasing the productivity of the extrusion process connected with the implementation of these new quality aluminum alloys are also an element of the struggle for the market and increase in the productivity of companies.

Author details

Rafał Hubicki[1]* and Maria Richert[2]

1 Grupa Kęty S.A., Poland

2 AGH University of Science and Technology, Krakow, Poland

*Address all correspondence to: h_rafal@hotmail.com

References

[1] Butu M, Moldovan P, Marcu FD, Ungureanu I. Thermodynamics of in situ production of aluminium matrix composites comparative analysis. Materiale Plastice. 2016;**53**(3):428-433

[2] Available from: https://www.azom.com/article.aspx?ArticleID=2863

[3] Vissers R, Van Huis MA, Jansen J, Mariora CD, Andersen SJ. The crystal structure of the β′ phase in Al–Mg–Si alloys. Acta Materialia. 2007;**55**(11):3815-3823

[4] Marioara CD, Andersen SJ, Jansen J, Zandbergen HW. Atomic model for GP-zones in a 6082 Al–Mg–Si system. Acta Materialia. 2001;**49**:321-328

[5] Andersen SJ, Zandbergen HW, Jansen J, Traeholt C, Tundal U, Reiso O. The crystal structure of the β″ phase in Al–Mg–Si alloys. Acta Materialia. 1998;**46**(9):3283-3298

[6] Derlet PM, Andersen SJ, Marioara CD, Froseth A. A first-principles study of the β″ phase in Al-Mg-Si alloys. Journal of Physics: Condensed Matter. 2002;**14**:4011-4024

[7] Froseth A. PhD thesis. Trondheim: Norwegian University of Science and Technology; 2003. pp. 63-111

[8] Andersen SJ, Marioara CD, Froseth A, Vissers R, Zandbergen HW. Crystal structure of the orthorhombic U2-$Al_4Mg_4Si_4$ precipitate in the Al–Mg–Si alloy system and its relation to the β′ and β″ phases. Materials Science Journal Engineering A. 2005;**390**(1-2):127-138

[9] Matsuda K, Sakaguchi Y, Miyata Y, Uetani Y, Sato T, Kamio A, et al. Precipitation sequence of various kinds of metastable phases in Al-1.0mass% Mg_2Si-0.4mass% Si alloy. Journal of Materials Science. 2000;**35**:179-189

[10] Madsuda K, Ikeno S, Sato T, Kamio A. A metastable phase having the orthorhombic crystal lattice in an Al-1.0mass% Mg_2Si-0.4mass% Si alloy. Scripta Materilia. 1996;**34**:1797-1802

[11] Froseth A, Hoier R, Derlet PM, Andersen SJ, Marioara CD. Bonding in MgSi and Al-Mg-Si compounds relevant to Al-Mg-Si alloys. Physical Reviews B. 2003;**67**:224-106

[12] Marioara CD, Andersen SJ, Zandbergen HW, Holmestad R. The influence of alloy composition on precipitates of the Al-Mg-Si system. Metallurgical and Materials Transactions A. 2005;**36A**:691-702

[13] Chakrabarti DJ, Peng Y, Laughlin DE. Precipitation in Al-Mg-Si alloys with Cu additions and the role of Q' and related phases. In: Materials Science Forum. Vol. 396-402. USA: Alcoa Technical Center; 2002. pp. 857-862

[14] Edwards GA, Stiller K, Dunlop GL, Couper MJ. The Precipitation Sequence in Al-Mg-Si Alloys. Acta Materialia. 1994;**46**:3893-3904

[15] Chakrabarti DJ, Cheong BK, Laughlin DE. Automotive Alloys. USA: TMS; 1998. p. 27

[16] Kuroda Y, Yoshino D, Lee SW, Ikeno S, Matsuda K. Microstructure of small amount of TM added Al-Mg-Si alloys with two step ageing. In: Proceedings of 11th Polish-Japanese Joint Seminar on Micro and Nano Analysis, Gniew, September 11-14. 2016. DOI: 10.12693/APhysPolA.131.1373

[17] Zandbergen MW, Cerezo A, Smith GDW. Study of precipitation in Al-Mg-Si alloys by atom probe tomography. II. Influence of

Cu additions. Acta Materialia. 2015;**101**:149-158

[18] Takeshi Saito S, Wenner A, Osmundsen CD, Marioara SJ, Andersen J, Royset W, et al. The effect of Zn on precipitation in Al-Mg-Si alloys. Philosophical Magazine. 2014;**94**:2410-2425. DOI: 10.1080/14786435.2014.913819

[19] Lee DH, Park JH, Nam SW. Enhancement of mechanical properties of Al-Mg-Si alloys by means of manganese dispersoids. Materials Science and Technology. 2013;**15**(4):450-455

[20] Chen JH, Constan E, van Huis MA, Xu Q, Zandbergen HW. Atomic pillar-based nanoprecipitates strengthen AlMgSi alloys. Science. 2006;**312**:416-419. DOI: 10.1126/science.1124199

[21] Ringdalen I, Wenner S, Friis J, Marian J. Dislocation dynamics study of precipitate hardening in Al-Mg-Si alloys with input from experimental characterization. MRS Communication. 2017;7(3):626-633. DOI: 10.1557/mrc.2017.78

[22] Richert J, Mroczkowski M, Gellner J. Nowy sposób starzenia stopów AlMgSi wyciskanych w stanie T5. Rudy i Metale Nieżelazne. 2006;**51**(12):747-753

[23] Ryen Ø, Holmedal B, Marthinsen K, Furu T. Precipitation, strength and work hardening of age hardened aluminium alloys. Materials Science and Engineering. 2015;**89**:012013. DOI: 10.1088/1757-899X/89/1/012013

[24] Andersen SJ, Marioara CD, Friis J, Wenner S, Holmestad R. Precipitates in aluminium alloys. Advances in Physics: X. 2018;**3**(1):1479984. DOI: 10.1080/23746149.2018.1479984

[25] Available from: https://www.bing.com/images/search?view=detailV2&ccid=V60%2B5rNV&id=019EC72102CAE67AB922781ADE5BCA40109097CB&thid=OIP.V60-5rNVOMKKmWU8zuw79gHaFf&q=mg2si+precipitations+in+6060+alloy&simid=608017416757841339&selectedindex=1&mode=overlay&first=1

[26] Grupa Kęty S.A. Photos by authors.

[27] Sheppard T. Temperature and speed effects in hot extrusion of aluminium alloys. Metals Technology. 1981;**8**(1):130-141. DOI: 10.1179/030716981803276009

[28] Li J, Liu B, Wang Y, Shan T, Liu Y, Lu X. A study on the Zener-Hollomon parameter and fracture toughness of an Nb-particles-toughened TiAl-Nb alloy. Metals. 2018;**8**(4):287. DOI: 10.3390/met8040287

[29] Quan G-z, Wang Y, Liu Y-Y, Zhou J. Effect of temperatures and strain rates on the average size of grains refined by dynamic recrystallization for as-extruded 42CrMo steel. Materials Research. 2013;**16**(5):1092-1105. DOI: 10.1590/S1516-14392013005000091

[30] McQeen HJ, Celliers OC. Application of hot workability studies to extrusion processing: Part II. Microstructure development and extrusion of Al, Al-Mg and Al-Mg-Mn alloys. Canadian Metallurgical Quarterly. 1996;**35**(4):305-319

Chapter 5

Corrosion Resistance of Precipitation-Hardened Al Alloys: A Comparison between New Generation Al-Cu-Li and Conventional Alloys

Uyime Donatus, Michael Oluwatosin Bodunrin, Ayotunde Olayinka, Mariana Xavier Milagre, Olamilekan Rasaq Oloyede, Sunday Aribo, João Victor de Sousa Araujo, Caruline de Souza Carvalho Machado and Isolda Costa

Abstract

The corrosion resistance of conventional (AA2024-T3, AA6082-T6 and AA7050-T7451) and the new generation (AA2050-T84, AA2098-T351, AA2198-T8, and AA2198-T851) precipitation-hardened alloys has been studied and compared using electrochemical and non-electrochemical approaches. The AA6082-T6 was the most resistant alloy followed by the new generation Al-Cu-Li alloys, except the AA2050-T84. All the alloys exhibited pseudo-passivity, except for the AA2024-T3 alloy which presented the highest number of pitting sites per cm^2 and also exhibited the most insidious form of corrosion amongst the alloys tested. However, the alloy with the highest corrosion depth was the AA2050-T84 alloy followed by the AA2024-T3 and AA7050-T7451 alloys. Intergranular corrosion was associated with rapid rates of penetration. In addition to the microstructural features of the alloys before corrosion, the modes of localized corrosion in the alloys were also influenced by evolving microstructural features (such as re-deposited Cu) during corrosion.

Keywords: wrought Al alloys, new-generation Al alloys, localized corrosion, microstructure, SEM

1. Introduction

New generation Al alloys are being developed to meet lightweight requirements in the automotive and aerospace sectors. For a successful introduction, the alloys must possess specific strengths, mechanical properties and corrosion resistance superior or at least equal to those exhibited by conventionally used Al alloys. Concerning corrosion, the currently used high strength 2xxx and 7xxx series alloys in the aerospace sector are highly susceptible to severe localized corrosion (SLC), but adequate care should be taken in ensuring that the proposed replacements (the new generation Al-Cu-Li alloys) exhibit better in-service corrosion performances.

There are reports in the literature [1–3] comparing the corrosion resistance of selected conventional alloys to that of the new generation Al-Cu-Li alloys, and most of these studies were based on the use of electrochemical techniques [1, 3]. However, electrochemical techniques alone cannot give enough information about the corrosion behavior of these alloys. This is because electrochemical techniques are largely designed to generate data from activities occurring on the surfaces of materials. Sub-surface details from tens to hundreds of microns beneath the surface are very difficult to obtain via electrochemical methods. Attacks in precipitation-hardened aluminum alloys can penetrate hundreds of microns beneath the surface as fissures with non-linear pathways that are difficult to follow from the surface. This is in addition to the fact that the attacks can also transit easily from one form to the other, and the active area of corrosion is very difficult to establish [4].

Thus, it is always important to examine the surfaces and cross-sections of the corroded alloys via microscopic techniques before concluding. In this regard, it is also important to mention that conclusions from nano to mesoscale microscopic approaches should be drawn with caution because even macro/microscale results can be very misleading if care is not taken.

In this work, the comparison between the corrosion resistance of the new generation Al-Cu-Li alloys and that of conventional aluminum alloys have been made using scanning electron microscopy with the results correlated with potentiodynamic polarization results. Alloys from all the precipitation-hardened series (2xxx, 6xxx and 7xxx) were selected. The selected alloys are industrial alloys in the common tempers in which they are being employed. These alloys derive their strengths from the formation of finely and uniformly distributed nano-sized phases in their matrix. To accomplish the precipitation of these phases, alloying elements with reducing solid solubility as temperature decreases are used for this purpose. Examples of elements that fall into this category include, copper, magnesium, zinc and, lithium. The potentials of the intermetallic particles formed by these elements are often different from that of the matrix (i.e. these particles are either cathodic or anodic to the matrix) when exposed to aggressive environments, and this results in the development of localized corrosion which compromises the integrity of the alloys in service. The form and extent of localized corrosion are alloy specific. Establishing the severity and how insidious the different forms of attacks in competing alloys are crucial to improving the performance of the components built from these alloys. Thus, in this work we have compared the forms and how insidious the corrosion attacks in selected industrially important alloys are, especially by contrasting between the attacks in new generation and conventional aluminum alloys in sea-water environment.

2. Experimental

The conventional alloys employed in this study are the AA2024-T3, AA6082-T6 and the AA7050-T7451 alloys, and the new generation alloys are the AA2050-T84, AA2098-T351, AA2198-T8 and AA2198-T851 alloys. The compositions of these alloys are as presented in **Table 1**. These alloys are commercial alloys in the tempers in which they are mostly employed.

Prior to the corrosion tests, the samples were sequentially polished to a 1 μm surface finish using SiC papers and diamond pastes.

	Cu	Li	Fe	Zr	Cr	Mg	Zn	Si	Mn	Ag
AA2024-T3	4.2		0.2		0.1	1.6	0.2	0.1	0.4	0.15
AA2050-T84	3.64	1.0	0.04	0.12		0.36	0.02		0.39	
AA2098-T351	3.4	1.0	0.04	0.4		0.3	0.02	0.05	0.003	0.3
AA2198-T8	3.32	0.96	0.005	0.51		0.31	0.004	0.004	0.002	0.26
AA2198-T851	3.31	0.96	0.004	0.4		0.31	0.01	0.03		0.25
AA6082-T6			0.33			0.74	0.05	0.71	0.40	
AA7050-T7451	2.15		0.04	0.14		1.53	6.80	0.08		

Table 1.
Composition (wt. %) of the alloys used in this study.

The main investigation in this work was based on optical and scanning electron microscopy of the surfaces and cross-sections of the alloys after a 72 h corrosion immersion test in 3.5% NaCl solution. Polished samples of the alloys were employed, and beeswax was used to expose an area of 1 cm^2 on the alloys.

Other corrosion tests employed in the investigation were potentiodynamic polarization tests, agar-gel visualization test and scanning vibrating electrode technique (SVET) measurements.

Potentiodynamic polarization curves of the alloys were obtained in the 3.5% NaCl solution. A three-electrode cell comprising the sample as the working electrode, an Ag/AgCl reference electrode and a platinum wire as a counter electrode was employed for the polarization tests. The scans were initiated at −100 mV of the open circuit potential (OCP) values to +800 mV of the OCP. OCP measurements were conducted for 90 min prior to the polarization measurements, and a scan rate of 1 m V/s was employed.

Details of the agar-gel visualization test are similar to those reported in previous work from the same group [5], except that a universal indicator was employed this time in the place of phenolphthalein.

For the scanning vibrating electrode technique (SVET) measurements, an Applicable Electronic Produced SVET machine with an ASET 4.0 software was employed using a 5 mM NaCl solution. SVET maps and optimal images were obtained every 2 h. Further details on the SVET procedures can be found in a previously published work from the same group [6].

Scanning electron microscopy (SEM) analysis was conducted using a JEOL JEM 6010 LA and TM 3000 microscopes equipped with an energy-dispersive x-ray spectroscopy (EDS) detector. Transmission electron microscopy (TEM) was conducted using a JEM-2100F microscope. The TEM samples were prepared by twin-jet electropolishing using 35% nitric acid in methanol after an initial thinning through grinding with SiC papers.

3. Results and discussion

The results of the corrosion studies are presented in two parts. Firstly, results from electrochemical analyses are presented and compared. Majorly, potentiodynamic polarization and SVET results are compared, while agar visualization test was employed to understand the pH variation around an SLC site further. Subsequently, SEM examination of the surfaces and cross-sections of the alloys are presented and compared.

3.1 Electrochemical tests

3.1.1 Potentiodynamic polarization

Presented in **Figure 1** are the potentiodynamic polarization results of the selected alloys in 3.5% NaCl solution. **Figure 1a** combines the potentiodynamic polarization curves of all the alloys investigated, while **Figure 1a, b** shows the curves of the new generation Al-Cu-Li alloys (AA2050-T84, AA2098-T351, AA2198-T8 and AA2198-T851) and the curves of the conventional alloys (AA2024-T3, AA6082-T6 and AA7050-T7451), respectively. **Figure 1d** is a plot of extrapolated corrosion potential (Ecorr) and pitting potential (Epit) values for the different alloys from the plots in **Figure 1a** (The Epit, in this case, is the potential beyond which there is a large increase in current density compared with the pseudopassive region just above the Ecorr). These results show that, amongst the new generation alloys, the AA2050-T84 alloy with the lowest Ecorr value (≈−0.82 V) exhibited the highest tendency to corrode, while the AA2098-T351 and AA2198-T8

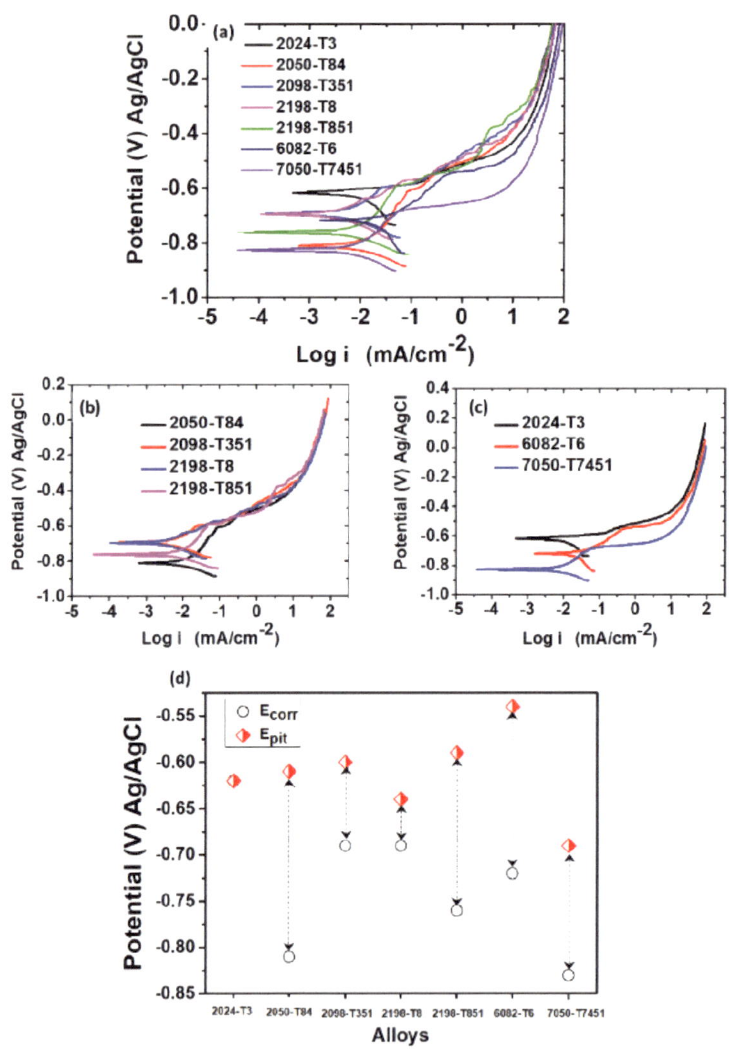

Figure 1.
Potentiodynamic polarization results of selected heat-treatable aluminum alloys in 3.5% NaCl solution.

alloys, with the highest Ecorr values (≈ -0.68 V), exhibited the lowest tendency to corrode. The Ecorr value (≈ -0.76 V) of the AA2198-T851 alloy was in between those of the AA2050-T84 and AA2098-T351/2198-T8 alloys indicating that the AA2198-T851 alloy has more corrosion tendency than the AA2098-T351/2198-T8 alloys but lower tendency than the AA2050-T84 alloy.

Nonetheless, all the new generation alloys exhibited pseudo-passive behavior under the conditions tested. The Epit – Ecorr difference shows the potential range for the active – pseudopassive behavior. Pseudopassivation occurs because of the formation of a non-protective oxide layer on the alloys [7, 8]. It should be noted that the Epits in these alloys are not the actual pitting potentials of the alloys. These alloys develop severe localized corrosion at OCP (i.e. at potentials lower than the pseudopassive range). The oxide formed after the active regions is only formed on non-pitting sites. And the contribution of the pitting sites to the total current is overshadowed by the current flowing from the larger surface with an oxide layer [8]. Thus, the pseudopassive current predominates at this potential range. However, after the potentials designated as Epits, the contribution of the pitting areas to the overall current flowing from the surface becomes significant [8] and superior to the pseudopassive current, and this leads to pronounced current density increase per potential.

For the conventional alloys, the AA7050-T8451 alloy, with an Ecorr value in the range of ≈ -0.84 V, exhibited the highest tendency to corrode, followed by the AA6082-T6 alloy with an Ecorr value of ≈ -0.72 V. The AA2024-T3 alloy, with an Ecorr value of ≈ -0.63 V, exhibited the lowest tendency to corrode. Amongst the three conventional alloys compared, the AA2024-T3 alloy did not show any pseudo-passive range in the condition tested. In fact, amongst all the alloys compared, the AA2024-T3 alloy was the only alloy that did not exhibit a pseudopassive behavior. This implies that the contribution of the pitting areas to the total current was significant (from potentials equal to or below the Ecorr) and swamped that from the oxide-covered surface. Thus, the AA2024-T3 alloy possibly presented a higher active pitting area compared with the other alloys.

Nonetheless, for all the alloys, AA2024-T3 > AA2098-T351/AA2198-T8 > AA608 2-T6 > AA2198-T851 > AA2050-T84 > AA7050-T7451 in terms of Ecorr values. Based on this, the AA7050-T7451 alloy exhibits the highest tendency to corrode and should be the most susceptible to corrosion amongst all the alloys. The AA7050-T7451 alloy also presented the lowest potential at which the current from the pitting areas contributes significantly to the total current flowing from its surface. Also, it is expected that the AA2024-T3 alloy should exhibit the least tendency to corrode in NaCl environment amongst the alloys investigated. Also, since the new generation alloys have lower Ecorr values compared with the AA2024-T3 alloy, the new generation Al-Cu-Li alloys should be more susceptible to corrosion compared with the AA2024-T3 alloy.

However, potentiodynamic polarization results are not sufficient to establish the corrosion resistance of these alloys, especially as it is difficult to rely on extrapolated current density values for aluminum alloys in near-neutral NaCl environments. One of the reasons being that it is difficult to establish the active corroding area [4]. For a quick comparison, SVET immersion tests were conducted on samples representing the Al-Cu-Mg, Al-Cu-Li, Al-Mg-Si and Al-Zn-Mg series as presented in the section below.

3.1.2 SVET immersion tests

Figures 2 and **3** present the SVET result of the AA2024-T3 (Al-Cu-Mg), AA2198-T851 (Al-Cu-Li), AA6082-T6 (Al-Mg-Si) and AA7050-T7451 (Al-Zn-Mg) alloys after 2 and 18 h of immersion (Two 2xxx series alloys were selected to have a conventional (AA2024-T3) and a new generation Al-Cu-Li alloy (AA2198-T851) representation. Also, the solution employed in this case was 5 mM NaCl solution. This solution is less aggressive compared with the 3.5% NaCl solution since it contains less chloride ions, and it was chosen to allow for easy monitoring of the in-situ corrosion activities on the alloys with time.)

Pronounced anodic activities were observed on the AA7050-T7451 and AA2024-T3 alloys within the first 2 h of immersion (**Figure 2**). The SLC sites were easily discernible on the AA2024-T3 alloy (**Figure 2b**) but difficult to find on the AA7050-T7451 alloy (**Figure 2h**) at macroscale because of the nature of pit covering and corrosion product formation on the later alloy (this is discussed further in the section below). For the AA2198-T851 alloy, the anodic activities were not that pronounced (compared with these two alloys), and only traces of SLC sites were observed (an example is indicated by the red arrow). Localized activities associated with SLC were not observed on the AA6082-T6 alloy. The localized activities observed in the early hours were transient and no stable SLC site was initiated on this alloy after 2 h.

In the later hours of the test, the anodic activities were reduced at the surfaces of the AA2024-T3 and AA7050-T7451 alloys were pronounced SLC activities were observed because corrosion products had covered the sites on the AA7050-T7451 and AA2024-T3 alloys. At this stage, visible SLC sites were evident on the AA2198-T851 alloy. However, corrosion products also formed on these sites and reduced the anodic activities recorded by the SVET. Again, for the AA6082-T6 alloy,

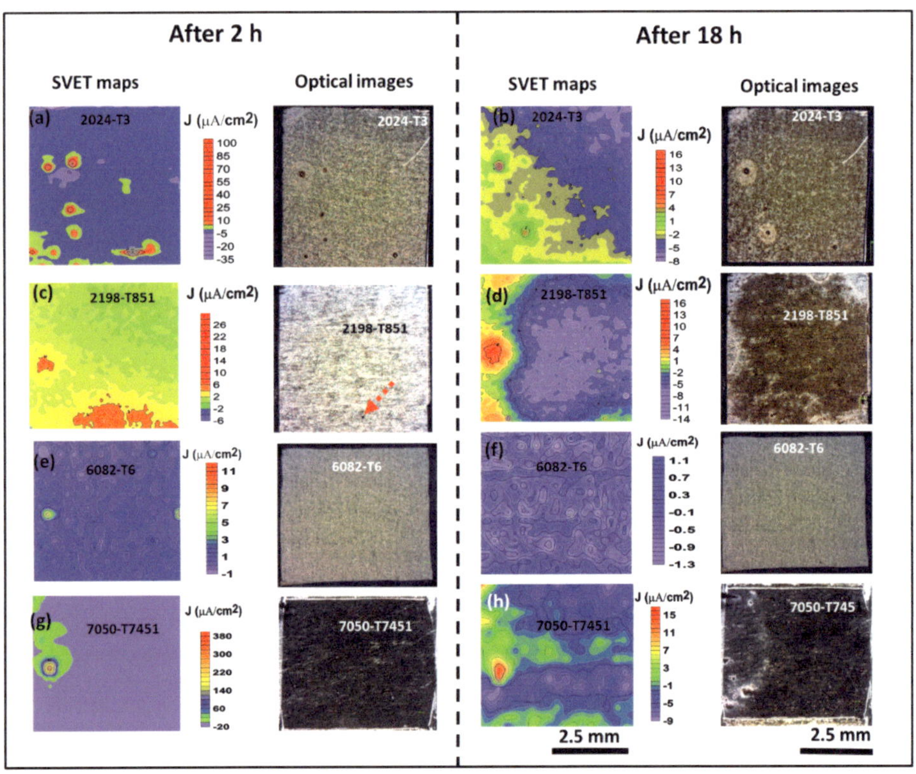

Figure 2.
SVET current density maps and optical images of the tested alloys in 5 mM NaCl solution after 2 and18 h of immersion.

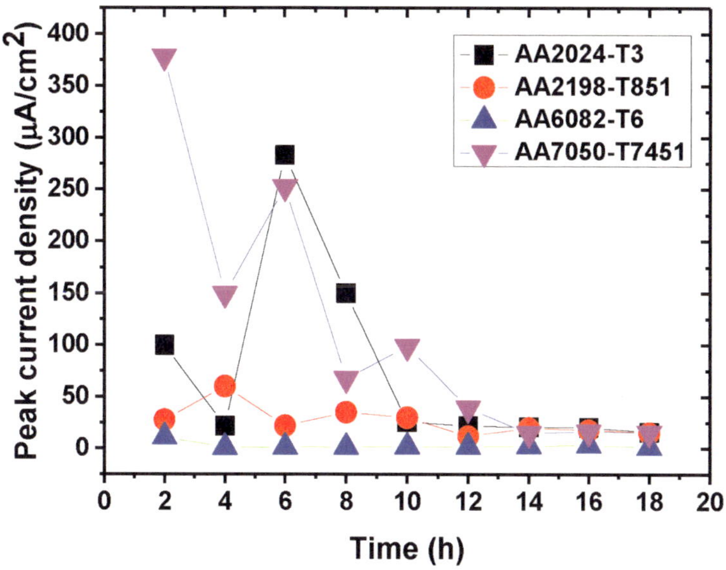

Figure 3.
Plots of peak current density values recorded on the alloys during the SVET immersion test.

no trace of SLC activity was recorded by the SVET and the optical macrograph also did not reveal any trace of SLC site.

Presented in **Figure 3** is a plot of the peak current density recorded on the alloys throughout the test. As evident, the highest peak current density values were recorded on the AA2024-T3 and AA7050-T7451 alloys. Peak current density values as high as 298.3 and 377.8 μA/cm^2 were recorded on both alloys, respectively. Lower peak current density values were recorded on the AA2198-T851 alloy with the highest being in the range of 60 μA/cm^2. For the AA6082-T6 alloy, the peak current density values were near zero with the highest being about 11 μA/cm^2. The average peak current density values recorded on these alloys were 114.49 μA/cm^2 for AA7050-T7451, 73.03 μA/cm^2 for AA2024-T3, 21.0 μA/cm^2 for AA2198-T851 and 2.38 μA/cm^2 for AA6082-T6. This implies that the AA7050-T7451 alloy was the most susceptible and the corrosion rate on the alloy was the highest. However, it should be noted that the peak current values on this alloy were emanating from a few SLC sites compared with the AA2024-T3 alloy. The number of SLC sites were highest on the AA2024-T3 alloy, and high current density values were emanating from multiple sites across the surface of the alloy. This possibly explains why no pseudo-passivity was observed on the AA2024-T3 alloy during potentiodynamic polarization, since the sum of the current from the pitting sites would be very significant, swamping the total current flowing from the oxide-covered area.

The SEM images of the surfaces of the alloys before the removal of corrosion products show SLC sites on the AA2024-T3, AA7050-T7551 and AA2198-T851 alloys but not on the AA6082-T6 alloy (not even trenching associated with the cathodic Al-Fe-Si rich phases were observed on the AA6082-T6 alloy). The reason for the immaculate corrosion resistance of the AA6082-T6 alloy in the test environment used may be because of the insignificant amount of Cu. Galvanic coupling activities associated with Cu-rich particles are often more pronounced than those associated with Fe and Si-rich particles. The AA6082 alloy is, however, susceptible in chloride environment when Mg_2Si particles are precipitated in the presence of precipitate free zones (PFZs) at the grain boundaries (GBs), and this is most common in the weld heat-affected zones and overaged temper (T7) of the alloy. For the other alloys, the microstructural factors associated with the formation of SLC are readily present in their microstructure in the present tempers. These factors include S-phase and Cu-rich particle clusters for the AA2024, η phase (and its variants) and PFZs for AA7050, and T1 particles for the AA2198 alloy.

The SLC sites on these susceptible alloys were covered with corrosion products, and these sites were found within corrosion rings (especially as showcased in the SEM image of the AA2198-T851). The observed corrosion rings resulted from pH difference between the regions around the SLC pits and the surroundings. The reduction reaction of dissolved oxygen occurs in the surrounding region, while H^+ ions are generated from the hydrolysis of Al^{3+} inside the pit. During the pitting process, the H^+ ions migrate from within the pits to the mouths of the pits due to electrostatic potential difference [9]. Thus, H^+ ions are present around the pit/SLC site and cause a local reduction in the pH around the pit mouths as evident in the agar-visualization result in **Figure 4a–d** (see the evolution of the sites labelled 1 and 2). Also, although the predominant reduction of H^+ to generate H^2 bubbles occurs inside the pits, some of the ions are reduced around the pit mouths. What is clear, as will be seen in the section below, is that there is a boundary between the low pH region around the SLC sites and the high pH region surrounding the sites, and this boundary defines the domain of the corrosion rings as clearly depicted by the black arrow of the site labelled 2 in **Figure 4a–d**. The site labelled 1.

SEM analysis was further carried out after the removal of corrosion products on the surfaces of the alloys that exhibited SLC. In agreement with the SVET peak current density values, the width and extent of the attack on an SLC site were most pronounced on the AA7050-T7451 alloy. However, those of the AA2024-T3 were not as pronounced as expected especially when compared with those on the AA2198-T851 alloy. Intergranular corrosion (IGC) expanding only within about 70 μm was observed on the surface of the AA2024-T3 alloy, whereas intragranular corrosion expanding beyond 100 μm was observed on the AA2198-T851 alloy. Also, it appears as if more

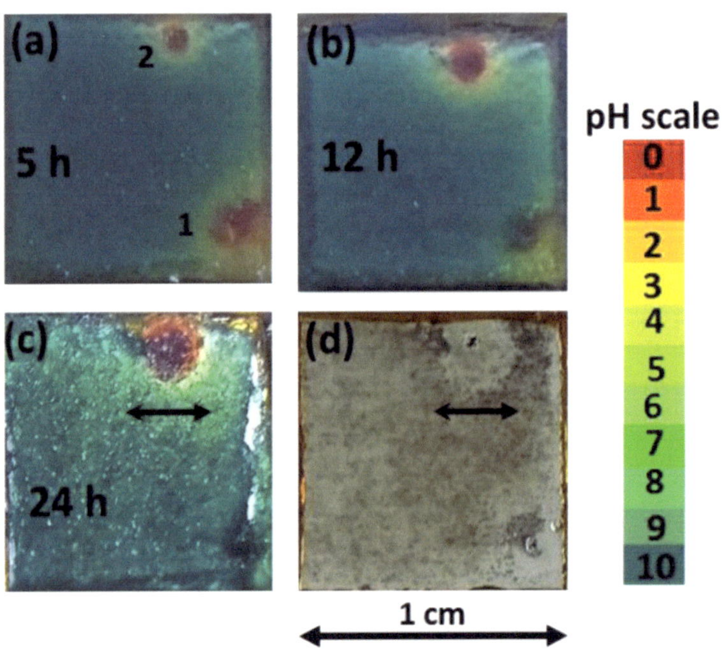

Figure 4.
Optical images showing the pH around SLC sites on AA2198-T851 alloy during agar visualization test and the corresponding surface after the removal of the gel.

materials were consumed on the AA2198-T851 alloy compared with the AA2024-T3 alloy. From these two alloys, the widths and the intensity of corrosion observed from the SEM images do not appear to correlate well with the current density values recorded during the SVET measurements. The peak current density values and the number of SLC sites indicate that the AA2024-T3 alloy was more prone to corrosion compared with the AA2198-T851 alloy. However, the SEM images of the surfaces after the corrosion test, tend to indicate otherwise. Also, the diameters of the corrosion rings were larger on the AA2198-T851 alloy compared with the AA2024-T3 alloy.

Nonetheless, the SVET results indicate that in order of rate of corrosion attack the AA7050 > AA2024 > AA2198-T851 > AA6082-T6 alloy. Thus, except for the AA6082-T6 alloy which exhibited no trace of corrosion during the test, the new generation AA2198-T851 Al-Cu-Li alloy is better than the conventional AA2024-T3 and AA7050-T7451 alloys. However, it is difficult to relate the observed SVET results to those of the potentiodynamic polarization curves of the alloys based on Ecorr values. The only relatable correlation is the magnitude of the current density values, which is quite high for the AA2024 alloy in agreement with the predominance of the pitting current over the pseudopassive current from OCP as observed on the potentiodynamic polarization curve. Thus, to get more details in a bid to establish the corrosion resistance of the alloys, a non-electrochemical approach is needed.

3.2 Non-electrochemical approach

The non-electrochemical approach employed in this study involves optical and scanning electron microscopy analyses of the surfaces of the selected alloys after a 72-h immersion test. The surfaces were examined before and after the removal of corrosion products. Following these, cross-sectional examinations of the corrosion attacks were then carried out.

Presented in **Figure 6** are the optical images of the surfaces of the alloys after the immersion test. The new generation Al-Cu-Li alloys are placed on top—**Figure 7a–d**, while the conventional alloys are placed below, **Figure 6e–g**. At this scale, discernible SLC pits were pronounced on the new generation Al-Cu-Li alloys with the AA2050-T84 alloy appearing to exhibit the most number of pitting sites. The conventional alloys did not exhibit pronounced discernible pitting sites except for the AA7050-T7451 alloy (as depicted by the arrow). However, similar to the SVET sample, the surface of the AA7050-T7451 alloy was glossy and did not show any trace of corrosion (except for the area depicted with red arrow). No trace of SLC was found on the AA6082-T6 alloy at this scale. The AA2024-T3 alloy also appeared to show no trace of SLC sites when wet. However, after the surface was dried under an air stream, multiple SLC sites appeared to be present on the surface of the alloy. Thus, the optical micrographs presented tend to show that the new generation Al-Cu-Li alloys are more prone to corrosion compared with the conventional alloys. Amongst the conventional alloys, the AA6082-T6 appeared to be the most resistant compared with the other two alloys.

However, polarization curves show that the AA6082-T6 and AA7050-T7451 alloy have high tendencies to corrode in the test environment. Thus, it is necessary to examine the extent of corrosion further at higher magnifications. In this regard, the surfaces of the corroded alloys and the cross-sections are examined using the SEM.

3.2.1 New generation Al-Cu-Li alloys

3.2.1.1 AA2050

Figure 5 presents the SEM images of the corroded surface and cross-section of the AA2050 alloy before and after the removal of corrosion products. The SLC sites were situated within corrosion rings. The number of SLC sites per cm^2 was about 14. The corrosion features observed on the surface of the alloy suggest that the attack was predominantly intergranular, and the attacks were aligned according to

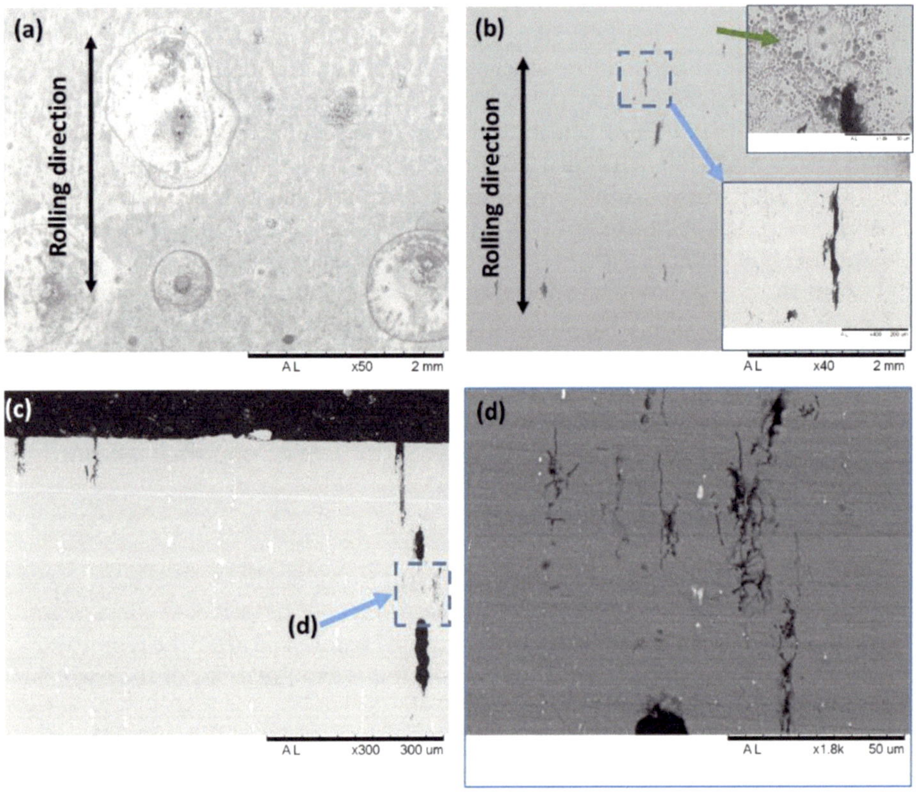

Figure 5.
SEM images of the corroded (a, b) surface of AA2050-T84 alloy ((a) before and (b) after the removal of corrosion products) and (c, d) cross-section showing the depth and different SLC types in the alloy.

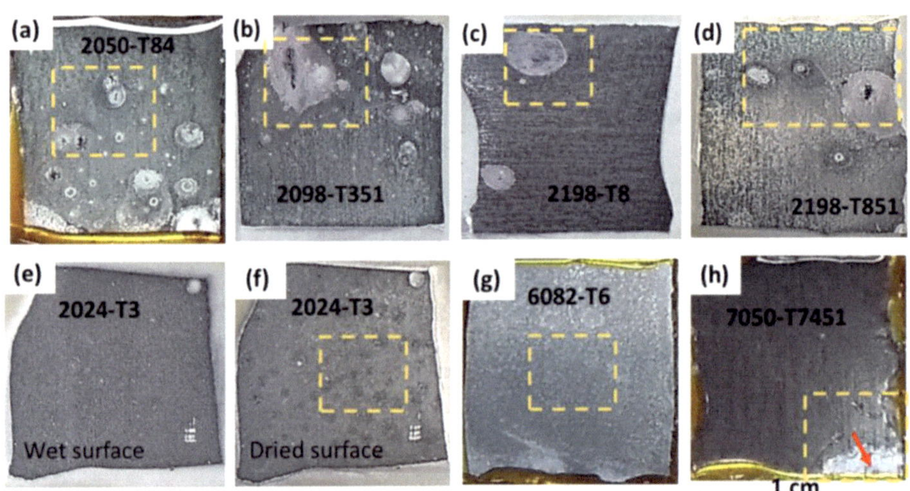

Figure 6.
Optical images showing the corroded surfaces of selected aluminum alloys after 72 h immersion in 3.5% NaCl solution at macroscopic scale.

the direction of deformation (see the inset from the region highlighted with blue rectangle in **Figure 5b**). In certain regions, superficial attacks were observed at the mouths of the pits (as indicated by the green arrow in the inset **Figure 5b**). It does appear as if the corrosion products from the pit preferentially etches the surface around the pit mouth. As earlier mentioned, H^+ ions migrate from within the pits to the mouths of the pits. This migration results in the decrease of the pH near the pit mouths. Thus, the local chemistry around the pit mouth is different from those in the surroundings. In this region, the solution can be aggressive owing to the reduced pH which can result in the mild attack of the surface of the alloy.

Cross-sectional images of the corroded AA2050-T84 alloy are presented in **Figure 5c, d** and **Figure 8a, b**. Attacks were observed penetrating as deep as 420 μm beneath the surface. Also, both intergranular and intragranular attacks were observed. The observed attack morphologies suggest that the attack initiated as IGC and then transited to intragranular. The magnified images in **Figure 8** show the typical progression of the attack. Cu-rich particles were observed to promote the dissolution of the adjacent matrix in the direction of corrosion propagation. Non-uniform precipitation was also observed as depicted in the images. The A-regions were richer in particles than the B-regions. This may affect the rapid propagation of the attack as observed since a galvanic cell will be most likely created between the particle-rich bands and the bands with lesser particles. Another interesting feature was the activities of redeposited Cu (**Figure 8b**). The re-deposited Cu promoted the dissolution of the matrix in a version similar to the Cu-rich particles. This sort of secondary attack caused the transition of the attack from intergranular to intragranular.

In the AA2050 alloy, initiation of IGC has been associated with Cu and Li enrichment of the GBs although reports are associating the corrosion susceptibility of the AA2050 alloy with the activities of T1 particles at the GBs [10, 11]. A recent detailed report by Yan et al. [12] has shown that IGC attacks are most likely due to Cu-Li enrichment or the presence of S-phase at the GBs. Also, Guerin et al. [13] showed that, even though the T1 precipitates populated the GBs in AA2050 alloy, the IGC observed in the T34 alloy was not due to the activities of the T1 phase. Other factors, such as high level of misorientation, were suggested to have more influence on IGC susceptibility. Thus, the attack observed in this work probably initiated at the GBs due to Cu and Li enrichment of the GBs, but the transition to intragranular corrosion occurred due to non-uniform precipitation, the presence of cathodic Cu-rich coarse particles in the corrosion paths, and the activities of re-deposited Cu which acted as local cathodic sites for the dissolution of the adjacent matrix.

3.2.1.2 AA2098-T351, AA2198-T8 and AA2198-T851

The corrosion behaviors of these alloys are very similar, and this is why they are grouped in this section. **Figure** 7 presents the SEM images of the surfaces and cross-sections of the AA2098-T351 alloy. The corrosion features in the three alloys are similar.

These alloys exhibited SLC sites formed within corrosion rings, and the attacks spread on the surfaces of the alloy with re-deposited Cu at the edges of the attacks. From the examination of the images of the surfaces after the removal of the corrosion products, it was observed that the attacks spread laterally with that of the AA2098-T351 alloy being the most pronounced (as much as 2 mm). In addition to the lateral spread, undercutting also occurred for the three alloys. Furthermore, the number of pits per area were significantly lower compared with the AA2050-T84 alloy - ranging between 3 and 8 per cm^2 for the three alloys. The AA2198-T8 alloy was the least susceptible to corrosion. It exhibited the lowest number of pits per

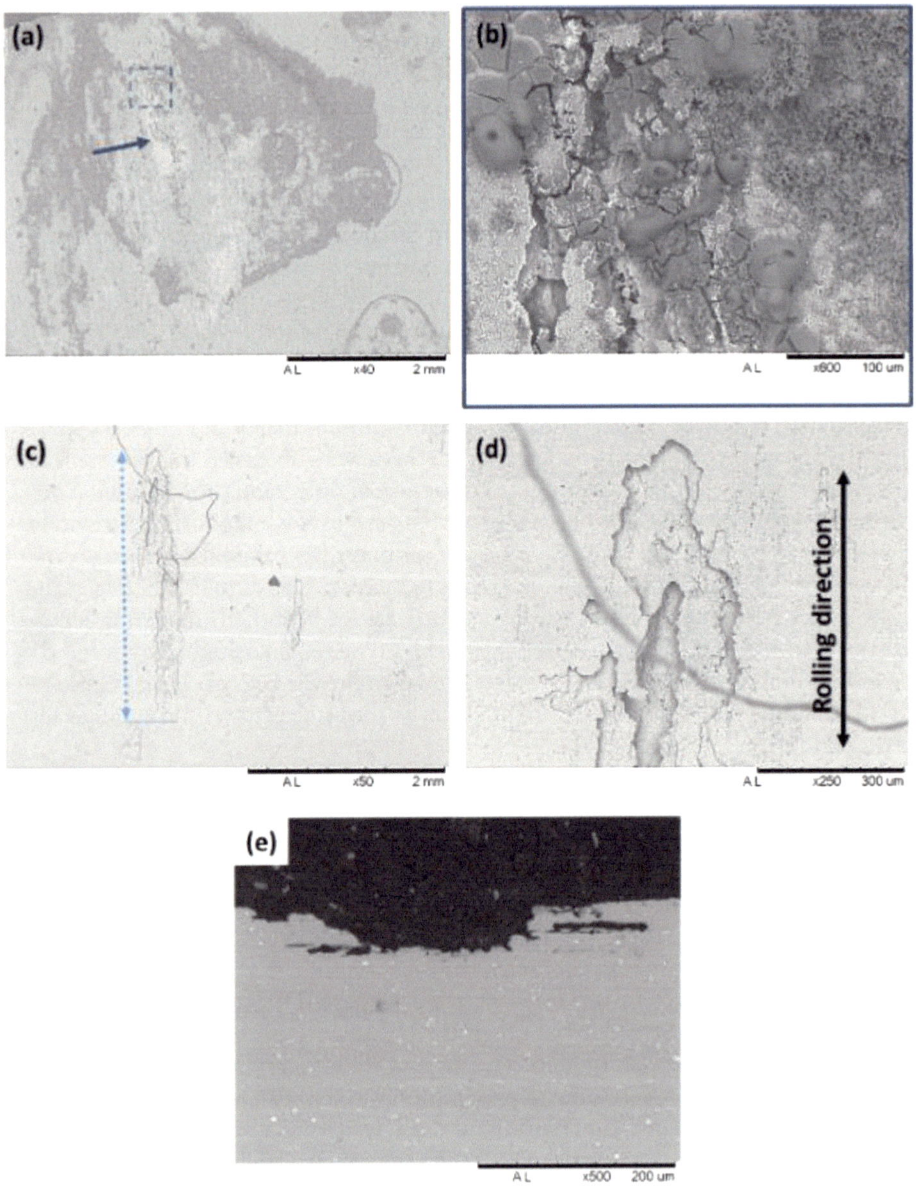

Figure 7.
SEM images of the corroded surface of AA2098-T351 alloy (a, b) before and (c, d) after the removal of corrosion products. (e) A cross-sectional view of the corrosion.

area and the depth of attack was only in the range of 50 μm. The depth on the AA2098-T351 alloy was around 70 μm. The most susceptible of the three alloys was the AA2198-T851 alloy. The attack depth on this alloy was as much as 110 μm. The reason for the increased susceptibilities of the AA2098-T351 and AA2198-T851 alloys to corrosion is because of the "51" temper treatment. This treatment involves an extra deformation process which introduces more dislocations into the alloys. Dislocations are preferred sites for the precipitation of the T1 phase which is the phase responsible for the formation of SLC in these alloys [6, 14–19]. Thus, the AA2198-T851 and AA2098-T351 alloys contain more T1 particles, particularly the AA2198-T851 alloy since it is artificially aged. The higher densities of the TI particles in these two alloys make them more susceptible compared with the AA2198-T8 alloy. And this is

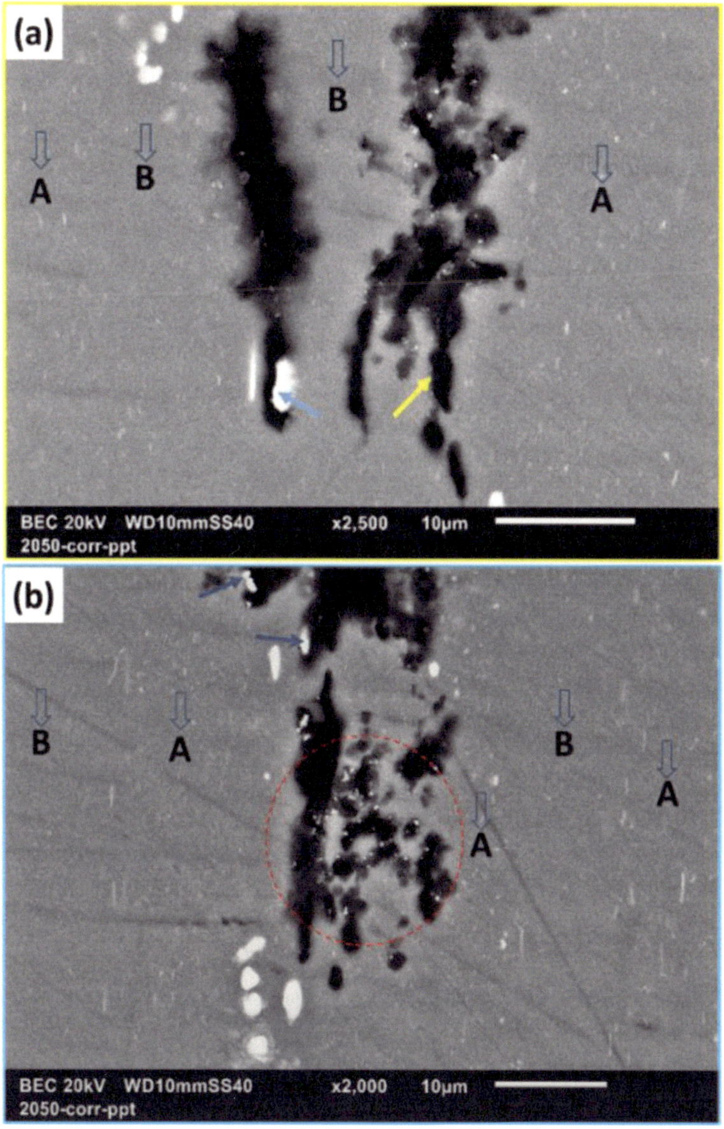

Figure 8.
SEM images of the cross-section of AA2050-T84 alloy showing SLC morphology and non-uniform precipitation in the alloy.

evident in the depths and spread of the attack. Nonetheless, the attacks on the three alloys spread more laterally and did not penetrate to very high depths as observed on the AA2050-T84 alloy. These three alloys, the AA2098-T351, AA2198-T8 and AA2198-T851, did not exhibit any form of intergranular corrosion. Also, it is important to note that the coarse intermetallic particles in these alloys are not associated with the initiation of SLC. The coarse particles cause the localized dissolution of the surrounding matrix, and, consequently are associated with the formation of trenches and cavities (micro-pits).

The increased depth in the AA2050-T84 alloy may be associated with the highly localized regions of attack with pronounced non-uniform precipitation in bands. Another important factor to note is that the SLC initiation in the AA2098-T351, AA2198-T8 and AA2198-T851 alloys is associated with the T1 phase which is present in the interiors of the grains as shown in **Figure 9**. This results in the intragranular attack observed. The GBs in these alloys are resistant to corrosion as shown in previous works from the same group [20, 21]. On the other hand, the SLC initiation in the AA2050 alloy is associated with Cu and Li enrichment at the GBs [12]. This results in intergranular corrosion which transits to intragranular (due to the effects of non-uniform precipitation, cathodic Cu-rich particles and Cu-redeposition) and penetrates very deep into the alloy. Thus, the corrosion behaviors of the new generation Al-Cu-Li alloys are not the same, the corrosion morphology and rate in the AA2050 alloys are very different from those of the other new-generation Al-Cu-Li alloys. For the later alloys, the attacks propagate laterally, predominantly. What is however common amongst the alloys is that irrespective of whether the attacks are penetrating deeply into the alloy or spreading laterally, the deformation the alloys were previously subjected to played a role in the propagation of attacks. In the AA2098 and AA2198 alloys, the attacks spread laterally according to the rolling direction. Also, in the AA2050 alloy, the attack spread and penetrated according to the rolling effect. Thus, in these alloys, there is a relationship between deformation and the propagation of SLC.

3.2.2 AA2024-T3 alloy

Figure 10 presents the SEM images of the surface of the AA2024-T3 alloy before and after the removal of corrosion products. Before the removal of corrosion products, it was difficult to locate the SLC sites. After the removal, small-sized SLC sites were observed all over the surface of the alloy. The number of SLC sites per area was more than 400. The cross-sectional images presented in **Figure 11** shows the attack depths and the corrosion morphologies. The region in brown square region in **Figure 11a** is further analyzed in **Figure 12**. **Figure 11c** and **11d** are magnified images of the blue and green square regions in **Figure 11a** and **11b**, respectively. The red arrows indicate intergranular attack, the blue arrows indicate trenched particles and the yellow arrows indicate regions that have been attacked intergranularly. Attacks were observed to have penetrated as deep as 220 μm. Thus, the attack on

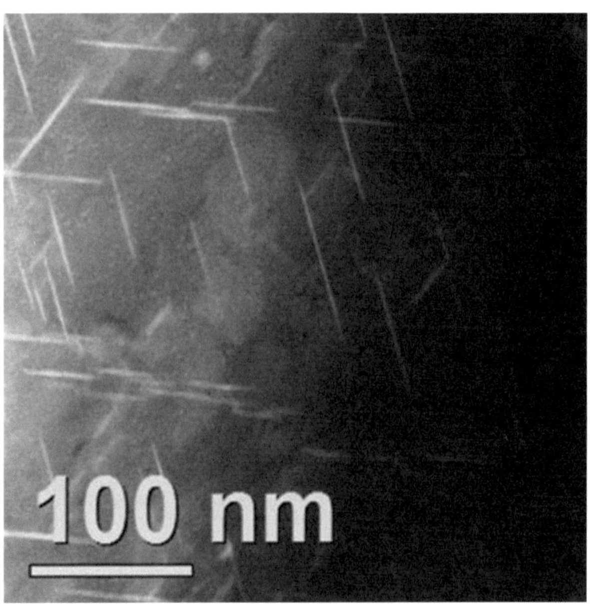

Figure 9.
HAADF image of the grain interior of AA2198-T851 alloy showing the typical hexagonal T1 precipitates present in new generation Al-Cu-Li alloys.

this alloy is very insidious since numerous small-sized attacks are penetrating very deep into the alloy without pronounced signs at the surface, especially during corrosion as shown by the differences between the wet and dried surface in **Figure 6**.

SLC attacks in AA2024 alloy are usually nucleated at regions with clusters of Cu-rich particles irrespective of whether they are S-phase particles or not [22–27]. However, in this work, most of the particles analyzed in the vicinity of attacks were predominantly S-phase particles. This is not surprising since the S-phase constitutes more than 60% of the coarse intermetallic particles present in the AA2024 alloy [28]. The S-phase associated attack resulted in pitting and transition to intergranular attack as the attack propagated. This is reflected in **Figure 11**, where trenching around cathodic particles are also revealed. Also, partial consumption of particles which resulted from the heterogeneities of the individual particles was observed (**Figure 12**).

With respect to the attack features, three types of coarse particles were observed (**Figure 12**): highly heterogeneous particles (HT); more homogeneous particles (HM); and high Cu-containing particles (HC). Partial dissolution of particles is associated with the HT particles. From **Figure 12**, it can be seen that the HT particles contain regions that are richer in Cu relative to the other regions of the particles. The EDX analysis of the observed HT particles revealed that they were predominantly Al-Cu-Mg particles with significant differences in the weight percentages of the three elements. For instance, in one of the particles, the Al, Cu, Mg weight percentages were 25.16, 63.69 and 1.23, respectively, in one region and 35.29, 49.02 and 12.59 in another region. The presence of multi-components

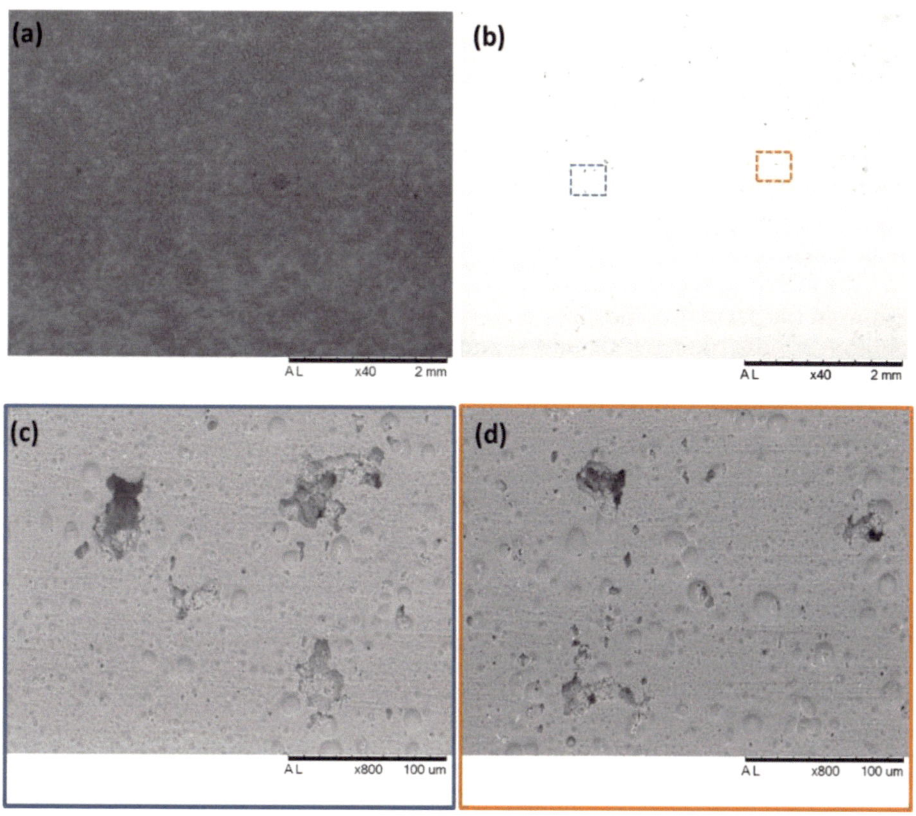

Figure 10.
SEM images of the corroded surface of AA2024-T3 alloy (a) before and (b–d) after the removal of corrosion products.

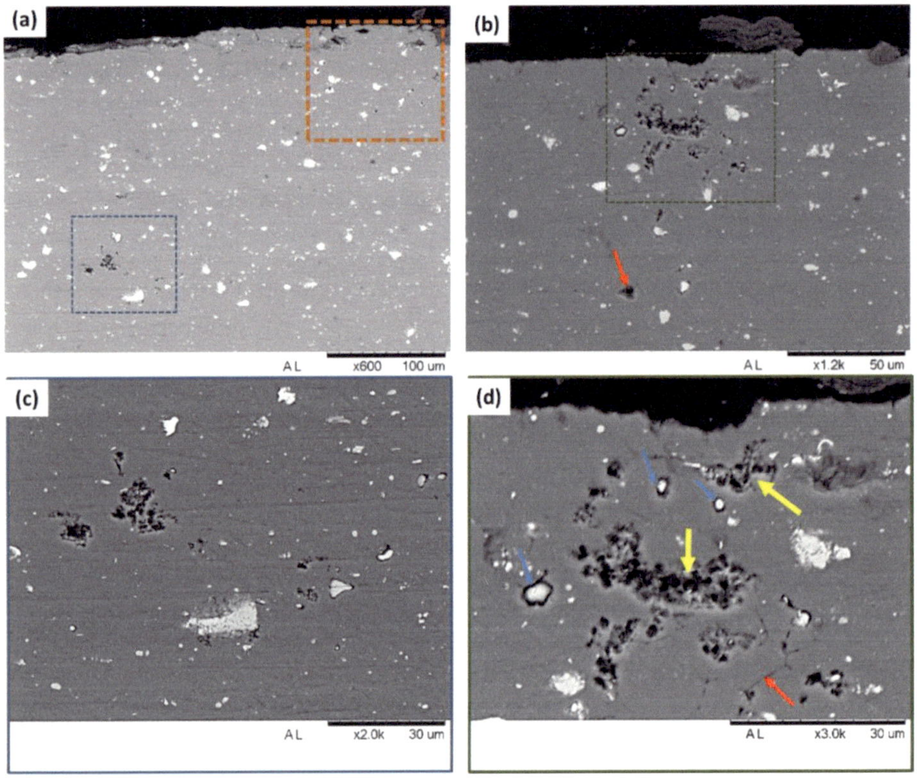

Figure 11.
SEM images of the cross-section of corroded AA2024-T3 alloy showing different depths and morphologies of attacks.

in a coarse particle has been previously reported in the literature [25, 29]. Micro-galvanic coupling can occur within HT particles [29], and the micro-galvanic interactions between the compositionally different domains in the particles (and the matrix) possibly resulted in the selective dissolution of the most active regions in the particles. These regions are most likely richer in Mg compared with the other regions, and are, therefore, anodic both to the matrix and the other regions of the particles. The HM particles were not found to be associated with any form of corrosion activities in this work. They are possibly (Al,Cu)x(Fe,Mn) ySi group of particles with lower Cu/Fe ratio as reported by Boag et al. [22] who also showed that trenches were not formed around these particles except after prolonged hours of exposure. The HC particles are Cu-enriched Al-Cu-Mg particles and Al-Cu-Fe-Mn particles. EDX analysis revealed that the Cu-enriched Al-Cu-Mg particles are dealloyed S-phase particles with Al and Mg contents in the range of 17 and 1.0 wt %, respectively, compared with 35.29 and 12.59 wt % of the HT particles. The selective leaching of the Al and Mg components resulted in the formation of Cu-rich remnants. The high Cu-content of the Al-Cu-Fe-Mn particles are associated with the re-deposition of Cu on these particles since they are cathodic particles [24]. The HC particles caused the dissolution of the adjacent matrix, and, hence, the formation of trenches.

From **Figure 11**, it is clear that there are links between the particle-associated attack and the GB attacks, and these links provide pathways for deep penetration into the alloy. However, it is important to note that the transition from pitting to intergranular corrosion is a typical corrosion characteristic of this alloy [30–32].

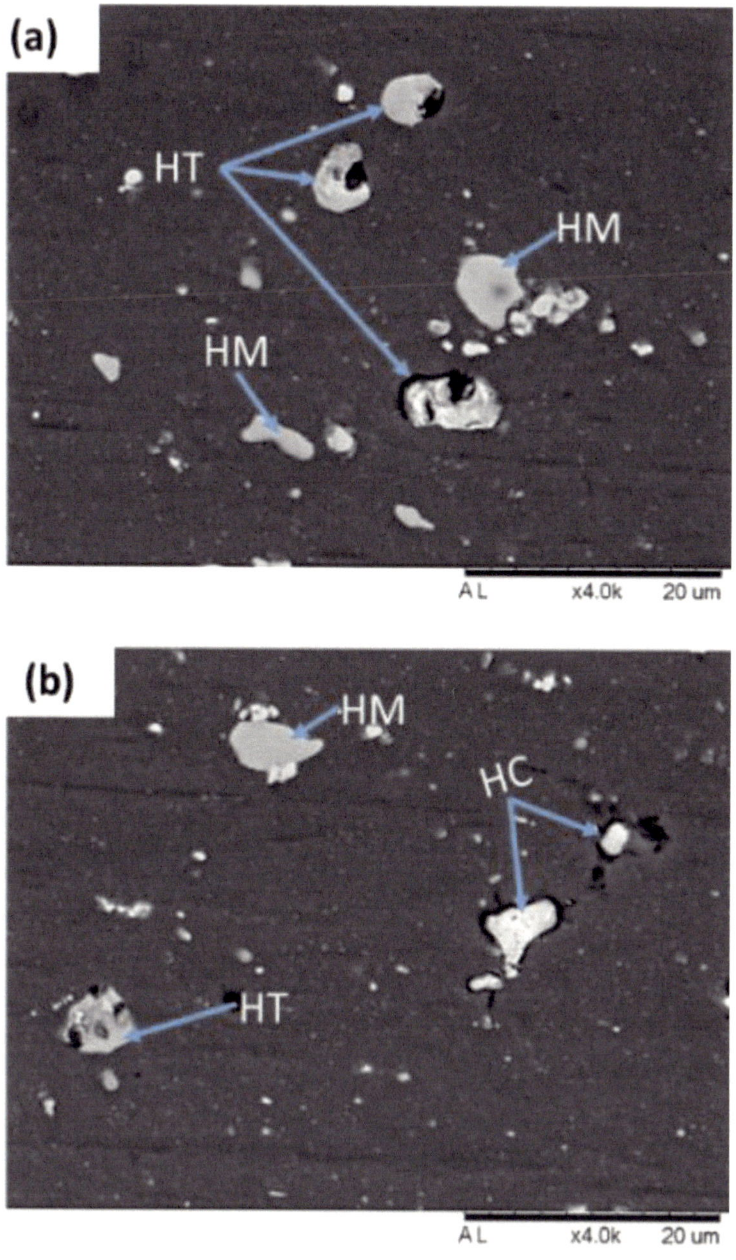

Figure 12.
SEM images of the cross-section of a corroded AA2024-T3 alloy showing different types of precipitates with respect to the corrosion activities. These images were obtained from areas within the brown square in ***Figure 11a****.*

Also, because of the non-linear form of these links, the attacks branch significantly such that it is often difficult to follow the attacks from the surface to regions far beneath the surface through cross-sectional examination.

Unlike the new generation alloys, especially the AA2098 and AA2198 alloys, the initiation of SLC in the AA2024-T3 alloy is associated with the coarse intermetallic phases, and the propagation of attack appears to have no relationship with the rolling direction. Also, far more SLC sites were observed on the AA2024-T3 alloy, and the attacks penetrated very deep into the alloy - twice as deep as those observed on the new generation alloys (except for the AA2050-T84 alloy). This shows why the total current from the pits swamps that of the pseudopassive as observed on the polarization curves. Thus, it can be argued that the new generation Al-Cu-Li alloys (except for the AA2050-T84 alloy) are more corrosion resistant compared with the AA2024-T3 alloy.

3.2.3 AA6082-T6

Presented in **Figure 13** are SEM images showing the corroded surfaces of the AA6082-T6 alloy before and after the removal of corrosion products and the cross-sectional image obtained afterwards.

In this alloy, the predominant form of corrosion was trenching, and cavity (micro-pit) formation and this was associated with the activities of the coarse Fe-rich particles. The Fe-rich particles were the predominant coarse particles on this alloy. The formation of SLC was very rare. In fact, only a site was found in the entire area exposed, and it was not pronounced. The observed SLC was intergranular (**Figure 13d**) and penetrated only as deep as 30 μm. The formation of IGC in this alloy is as a result of the formation of Mg_2Si particles at the GBs with widened precipitate free zones (PFZs) [33] as typified in the TEM image in **Figure 14a**. The Mg_2Si (β) phase is highly anodic to the Al matrix [34, 35]. The absence of precipitates at regions immediately adjacent to the GBs promotes galvanic interactions between a highly anodic Mg_2Si phase and the PFZs. Upon exposure, the Mg component of the Mg_2Si phase is selectively dissolved leaving behind a Si-enriched particle

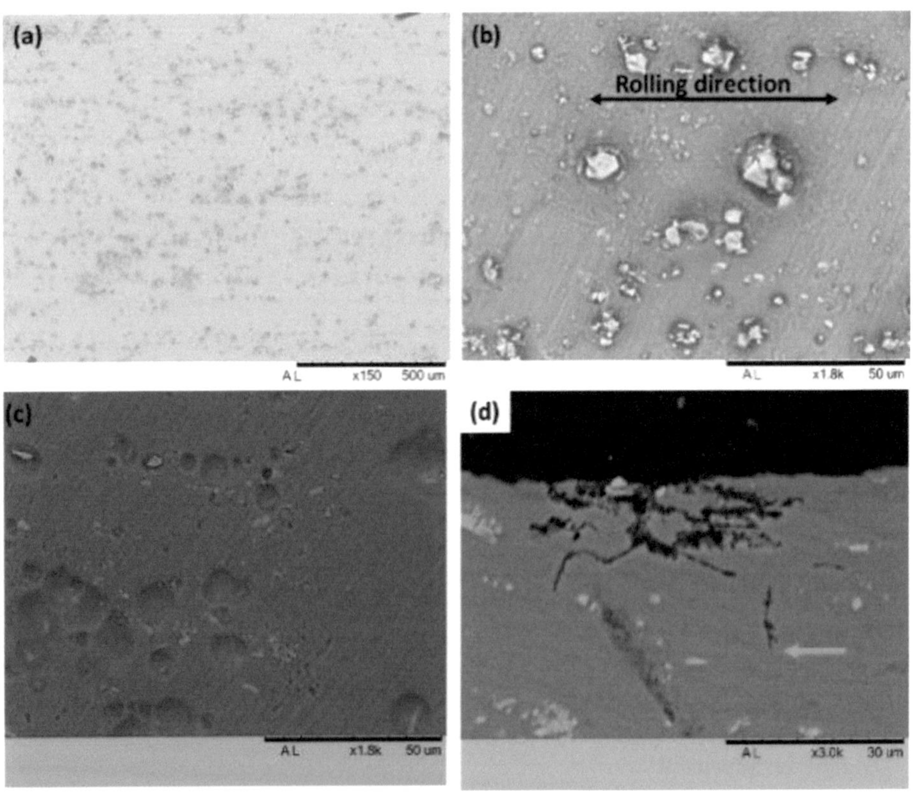

Figure 13.
SEM images of the corroded surface of AA6082-T6 alloy (a, b) before and (c) after the removal of corrosion products. (d) Cross-section of the corroded alloy showing the intergranular attack.

remnant. Although some authors have argued that Si-enriched particles rapidly form SiO_2 in the presence of water and are therefore not effective cathodes [36–38], it is believed that this remnant is cathodic and causes the dissolution of the adjacent PFZs [33, 39]. An example of the activities of the Si-enriched Mg_2Si particles can be seen in **Figure 14b, c**. The images provided were obtained from a mildly corroded region of an AA6082-T6 sample that was exposed to 3.5% NaCl solution for 7 days. The corrosion products were not removed before obtaining the images. From these images, it can be seen that the Si-enriched particles cause the dissolution of the adjacent matrix leading to the selective dissolution of the GB.

Nonetheless, the AA6082-T6 alloy is the most resistant to corrosion amongst the selected alloys compared—the number of SLC sites per area and the depth of SLC penetration were the lowest. This agrees with the SVET result.

3.2.4 AA7050-T7451

Figures 15 and **16** present the SEM images of the corroded surfaces (before and after the removal of corrosion products) and the cross-section of the AA7050-T7451 after the 72-h immersion test. **Figure 15b** is a magnified image of the marked region in **Figure 15a**. As with the other alloys, corrosion rings were also formed around SLC sites. However, except for the highly pronounced region of attack, the corrosion products blended well with the surface and covered

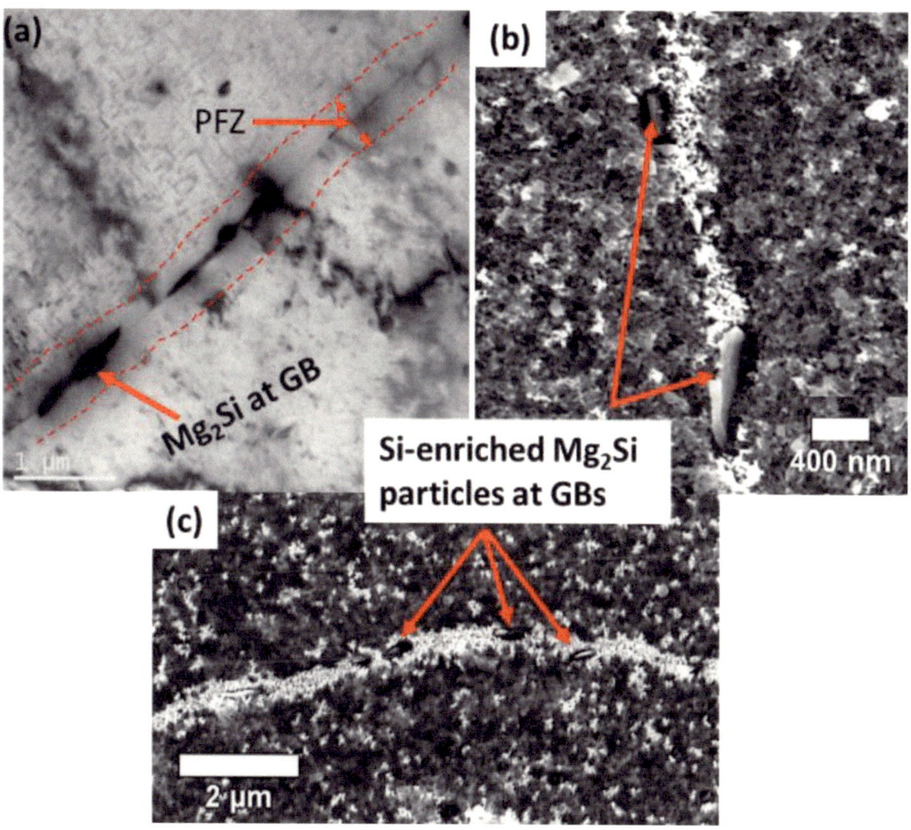

Figure 14.
(a) TEM bright-field image showing widened PFZ and Mg_2Si precipitates at the grain boundary of the AA6082-T6 alloy. (b) SEM images showing Si-enriched remnants at the GB of the corroded surface of AA6082-T6 alloy.

pitting sites smoothly. And this is the reason behind the glossy appearance of the surface at macro-scale (**Figure 6h**). After the removal of the corrosion products, pronounced SLC sites numbering up to 24 per cm^2 were revealed. The SLC was predominantly IGC, but other forms of intragranular corrosion were also observed (see **Figure 16c** and the region labeled z in **Figure 16d**). Furthermore, superficial IGC were observed across the entire surface examined (**Figure 16e**). It is clear from the images in Figure that the selective dissolution of the particles precipitated at the GBs plays an important role in the IGC attack. In the 7xxx series alloys, the $\eta(Zn_2Mg)$ phase and its variants are usually associated with IGC [40–42]. This phase is highly active compared with the matrix of the alloys and preferentially dissolves upon exposure to aggressive media. The potential difference between a widened PFZ and the grain interior also plays a role in the selective dissolution of the GBs in 7xxx series alloys [43].

The SLC attacks propagated according to the elongation induced by the prior deformation process. The SLC attacks penetrated as deep as 143 μm into the alloy (**Figure 16f**). The SEM image in **Figure 16c** shows that the propagation of the SLC was also affected by grain specific bands similar to the grain features revealed by Donatus et al. [44, 45]. Again, superficial dissolution of regions around some pit mouths, similar to what was revealed on the AA2050 alloy, were observed. This further shows the effect of the local chemistry changes around the mouths of the pits.

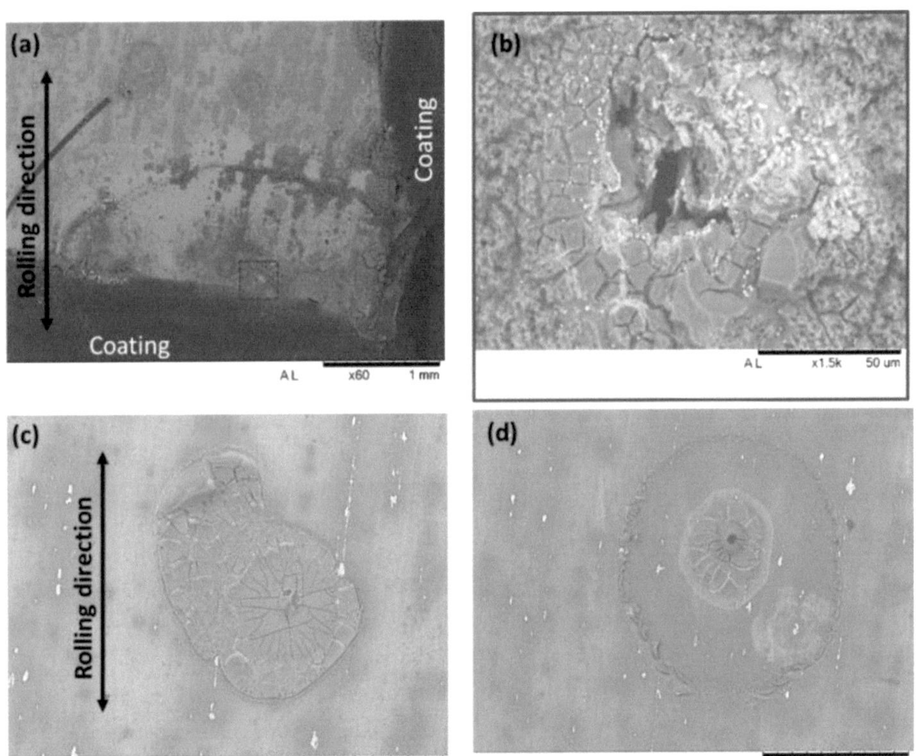

Figure 15.
SEM images of the corroded surface of AA7050-T7451 alloy before the removal of corrosion products showing different types of SLC sites on the alloy.

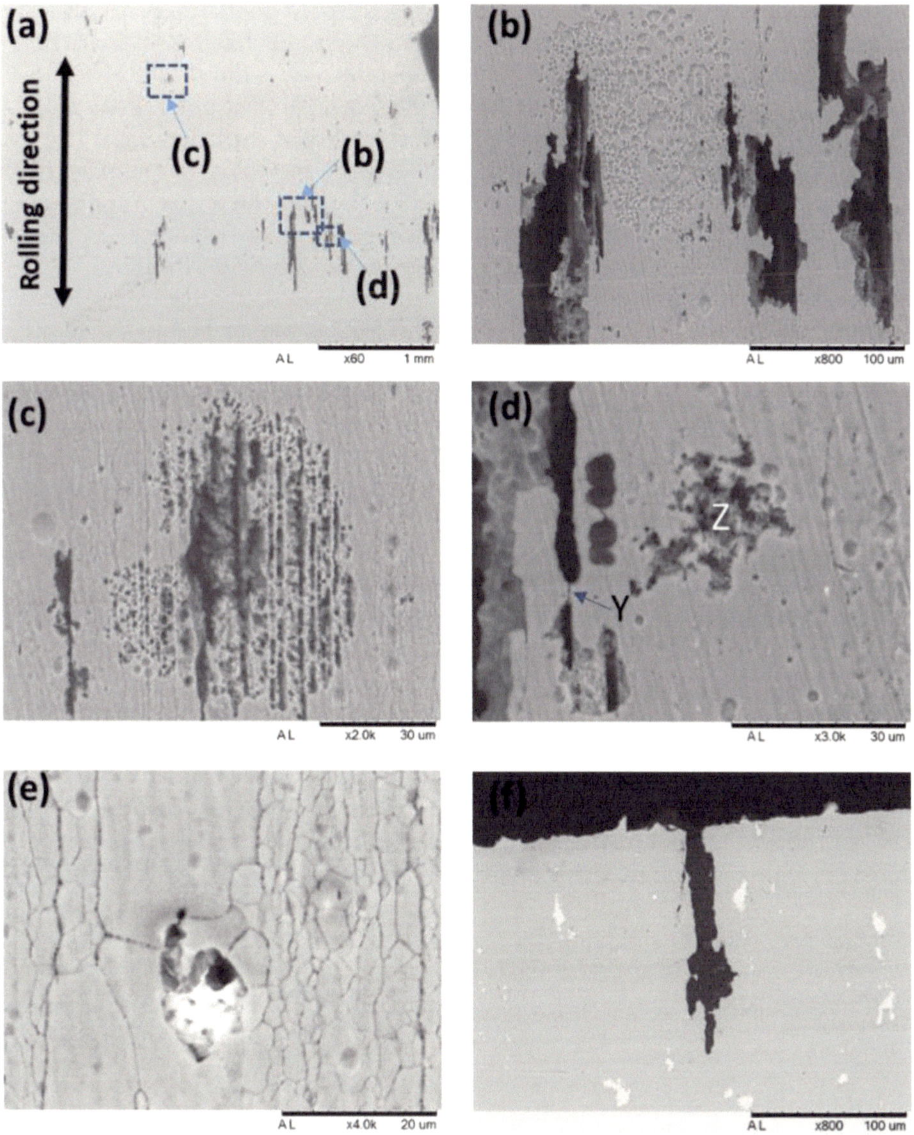

Figure 16.
SEM images of the corroded (a–e) surface and (f) cross-section of AA7050 alloy after the removal of corrosion products. Different forms of corrosion including a superficial form of intergranular attack (e) were observed.

Compared with the new-generation Al-Cu-Li alloys, the AA7050-T7451 alloy is far more susceptible to corrosion both in terms of depth of penetration (except for the AA2050-T84 alloy) and the number of attacks per cm^2. Also, the attack on AA7050-T7451 is far more insidious compared with all the alloys tested since it is very difficult to detect at macro-scale.

4. Summary

Table 2 gives a summary of the corrosion type, morphology and the microstructural features associated with the corrosion of the selected aluminum alloys.

Presented in **Figure 17a** are curves showing the peak depth of corrosion attack and the number of SLC sites per cm^2 for the selected alloys. In terms of depth of penetration, the most corrosion-resistant alloy is the AA6082-T6 alloy followed by the new-generation AA2098-T351, AA2198-T8 and AA2198-T851 alloys. The "51" treatment increased the susceptibility of the later alloys. The corrosion rate of the AA2050-T84 is the highest with the attack penetrating twice as deep as that of the nearest alloy (the AA2024-T3) alloy. In terms of the number of pits per cm^2, the AA2024-T3 presented the highest number followed by the AA7050-T7451 alloys. These two alloys, alongside the AA2050-T84 alloy, are the least corrosion-resistant alloys.

Figure 17b shows the length of the spread of SLC on the surfaces of the alloys. Although the new generation Al alloys are more resistant in terms of corrosion penetration and the number of pits per cm^2, these alloys presented the largest lengths of SLC attacks on the average. This was most evident in the AA2098-T351 alloy. These alloys only exhibited an intragranular form of attack and this form of attack only propagated and spread laterally, but predominantly according to the direction of deformation.

It is also important to note that the corrosion behaviors of the Al-Cu-Li alloys are not the same. Different factors trigger and promote SLC in these alloys. The AA2050-T84 alloy presented a very different degree of susceptibility and morphology in comparison with the other third-generation Al-Cu-Li alloys namely, the AA2098-T351, AA2198-T8 and the AA2198-T851 alloys. However, it is important to state that the AA2098 alloy is a precursor to the AA2198 alloy [46]. This probably explains why their corrosion behaviors are similar. It is also very evident from the results that, except for the AA6082-T6 alloy, every other alloy that exhibited IGC presented very high rates of corrosion attack penetration. The new generation Al-Cu-Li alloys (except for AA2050-T84) that exhibited only intragranular corrosion were more resistant to corrosion penetration.

Alloy	SLC type and morphology	Associated microstructural features/ phases
AA2050-T84	Predominantly IGC and pitting developed from IGC Deeply penetrating and less branched attack	GB enrichment [12, 13]. Redeposited Cu. Non-uniform precipitation of particles.
AA2098-T351	Intragranular attack (no IGC) Exfoliating layers in attacked grains. Shallow and laterally spreading attack.	Intragranular T1 particles.
AA2198-T8	Same as AA2098-T351.	Same as AA2098-T351.
AA2198-T851	Same as AA2098-T351 + band-like attack.	Same as AA2098-T351.
AA2024-T3	Predominantly IGC and pitting with particle consumption, and particle-GB linked attack. Deeply penetrating and highly branched attack.	S-phase and heterogeneous Cu-rich particles.
AA6082-T6	IGC Less pronounced penetration.	Mg_2Si (β) particles and precipitate free zones at GBs.
AA7050-T7451	IGC, pitting and intragranular band-like attack Deeply penetrating and less branched attack.	Zn_2Mg (η) and η-phase variants at GBs [40–42], and the presence of PFZs [43].

Table 2.
Summary of SLC type and morphology and associated microstructural features and phases in the alloys investigated.

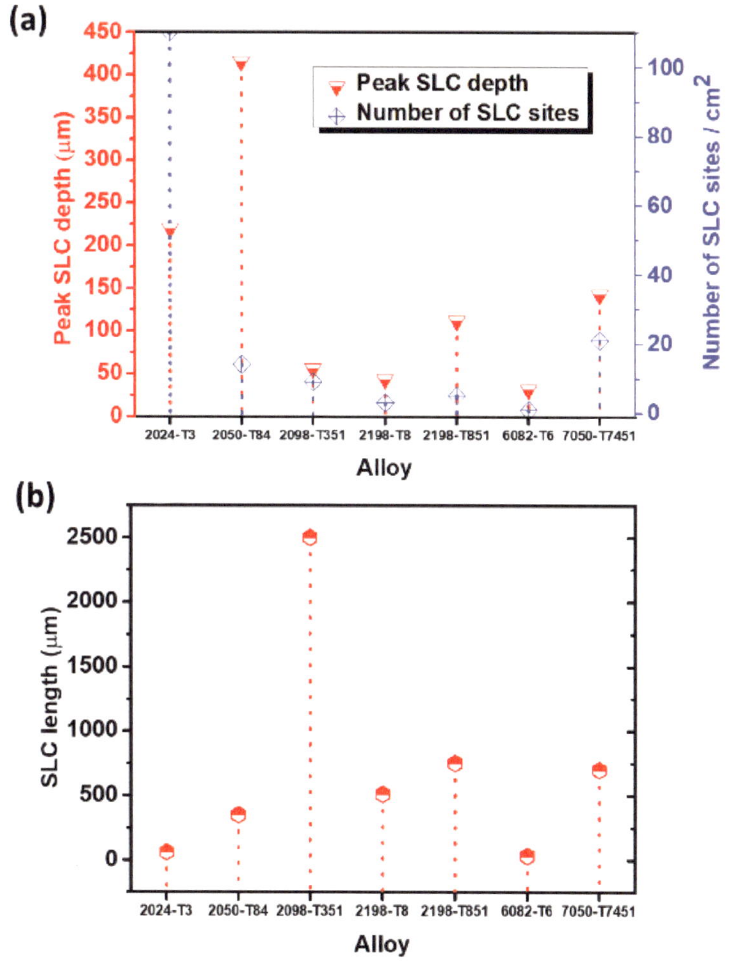

Figure 17.
Plots showing the observed (a) peak SLC depths and number of SLC sites per cm² and (b) peak lengths of SLC on the selected aluminum alloys.

Furthermore, there seems to be a strong relationship between the direction of attack propagation and spread with the direction of deformation. This was evident in at least five out of the seven alloys investigated. Also, all the alloys exhibited trenching and the formation of cavities (micro-pits). These types of attack are associated with the activities of cathodic coarse intermetallic particles.

In conclusion, the new generation aerospace alloys (except for the AA2050-T84) are more resistant to corrosion than the conventional aerospace alloys (AA2034-T3 and AA7050-T7451) but are less resistant compared with the AA6082-T6 alloy. It was difficult to fully establish these differences from electrochemical approaches (especially from the potentiodynamic polarization technique). Thus, it is advisable always to employ a non-electrochemical approach when the corrosion resistance of Al alloys are to be compared, and this can be combined with electrochemical techniques to gain more insight.

Acknowledgements

The authors appreciate FAPESP for financial support through the grant Proc. 2013/13235-6 and for sponsoring the postdoctoral fellowships of Dr. Uyime

Donatus (Process 2017/03095-3). The authors also wish to acknowledge the African Academy of Sciences Grant No [ARPDF 18-03], for providing financial support to carry out part of this research.

Author details

Uyime Donatus[1*], Michael Oluwatosin Bodunrin[2], Ayotunde Olayinka[3], Mariana Xavier Milagre[1], Olamilekan Rasaq Oloyede[4], Sunday Aribo[5], João Victor de Sousa Araujo[1], Caruline de Souza Carvalho Machado[1] and Isolda Costa[1]

1 Nuclear and Energy Research Institute, Sao Paulo, Brazil

2 School of Chemical and Metallurgical Engineering, University of the Witwatersrand, Johannesburg, South Africa

3 Louisiana Advanced Metal Processing Laboratory, Mechanical Engineering, University of Louisiana at Lafayette, Lafayette, Louisiana, USA

4 Institute for Materials Research, School of Chemical and Process Engineering, University of Leeds, UK

5 Research and Product Development, Evraz North America, Regina, SK, Canada

*Address all correspondence to: uyimedonatus@yahoo.com

References

[1] Moreto JA, Marino CEB, Bose Filho WW, Rocha LA, Fernandes JCS. SVET, SKP and EIS study of the corrosion behaviour of high strength Al and Al–Li alloys used in aircraft fabrication. Corrosion Science. 2014;**84**:30-41

[2] Moreto JA, Gamboni OC, Marino B, Bose Filho W, Fernandes S, Rocha LA. Corrosion behaviour of Al and Al-Li alloys used as aircraft materials. Corrosion and Protection of Materials. 2012;**31**(3/4):60-64. Available from: http://www.scielo.mec.pt/pdf/cpm/v31n3-4/v31n3-4a01.pdf [cited 20 July 2017]

[3] Garrard WN. Corrosion behavior of aluminum-lithium alloys. Corrosion. 1994;**50**(3):215-225. DOI: 10.5006/1.3293513

[4] Frankel G. The effects of microstructure and composition on Al alloy corrosion. Corrosion. 2015;**71**(11):1308-1320

[5] Donatus U, Terada M, Ospina CR, Queiroz FM, Fatima Santos Bugarin A, Costa I. On the AA2198-T851 alloy microstructure and its correlation with localized corrosion behaviour. Corrosion Science. 2018;**131**:300-309. DOI: 10.1016/j.corsci.2017.12.001

[6] de Sousa Araujo JV, Donatus U, Queiroz FM, Terada M, Milagre MX, de Alencar MC, et al. On the severe localized corrosion susceptibility of the AA2198-T851 alloy. Corrosion Science. 2018;**133**:132-140. DOI: 10.1016/j.corsci.2018.01.028

[7] Viramontes-Gamboa G, Rivera-Vasquez BF, Dixon DG. The active-passive behavior of chalcopyrite. Journal of the Electrochemical Society. 2007;**154**(6):C299-C311. DOI: 10.1149/1.2721782

[8] Burstein GT, Salter VCC, Ball J, Sharman JD. In: Datta M, MacDougall BR, Fenton JM, editors. The Growth of Pits during the Electrograining of Aluminium. Chicago, Illinois, USA: Electrochemical Society; 1996. Available from: https://books.google.com.br/books?id=P4TppG5uByoC&pg=PA133&lpg=PA133&dq=pseudopassive+behaviour+in+Al+alloys&source=bl&ots=FtQLepZJ2W&sig=ffZrJJn4a-KCydHErhyqy5cXy2o&hl=en&sa=X&ved=2ahUKEwjp1fmAhsDfAhVMCpAKHdKmCxsQ6AEwDXoECAkQAQ#v=onepage&q=pseudopassive [cited 26 December 2018]

[9] Tsuyuki C, Yamanaka A, Ogimoto Y. Phase-field modeling for pH-dependent general and pitting corrosion of iron. Scientific Reports. 2018;**8**(1):12777. Available from: http://www.nature.com/articles/s41598-018-31145-7

[10] Proton V, Alexis J, Andrieu E, Delfosse J, Deschamps A, De Geuser F, et al. The influence of artificial ageing on the corrosion behaviour of a 2050 aluminium-copper-lithium alloy. Corrosion Science. 2014;**80**:494-502

[11] Proton V, Alexis J, Andrieu E, Blanc C, Delfosse J, Lacroix L, et al. Influence of post-welding heat treatment on the corrosion behavior of a 2050-T3 aluminum-copper-lithium alloy friction stir welding joint. Journal of the Electrochemical Society. 2011;**158**(5):C139-C147

[12] Yan Y, Peguet L, Gharbi O, Deschamps A, Hutchinson CR, Kairy SK, et al. On the corrosion, electrochemistry and microstructure of Al-Cu-Li alloy AA2050 as a function of ageing. Materialia. 2018;**1**:25-36. Available from: http://linkinghub.elsevier.com/retrieve/pii/S2589152918300115

[13] Guérin M, Alexis J, Andrieu E, Laffont L, Lefebvre W, Odemer G, et al. Identification of the metallurgical parameters explaining the corrosion susceptibility in a 2050 aluminium alloy. Corrosion Science. 2016;**102**:291-300

[14] Buchheit RG, Moran JP, Stoner GE. Electrochemical behavior of the T1 (Al2CuLi) intermetallic compound and its role in localized corrosion of Al-2%Li-3%Cu alloys. Corrosion. 1994;**50**(2):120-130. DOI: 10.5006/1.3293500

[15] Luo C, Albu SP, Zhou X, Sun Z, Zhang X, Tang Z, et al. Continuous and discontinuous localized corrosion of a 2xxx aluminium-copper-lithium alloy in sodium chloride solution. Journal of Alloys and Compounds. 2016;**658**:61-70. DOI: 10.1016/j.jallcom.2015.10.185

[16] Ma ZY, Zhou X, Liao Y, Yi Y, Wu H, Wang Z, et al. Localised corrosion in AA 2099-T83 aluminium-lithium alloy: The role of grain orientation. Corrosion Science. 2015;**107**:41-48

[17] Donatus U, Klumpp RE, Mogili NVV, Aletobelli Antunes R, Milagre MX, Costa I. The effect of surface pretreatment on the corrosion behaviour of silanated AA2198-T851 Al-Cu-Li alloy. Surface and Interface Analysis. 2019;**51**(2):275-289

[18] Zhang X, Zhou X, Hashimoto T, Liu B, Luo C, Sun Z, et al. Corrosion behaviour of 2A97-T6 Al-Cu-Li alloy: The influence of non-uniform precipitation. Corrosion Science. 2018;**132**:1-8. Available from: https://www.sciencedirect.com/science/article/pii/S0010938X1730896X [cited 20 January 2018]

[19] Donatus U, Ferreira RO, Vardan Mogili NV, Gonçalves de Viveiros BV, Milagre MX, Costa I. Corrosion and anodizing behaviour of friction stir weldment of AA2198-T851 Al-Cu-Li alloy. Materials Chemistry and Physics. 2018;**219**(August):493-511. Available from: https://linkinghub.elsevier.com/retrieve/pii/S0254058418307193

[20] Donatus U, de Sousa Araujo JV, de Souza Carvalho Machado C, Vardhan Mogili NV, Antunes RA, Costa I. The effect of manufacturing process induced near-surface deformed layer on the corrosion behaviour of AA2198-T851 Al–Cu–Li alloy. Corrosion Engineering, Science and Technology. 2018;**2782**:1-11. DOI: 10.1080/1478422X.2018.1558932

[21] Donatus U, de Viveiros BVG, de Alencar MC, Ferreira RO, Milagre MX, Costa I, et al. Correlation between corrosion resistance, anodic hydrogen evolution and microhardness in friction stir weldment of AA2198 alloy. Materials Characterization. 2018;**144**:99-112. Available from: https://linkinghub.elsevier.com/retrieve/pii/S1044580318317546

[22] Boag A, Hughes A, Glenn A. Corrosion of AA2024-T3 part I: Localised corrosion of isolated IM particles. Corrosion. 2011;**53**(1):17-26. DOI: 10.1016/j.corsci.2010.09.009

[23] Glenn A, Muster T, Luo C, Zhou X, Thompson GE, Boag A, et al. Corrosion of AA2024-T3. Part III: Propagation. Corrosion Science. 2011;**53**(1):40-50. DOI: 10.1016/j.corsci.2010.09.035

[24] Chen GS, Gao M, Wei RP. Microconstituent-induced pitting corrosion in aluminum alloy 2024-T3. Corrosion. 1996;**52**(1):8-15

[25] Schneider O, Ilevbare GO, Scully JR, Kelly RG, Schneider O, Kelly RG, et al. In situ confocal laser scanning microscopy of AA 2024-T3 corrosion metrology. Journal of the Electrochemical Society. 2004;**151**(8):B465. DOI: 10.1149/1.1764781

[26] Zhou X, Luo C, Hashimoto T, Hughes AE, Thompson GE. Study of localized corrosion in AA2024 aluminium alloy using electron tomography. Corrosion Science. 2012;**58**:299-306. Available from: http://www.sciencedirect.com/science/article/pii/S0010938X12000807 [cited 18 March 2015]

[27] Hughes AE, Parvizi R, Forsyth M. Microstructure and corrosion of AA2024. Corrosion Reviews. 2015;**33**(1-2):1-30

[28] Dimitrov N, Mann JA, Vukmirovic M, Sieradzki K. Dealloying of Al_2CuMg in alkaline media. Journal of the Electrochemical Society. 2000;**147**(9):3283. DOI: 10.1149/1.1393896

[29] Queiroz F, Magnani M, Costa I, De MH. Investigation of the corrosion behaviour of AA 2024-T3 in low concentrated chloride media. Corrosion Science. 2008;**50**:2646-2657. Available from: http://www.sciencedirect.com/science/article/pii/S0010938X08002278 [cited 26 May 2013]

[30] Zhang W, Frankel G. Localized corrosion growth kinetics in AA2024 alloys. Journal of the Electrochemical Society. 2002;**149**(11):1-11. Available from: http://jes.ecsdl.org/content/149/11/B510.short [cited 09 May 2013]

[31] Zhang W, Frankel G. Anisotropy of localized corrosion in AA2024-T3. Electrochemical and Solid-State Letters. 2000;**3**(6):268. Available from: http://esl.ecsdl.org/content/3/6/268.short [cited 08 November 2012]

[32] Zhang W, Frankel G. Transitions between pitting and intergranular corrosion in AA2024. Electrochimica Acta. 2003;**48**(9):1193-1210. Available from: http://linkinghub.elsevier.com/retrieve/pii/S0013468602008289 [cited 28 October 2012]

[33] Donatus U, Thompson GE, Zhou X, Wang J, Cassell A, Beamish K. Corrosion susceptibility of dissimilar friction stir welds of AA5083 and AA6082 alloys. Materials Characterization. 2015;**107**:85-97. Available from: http://linkinghub.elsevier.com/retrieve/pii/S1044580315002491 [cited 10 July 2015]

[34] BirbilisN,BuchheitRG.Electrochemical characteristics of intermetallic phases in aluminum alloys an experimental survey and discussion. Journal of the Electrochemical Society. 2005;**152**(4):B140-B151. Available from: http://jes.ecsdl.org/content/152/4/B140%5Cn; http://jes.ecsdl.org/content/152/4/B140.full.pdf%5Cn; http://jes.ecsdl.org/content/152/4/B140.short

[35] Gupta RK, Sukiman NL, Fleming KM, Gibson MA, Birbilis N. Electrochemical behavior and localized corrosion associated with Mg_2Si particles in Al and Mg alloys. ECS Electrochemistry Letters. 2012;**1**(1): C1-C3. DOI: 10.1149/2.002201eel

[36] Svenningsen G, Larsen MH, Walmsley JC, Nordlien JH, Nisancioglu K. Effect of artificial aging on intergranular corrosion of extruded AlMgSi alloy with small Cu content. Corrosion Science. 2006;**48**(6):1528-1543

[37] Svenningsen G, Larsen MH, Nordlien JH, Nisancioglu K. Effect of thermomechanical history on intergranular corrosion of extruded AlMgSi(Cu) model alloy. Corrosion Science. 2006;**48**(12):3969-3987. Available from: http://linkinghub.elsevier.com/retrieve/pii/S0010938X06000989 [cited 04 March 2015]

[38] Liang WJ, Rometsch PA, Cao LF, Birbilis N. General aspects related to the corrosion of 6xxx series aluminium alloys: Exploring the influence of Mg/

Si ratio and Cu. Corrosion Science. 2013;**76**:119-128. DOI: 10.1016/j.corsci.2013.06.035

[39] Eckermann F, Suter T, Uggowitzer PJ, Afseth A, Schmutz P. The influence of MgSi particle reactivity and dissolution processes on corrosion in Al-Mg-Si alloys. Electrochimica Acta. 2008;**54**(2):844-855

[40] Wadeson DA, Zhou X, Thompson GE, Skeldon P, Oosterkamp LD, Scamans G. Corrosion behaviour of friction stir welded AA7108 T79 aluminium alloy. Corrosion Science. 2006;**48**(4):887-897. Available from: http://www.scopus.com/inward/record.url?eid=2-s2.0-33645025920&partnerID=tZOtx3y1 [cited 13 November 2014]

[41] Huang TS, Frankel GS. Kinetics of sharp intergranular corrosion fissures in AA7178. Corrosion Science. 2007;**49**(2):858-876

[42] Lu X, Han X, Du Z, Wang G, Lu L, Lei J, et al. Effect of microstructure on exfoliation corrosion resistance in an Al-Zn-Mg alloy. Materials Characterization. 2018;**135**:167-174. DOI: 10.1016/j.matchar.2017.11.029

[43] Marlaud T, Malki B, Henon C, Deschamps A, Baroux B. Relationship between alloy composition, microstructure and exfoliation corrosion in Al-Zn-Mg-Cu alloys. Corrosion Science. 2011;**53**(10):3139-3149. DOI: 10.1016/j.corsci.2011.05.057

[44] Donatus U, Thompson GE, Zhou X, Alias J, Tsai I-LL. Grain distinct stratified nanolayers in aluminium alloys. Materials Chemistry and Physics. 2017;**188**:109-114. DOI: 10.1016/j.matchemphys.2016.12.021

[45] Donatus U, Thompson GE, Elabar D, Hashimoto T, Morsch S. Features in aluminium alloy grains and their effects on anodizing and corrosion. Surface and Coatings Technology. 2015;**277**:91-98. Available from: http://linkinghub.elsevier.com/retrieve/pii/S025789721530147X [cited 25 July 2015]

[46] Fei ZS, Dong ZW, Hua YW, Ling SC, Jun WH. Ageing response of a Al-Cu-Li 2198 alloy. Materials and Design. 2014;**63**(April):368-374. DOI: 10.1016/j.matdes.2014.04.063

Chapter 6

Machining of Al-Cu and Al-Zn Alloys for Aeronautical Components

Jorge Salguero, Irene Del Sol, Alvaro Gomez-Parra and Moises Batista

Abstract

Machining operations are chosen by aircraft manufacturers worldwide to process light aluminum alloys. This type of materials presents good characteristics in terms of weight and physicochemical properties, which combined with a low cost ratio making them irreplaceable in aircraft elements with a high structural commitment. Conventional machining processes such as drilling, milling and turning are widely used for aeronautical parts manufacturing. High quality requirements are usually demanded for these kinds of components but aluminum alloys may present some machinability issues, basically associated to the heat generated during the process. Among others, surface quality and geometrical deviations are highly influenced by the condition of the cutting-tool, its wear and the cutting parameters. Consequently, the understanding of the relationship among the process parameters, the quality features and the main wear mechanism is a key factor for the improvement in the productivity. In this chapter, the fundamental issues of drilling, milling and turning are addressed, dealing with the relationship between cutting parameters, wear phenomena and micro and macro geometrical deviations.

Keywords: aluminum, drilling, milling, turning, cutting tool, tool wear

1. Introduction

Aluminum is considered a valuable material thanks to its lightness (around a third of copper and steel density), its mechanical properties and its resistance to corrosion. This highly malleable material presents an excellent thermal and electrical conductivity. It is also a magnificent light reflector that gives to it an attractive natural appearance. Furthermore, neither magnetic nor toxic, this metal is 100% recyclable, increasing its value even as a waste. In fact, its recycling offers powerful economic incentives [1, 2]. It should be noted that around 70% of the 761 million of tons of aluminum produced since 1886 continue in use [3].

The large number of chemical elements that can be alloyed to pure aluminum allow to find a suitable aluminum alloy for every industrial case. The different compositions help to enhance some of the mechanical properties, as is shown in **Table 1**.

Nonetheless, these alloys easily respond to hardening mechanisms, reaching strengths up to 30 times higher than the pure aluminum strength [4, 5]. For these

Al +	Cu	• Increased strength and hardness • Combined with Mg produces a heat treatable alloy
	Zn	• Increased strength and hardness • Possibility of stress corrosion • Combined with Mg produces a heat treatable alloy
	Si	• Good corrosion resistance • Combined with Mg produces a heat treatable alloy
	Mn	• Increased resistance to corrosion
	Mg	• Increases strength and hardness • Good corrosion resistance • Increased weldability

Table 1.
Main alloying elements used in aluminum alloys.

reasons, the variety of applications for aluminum is constantly increasing, being an essential part of our life.

Particularly, transport industry absorbs more than a quarter of the produced aluminum, being applied to any type of transport. Cast alloys are widely used in the automotive industry while forging alloys predominate in the aeronautic field.

This material is indispensable for the aeronautic industry since its origin, due to the high influence on the total weight of the aircraft. It is estimated that each ton of weight reduction in the structure of an airplane, considering an average weight of 80 kg per passenger, luggage on board included, results in an increase of 12 passengers per flight [6]. Likewise, the weight of the aircraft is directly proportional to fuel consumption as well as operational and maintenance costs, especially in landing gear consumable elements [7].

Consequently, light materials such as titanium-based alloys, composite materials and aluminum alloys are the most common choices for the structural elements of this sector. Despite nowadays it seems to be a trend trying to replace aluminum alloys by composite materials, forged aluminum alloys or some of its variants 2xxx (Al-Cu) and 7xxx (Al-Zn) series are still strategic materials for most of the structural parts (**Figure 1**) [8]. Their choice ensures a wide scope and a predictable in-service behavior [6]. That's why, 82% of the structure of a Boeing 747 aircraft and 70% of a Boeing 777 are made of this type material [1, 4].

Aeronautical parts have specific characteristics. They are designed to increase their strength and reduce their weight as well as to be integrated in the aircraft using assembly operations.

In the case of aluminum machining, the cost is significantly influenced by the machinability problems, which are basically associated with the heat generated during machining due to the deformation of the crystal lattice and the friction between the chip and cutting-tool [9].

The study of machining processes and the theory of metal cutting dates back to F.W. Taylor, who published "On the Art of Metal Cutting" in the early 20th century [10]. Since then, the scientific/technical advances have been spectacular, among the most noteworthy milestones are: new materials and tool coatings, automation of the machine tool, increasing the process accuracy and its monitoring, among others.

The most common operations are milling and turning to shape the component, while drilling is mainly used to prepare assembly operations. These processes keep in common the material removal in order to give the desired shape and dimensions, that is, to add value to the workpiece.

Due to the widespread use of forging alloys, it is common to find parts with a high Buy to Fly ratio (BtF). The BtF defines the relation between the weight of the

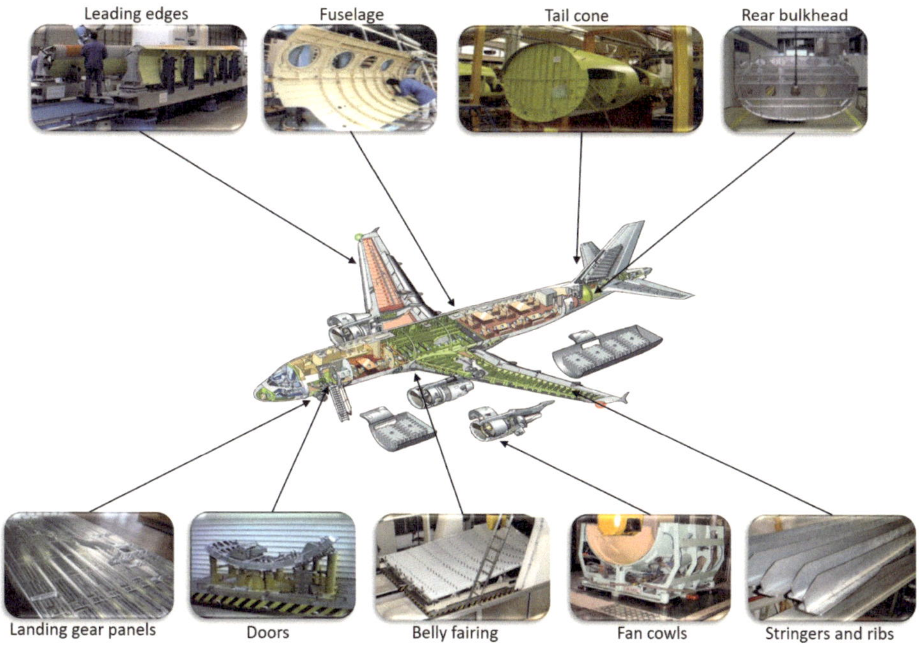

Figure 1.
Structural parts manufactured in aluminum in an A319 aircraft.

purchased raw material and the weight of the part that finally flies [8, 11], which means that most of the raw material is removed transforming it into chips during the machining operation. In fact, milling of monolithic parts can achieve up to 12: 1 BtF. Such high BtF factors will affect the primarily cost, weight, and performance of the part.

In summary, the aeronautical components start from a large volume of raw material, in which is necessary to remove the excess of material. Depending on the geometry and function of the component, various machining processes can be used, standing out the drilling, milling and turning processes. The correct control of these machining processes allows to obtain high quality parts, as those demanded in aeronautics.

1.1 Chapter scope

In the following sections, the main characteristics of these machining processes are presented based on a literature review. The text analyzes when and how each machining is used, covering the parameters and the cutting-tool wear effects on the quality of the produced components. This text is illustrated with examples extracted from experimental campaigns performed by the authors of this chapter.

All the experimental work and most of the literature find is focused on the two main Al series which characteristics are shown in **Table 2**.

2. Drilling of aerospace aluminum alloys

Aircrafts are subjected to a wide temperature ranges during their service, being able to reach up to 40°C while operating in airports and temperatures below -50°C when flying. This wide gradient implies that the structural joints must be designed to withstand stresses in a wide thermal gradient. Therefore, the joint

Identification	Aluminum Association	AA2024	AA7475	AA7050	AA7075
	UNS	A92024	A97475	A97050	A97075
	ISO	AlCu4Mg1	AlZn5.5MgCu(A)	AlZn6CuMgZr	AlZn5.5MgCu
Composition	Si	≤0.5	≤0.1	≤0.12	≤0.4
	Fe	≤0.5	≤0.12	≤0.15	≤0.5
	Cu	3.8–4.9	1.2–1.9	1.9–2.5	1.2–2.0
	Mn	0.3–0.9	≤0.06	≤0.1	≤0.3
	Mg	1.2–1.8	1.9–2.6	2.0–2.7	2.1–2.9
	Cr	≤0.1	0.18–0.25	≤0.04	0.18–0.28
	Zn	≤0.25	5.2–6.2	5.9–6.9	5.1–6.1
	Ti	≤0.15	≤0.05	≤0.06	≤0.2
	Al	Rem.	Rem.	Rem.	Rem.
Properties	Density (kg/m^3)	2.78	2.81	2.83	2.81
	Melting point (°C)	500–638	477–635	490–630	475–635
	Thermal conductivity (W/m°C)	121–151	163	157	130
	Thermal expansion (um/m°C)	23.2	23.2	24.1	23.6
	Young's Modulus (GPa)	73	72	72	72
	Percent elongation (%)	6–20	12	10	11
	Ultimate tensile strength-UTS (MPa)	440–495	531	495–550	525–570
	Heat treatment	T3, T4, T361, T6, T81,T861	T7651	T74	T6, T651, T73

Table 2.
Most frequent aeronautical aluminum alloys designation, compositions and technical properties [5].

must be made with rivets, while the use of welded joints is not certified. Because of this, drilling is one of the prior machining operations in the assembly of aerospace components [11, 12].

Drilling operation have a direct impact on the performance of the riveted joint, affecting mainly its dimensional compatibility with the part to join and the joint fatigue behavior, being particularly important in airframes parts. For this reason, drilling is monitored measuring the thrust force and the torque produced during the machining. Afterwards, some quality parameters such as roughness, burr formation or roundness, are usually measured to ensure the quality of the obtained hole [13–15].

This quality is usually affected by different factors, including the correct the selection of the cutting parameters, movements of the operator during the machining, incorrect chip removal, marks produced by the cutting-tool due to uncontrolled vibrations, as well as imperfections in the drilling angles [16].

Uncontrolled vibrations can be reduced by using automatic drilling machines or by reducing the length of the cutting-tool, decreasing the deflection of the tip and improving the roundness cylindricity of the hole [16]. However, the main problems of aluminum drilling are roughness, burr formation and cutting-tool wear, which are related mainly to the machining parameters [17].

Additionally, fatigue behavior is highly influenced by the roughness, even more than by residual stresses, so their control becomes a critical task. A higher roughness on the hole surface reflects deeper machining marks, scratches and gauges that may work as stress concentration points, increasing the possibilities of crack propagation by magnifying the stress on the part at the assembly point [18].

2.1 Influence of the drilling parameters

Drilling is an axial machining operation where the cutting-tool rotates and penetrates perpendicular to the surface to be drilled at the same time, **Figure 2**.

It is governed by two main parameters: tangential cutting speed (V) and linear feed rate (F). V is usually provided by the tool manufacturer, in function of the cutting-tool material. From V and the tool diameter (D), the spindle speed (S) can be calculated by using Eq. (1):

$$S = \frac{V}{\pi \cdot D} \tag{1}$$

Similarly, F depends on the feed per tooth (f_z), the number of cutting tooth (Z) and S. This parameter can be calculated using Eq. (2):

$$F = f_z \cdot Z \cdot S \tag{2}$$

Both parameters have a direct impact on the quality of the hole, and though on the quality of the joint.

In a general way, lower values of roughness are obtained when low cutting speed and feed-rate are used [17, 19]. Higher spindle speed brings longer chips that

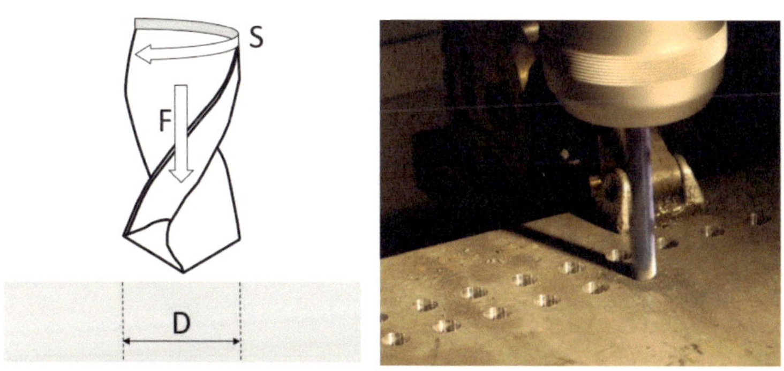

Figure 2.
Drilling scheme and image of a drilling process.

curl inside the hole, producing marks on its surface [19]. Other way to improve the roughness results it to use high point and helix angles [20], but in this case, diamond coatings should be considered to increase wear resistance [17]. Similarly, higher-feed rates increase thrust force and though the wear behavior, while an increase of the cutting speed slightly reduces the thrust forces.

Cylindricity and perpendicularity errors rise by increasing feed-rate and depth of cut, but cutting speed have a different effect in both of them. Low cutting speed reduces roundness error but it also peaks the perpendicularity error, so the optimum value should be selected depending on the part requirements [21].

Probably, the most relevant phenomenon in the drilling of aluminum alloys is the burr formation. Burrs in aluminum are usually classified as type "A", uniform along the hole [22]. They are related to the aluminum ductility (10–12% elongation), affected by the drill geometry (point angle, helix angle, diameter, web thickness and chisel edge) and the process parameters (F, S). Low feed-rates are recommended to reduce burr height, especially near the exit of the cutting tool [22], while a proper cutting speed can interact on the burr thickness. Additionally, large point angles minimize burr height [19]. If the selection of the suitable parameters cannot avoid burr appearance, deburring operations before riveting and assembly are performed. They depend on the burr height, being mandatory when it excess 0.3 mm.

Finally, the dimensional accuracy is mainly related to the helix angle. Larger helix angles increase it [20]. However, it should be considered that oversized holes are common and low cutting speed and high-feed rates can also increase its deviation from the nominal value [19], mainly due to adhesion wear mechanism.

Aluminum alloys are usually machined under dry conditions using tungsten carbide (WC-Co) tools but more aggressive parameters can be selected depending on the lubricating conditions. Carbide tools increase the process efficiency in terms of wear behavior [14] but they may have an impact on the hardness and the cylindricity whereas parameters need to be carefully selected to not increase the tool wear due to thermal effects.

When liquid lubricants are used, they should be placed inside the working area, but as drilling edges work inside the material, the chips produced inside it are constantly removed upward, forcing the lubricant to be evacuated from the place of action and reducing its cooling effect [16]. For this reason, the most adopted lubricating option to avoid this problem it to project the cutting fluids trough the cutting-tool, creating an internal lubrication.

Wear control is essential to obtain the expected quality parameters, affecting the diameter, the roughness and burr height. As example of the aforementioned, **Figure 3** shows different graphs analyzing the diameter, the roughness, the burr and cutting forces evolution in the dry drilling of AA7475 alloy.

It can be observed a diameter reduction produced by the heat effect on the expansion of the hole during the drilling operation. A slight opposite trend can be found for the thrust force due to the loss of cutting edge capacity, reduced by wear effects. Similarly, burr height easily grows over 0.3 mm, the maximum value permitted, forcing deburring operations as long as the tool wear increase. At last, Ra values present a high variability, indicating the presence of dynamic problems or a poor chip evacuation, as well as the alternate effects of the secondary adhesion wear mechanism. However, they are far from 3.2 μm, the maximum value allowed for metal alloys.

2.2 Advanced drilling techniques

Drilling of aluminum components is sometimes performed in multiple materials at the same time such as stacks or laminates. For this reason, different advanced

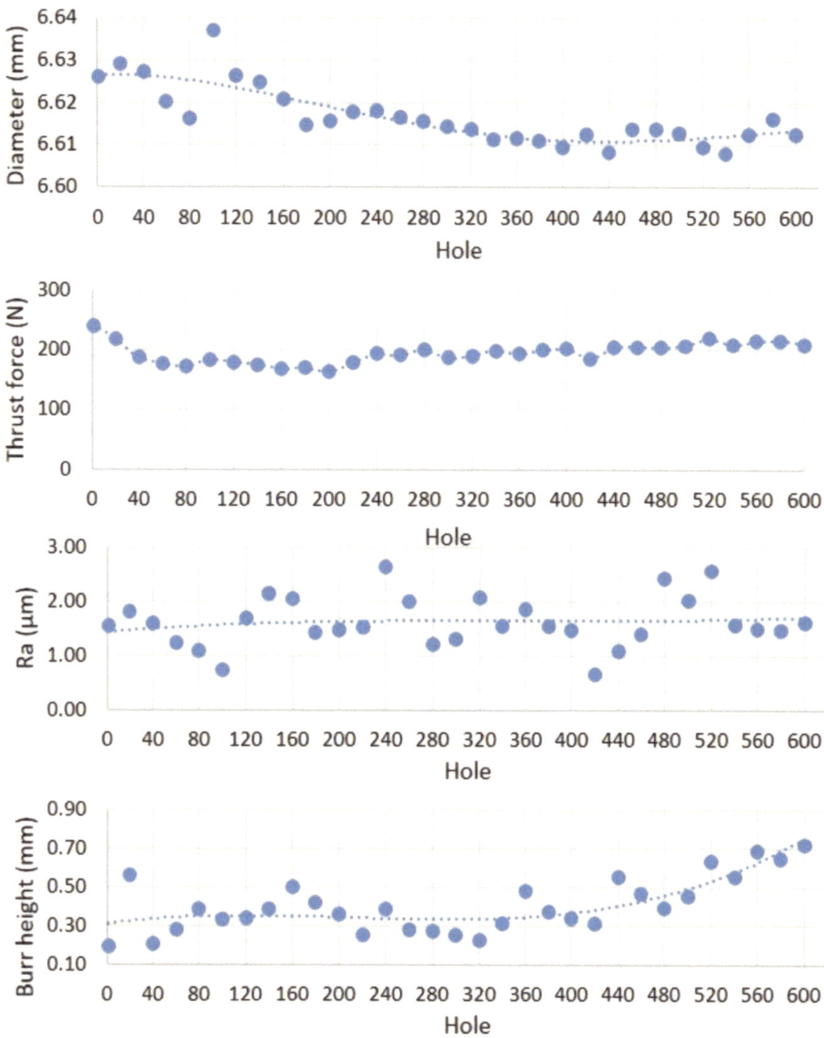

Figure 3.
Diameter, average roughness, burr and cutting forces evolution in the dry drilling of a AA7475 alloy drilled using a double-lip 6.6 mm diameter cutting-tool, S = 4800 rev/min and F = 1085 mm/min.

drilling techniques are used to improve the quality of the hole and avoid possible defects derived from them.

- Orbital Drilling (OD) is a technique where the hole is produced using a milling-tool instead of a drill bit. The cutting tool generates an orbital path to create the hole instead of an axial one. This technique is usually limited to 40 mm depth to reduce the possible vibration produced during the operation, which may decrease the hole quality as well as the cutting-tool life [23]. Moreover, cutting forces are lower than the obtained in conventional drilling, increasing the option of robots and machines that can be used in this operation [24]. It also drops the thermal effect of the machining, by its discontinuous cut and the short chip produced, that is also continuously removed reducing the need of lubricants [25]. This technique is quite useful when stacks of aluminum and titanium are drilled.

- Vibration Assisted Drilling (VAS) is a technique that combines the drilling operation with an impose high frequency vibration on the tool. It reduces the cutting forces, the burr formation and it increases the breakability of the chip, which increases the surface quality and reduces the dimensional error and the wear behavior [26, 27]. This technique includes peck drilling, where the amplitude and the frequency are higher and created by the alternative axial displacement in the machining center. VAS is a common technique for aluminum and fiber composite stacks drilling.

3. Milling of aerospace aluminum alloys

Milling of aluminum in the aeronautical industry is used to produce light components with a high accurate dimension. This operation creates slim parts or monolithic parts. Peripheral milling is performed in the first type to adjust the final dimensions of preformed sheets. Otherwise, monolithic parts come from a rough stock of aluminum and up to the 95% of the material can be removed [28], as it was mentioned before. For this reason, milling of aeronautic alloys selects the most aggressive parameters, so the material removal rate is as high as possible.

The quality of these type of parts is verified trough the surface quality and the dimensional accuracy [29] and both types involve at some point of the operation, a low stiffness situation were deflection and vibration, including chatter issues, may appear severally affecting their final quality. Consequently, low stiffness becomes determinant to select the machining parameters and reduce wear behavior. Additionally, this process has to consider residual stresses produced and released during the milling as well as the temperature achieved, especially to meet the precision targets [30].

In contrast to drilling, the continuous release of chip and cooling effect of inconstant contact favor manufactures to recommend high speed machining parameters. This option reduces the wear mechanism and increase the process efficiency. Nevertheless, high speed combined with low rigidity makes easier the appearance of chatter. These vibrations can arise due to the system excitation at the natural frequency response of the cutting-tool or the workpiece or due to the amplification of the displacements caused by the forces and the lack of stiffness [31, 32]. In these cases, the cut is unstable creating an un-constant chip thickness, which is afterwards reflected on the surface quality [33, 34]. Similarly, the static deflection produced by the forces involved in the process leads to over or under cuts, affecting the final dimension of the parts [35, 36]. These facts enhance the importance of the workholding to ensure the final quality of the parts.

Therefore, different workholding and fixture devices had being designed to increase the part stiffness. Most of them change their position during the operation to ensure the maximum rigidity of the complete system all over the operation [37]. It is common to combine them with active damping actuators to attenuate vibrations [38]. The better clamping system, the more aggressive the parameters will be and the higher the efficiency of the process.

Regarding the parameters and tool path selection, analytical approaches can be useful to reduce defects on the part and in-process issues. Simulations must include an accurate model of the material and a system that allows them to consider the continuous material removal, which will update the rigidity behavior of the part [39].

For the particular case of tool path and strategies, virtual twins' developments are common since they allow to predict the part behavior and improve the operation [40].

3.1 Influence of the milling parameters

Milling is a machining process where a rotary cutting-tools is used to remove material. The tool advance into a work piece varying the direction on one or several axes, **Figure 4**. This operation is defined with the same cutting parameters proposed in the expressions that govern drilling operation, Eqs. (1) and (2), including a radial feed rate.

Conventional milling strategies are prone to generate undercut while climb milling is usually related to overcut [37]. Symmetric tool paths are selected to compensate the effect of residual stresses in the part deformation [41] as well as to reduce in-process deflection. Similarly, particular tool paths may be designed to increase the part stiffness during the cutting operation [42].

In terms of parameter selection, higher cutting speeds and lower axial depths of cut reduce the cutting forces and though the deflection [35, 36]. High cutting speeds also have an impact on the temperature of the process. When high cutting speeds are used, the chip formation mechanism changes to a near to adiabatic process. In this situation, the chip takes a relevant role as a heat exchanger, evacuating most of the heat generated and keeping the workpiece and cutting-tool relatively cold. This fact directly influences the cutting force components, as is shown in **Figure 5**. The force in the feed rate direction is reduced up to 50% from 600 to 750 m/min, the high speed machining range for aluminum. The other force component is kept almost constant and proportional to the feed rate.

Roughness is also proportional to the feed-rate, directly increasing with it. Nevertheless, the adhesion wear mechanism can produce alterations in the geometry of the cutting-tools, improving the surface quality with the machining time, as is shown in **Figure 6**.

Milling operations are usually performed in different steps, roughing and finishing, as a consequence burr formation is less significant than in drilling operations. However, deburring operations may be included in the same process if the tool wear considerably increases, having direct impact on the burr height, **Figure 7**.

The cutting tools recommended for aluminum milling have the following characteristics. Very sharp edges to minimize adhesion and to perform a smooth cut. Two-lip cutting tools, with low helix angles (25°-30°) and long pitches to facilitate the evacuation of large chip flow rates [43, 44]. In order to reduce dynamic instabilities, lower than 5 length-to-diameter ratios (L/D < 5) are recommended.

Whereas the choice between integral or insert tools depends on the application, rounded inserts are more stable if they work with a considerably larger radial (20–60% diameter) than axial paths (2–8% diameter) [45] while integral

Figure 4.
Milling scheme and image of milling processes.

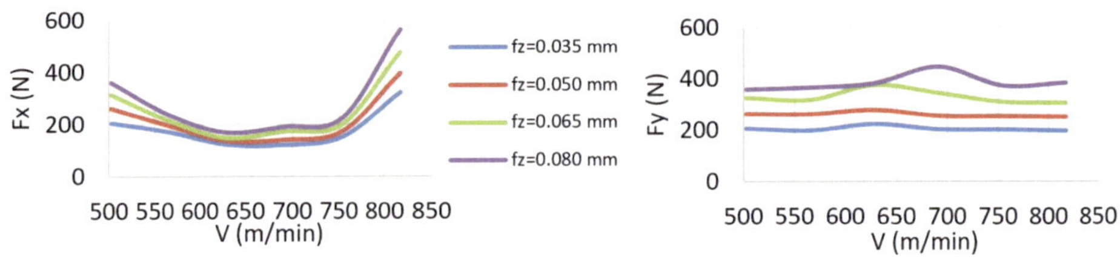

Figure 5.
Evolution of the cutting force components, obtained during the milling of a AA2024-T3 alloy, as a function of the cutting speed and the feed rate. Axial depth of cut is kept constant at 10 mm.

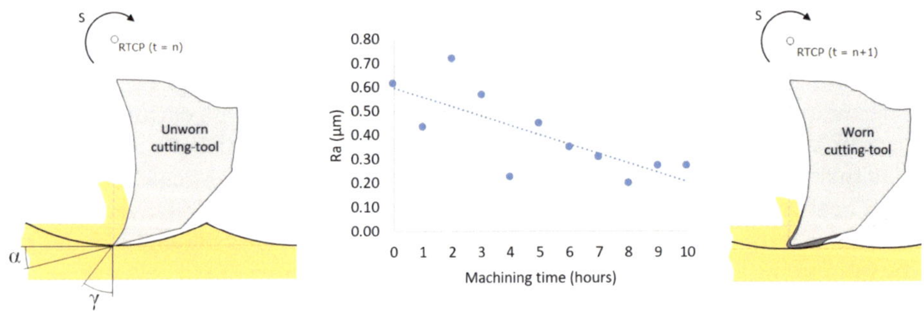

Figure 6.
Roughness average evolution during the milling of an AA2024-T3 alloy.

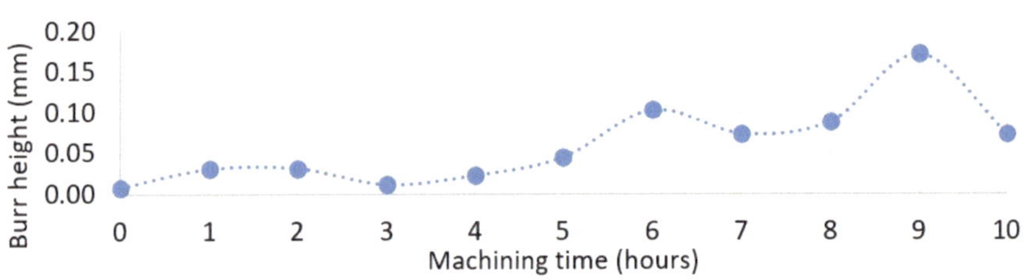

Figure 7.
Burr height evolution during the milling of an AA2024-T3 alloy.

cutting-tools, with both flat and toroidal tips, work much better laterally, with axial depth of cut between (50–150% diameter). However, these depths of cut also depend on the conditions described in the previous paragraph [45].

3.2 Advanced milling techniques

Nowadays there is a great interest in the machine intelligence. In this sense, different monitoring solutions are available to control the process. This measure allows to control the system state and with an adaptive control system, auto-regulate the cutting parameters.

For instance, an increase of the part vibration can be detected through the cutting forces or the acoustic emission frequency analysis will lead to a decrease of the surface quality, if it is detected on time and there is a model governing the case, the correct parameters can be changed so the situation is inverted [46].

Similarly, distance sensors are used to act on the depth of cut, so when the distance recorded is not in the expected range the system automatically modifies the depth of cut improving the dimensional accuracy [47, 48].

Finally, the cutting power and the cutting force signals can provide information about the tool wear or the process temperature improving the dimensional accuracy. It should also be considered that the variation on the cutting parameters can extend tool life slightly decreasing the material removal rate but increasing the process efficiency.

4. Turning of aerospace aluminum alloys

Despite turning processes are mainly applied in aerospace to critical elements, such as connecting bolts for gates and actuators built of titanium alloys, several non-critical elements are made of aluminum, as shafts, fasteners and spacers. These parts are also evaluated in terms of roughness, roundness, parallelism deviation and residual stresses to define their in-service behavior, but their requirements are less stringent than those for critical structural parts.

Turning is the simplest machining process, so its use is also essential to obtain preliminary results that may give an initial approximation to more complex processes like drilling and milling. That's why, this operation is commonly used to define the machinability of the alloys as well as the tool wear behavior in orthogonal or oblique cutting configurations, therefore its importance.

Several studies about the turning of aluminum alloys identify the importance of cutting parameters on the micro geometrical properties of the generated surfaces, evaluating them in terms of roughness average (Ra) [49]. Few of them correlate the residual stresses and machining of aluminum alloys [50], but the induced residual stresses are in all cases compressive not having a negative effect on the part. Finally, the roundness is usually measured from the parallelism deviation, since it is a relevant feature that can affect the in-service behavior.

4.1 Influence of the turning parameters

Turning is the most suitable machining process to create revolution surfaces by using a cutting-tool. This operation has two main movements to set the dimensions, one along the Z axis of the stock (F) and another, along the X axis, where the depth of cut (d) is set. At the same time, the tangential cutting speed is produced by the rotation of the part (S) cutting the part. These three actions are represented in **Figure 8**.

Regarding the cutting parameters, they are defined by the same expressions that govern drilling and milling operations, but in this case, D is the diameter of the cylindrical part. Turning cutting-tools work with just one cutting-edge, simplifying Eq. (2) in Eq. (3), where f is the feed per revolution:

$$F = f \cdot S \tag{3}$$

These parameters have a direct impact on the micro and macro geometrical deviations as well as on the tool wear. Generally, better Ra results are achieved when low feed-rates and high cutting speeds are applied during short machining times (**Figure 9**).

Ra is also affected by the machining time (**Figure 10**). It gradually decreases due to the adhesive wear [51]. This mechanism modifies the initial tool geometry, due to the material adhered to the rake and clearance faces, that reduces the edge position angle [52]. This fact decreases the height of the peaks created in each step, producing a smoother surface, in a similar way that was exposed for milling.

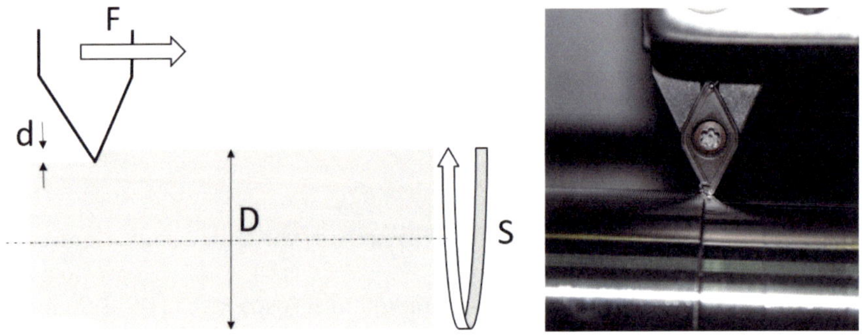

Figure 8.
Turning scheme and image of a turning process.

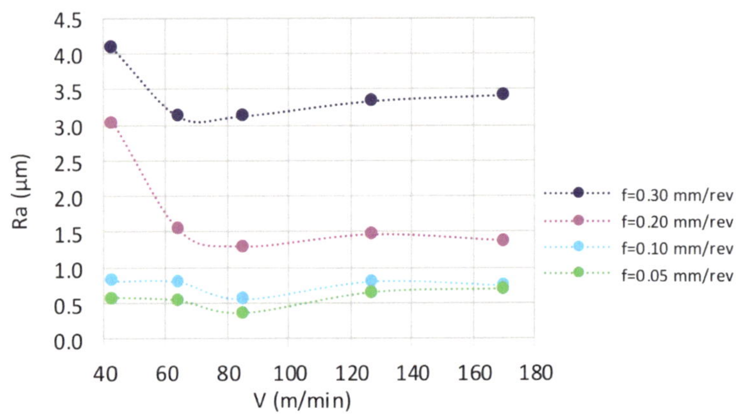

Figure 9.
Ra as a function of the cutting speed in the dry turning of AA2024-T3, depth of cut 1 mm and machining time 10 s.

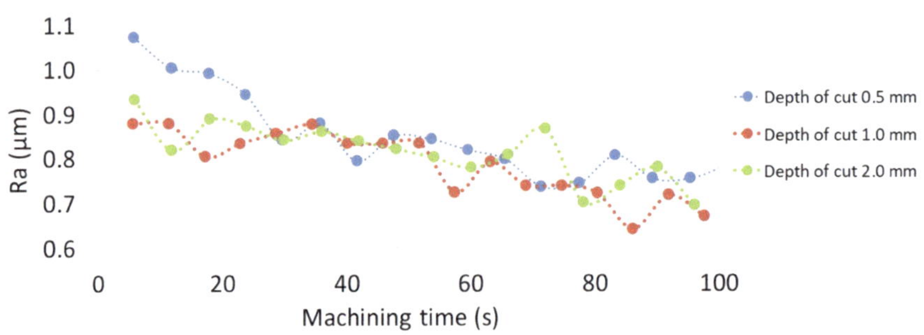

Figure 10.
Roughness average evolution in the turning of AA2024 as a function of the depth of cut.

Shape quality criteria for turning parts include roundness and parallelism deviations (*PD*). However, *PD* is easier to measure so its analysis is more usual [53, 54]. This criterion is also affected by the cutting parameters as is shown in **Figure 11** [55]. High cutting speeds achieve higher precision, while the feed rate has a combined effect in the deviations. The alloy used affects the machine dynamics, for instance the UTS of AA7475 (531 MPa) compared to AA2024 UTS (440 MPa) increase the deviation. For this reason, parametric surfaces are used to find the minimum *PD* by selecting the better cutting parameters.

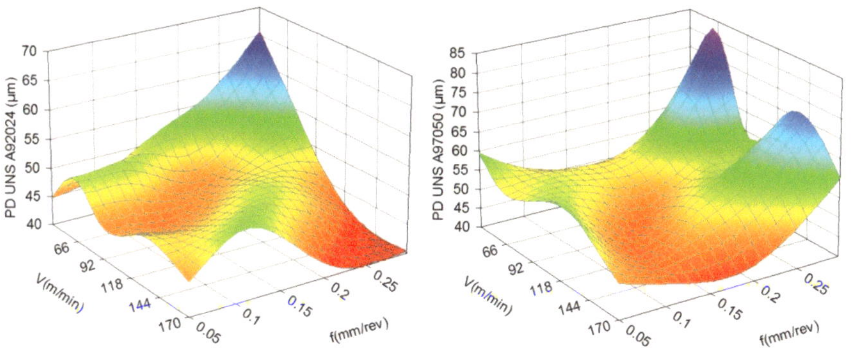

Figure 11.
Parametric surface PD (f,v) for AA2024 (left) and AA7050 (right).

Finally, as in the previous machining processes, the current trend in the machining of Al alloys is to minimize or eliminate the environmental impact reducing or avoiding the use of cutting fluids (dry turning) [56–58].

However, both turning and dry turning can have a negative impact on the in-service behavior of the parts or components manufactured, by reducing the functional performance of the process through the loss of the quality or the surface integrity [51, 56, 59]. Dry machining also has an impact on the wear behavior affecting the macro geometrical properties of the machined elements, in terms of dimensions or shape tolerances.

5. Cutting-tool wear in the machining of aluminum alloys

In the previous sections, the main machining techniques applied in the machining of aluminum aerospace alloys have been explained. For each machining process, it was indicated the influence of cutting-tool wear in the quality features of the machined parts. For this reason, it is relevant to explain the wear mechanisms that take place in the machining of aluminum alloys.

When the cutting-tool penetrates the part, it causes a compressive plastic deformation, which intensity can exceed the bond energy in certain planes, leading to shearing or sliding elements along planes. At the same time, the elastic recovery of the chip and the tribological interaction between the part-chip-tool provoke an exchange of heat that may thermally affect the tool properties. This change of properties or tool wear may be produced by different wear mechanism but all of them lead to possible changes in the cutting forces or in the dynamic stability of the process modifying, as a consequence, the properties of the surface generated [60–63].

The most common wear mechanism of aluminum alloys machining is secondary adhesion. This phenomenon takes place due to the temperature achieved in the process, the thermal conductivity (between 120 and 165 W/m°C) of the part-tool combination and the selected cutting speed. This mechanism, as well as the temperature and parameters associated, has been deeply studied for ferrous materials. Nevertheless, these studies are not directly applicable for softer materials such as aluminum. The high plasticity of this material favor chipping or notch wear for low cutting speeds [61, 62, 64].

Secondary adhesion appears divided in to two well-located phenomena, the Built-Up Edge (BUE) is located close to the cutting edge and the Built-Up Layer (BUL) is placed on the rake face [60, 65], as it is shown in **Figure 12**.

This adhesion process appears in different steps, which are represented in **Figure 13**. At the beginning of the machining process, a layer of material is adhered on the rake face of the cutting-tool, creating a BUL due to a mechanic-thermic effect of the cutting mechanism. Once it is formed, the cutting-tool geometry changes, promoting the growth of the adhered material over the cutting edge (BUE), which grows up to a critical thickness. Once this critical thickness is reached, BUE is mechanically extruded along the rake face, increasing the thickness of BUL and forming an adhered multilayer of material [66, 67].

Both the BUL and the BUE can disappear, be detached and be rebuilt, causing the gradually breaking of cutting-tool particles that are removed by the chip flow. This is therefore, a dynamic mechanism with successive layers of chip material welded and hardened. This cyclic behavior may change a gradual wear into a full weakening and even into the complete fracture of the tool [66, 67]. **Figure 14** shows the previous instants to the detachment of the adhered material, enriched with cutting-tool (WC-Co) elements in the machining of AA2024 alloy. This fact may also be favored by a weak edge or other types of tool wear, such as abrasion and diffusion.

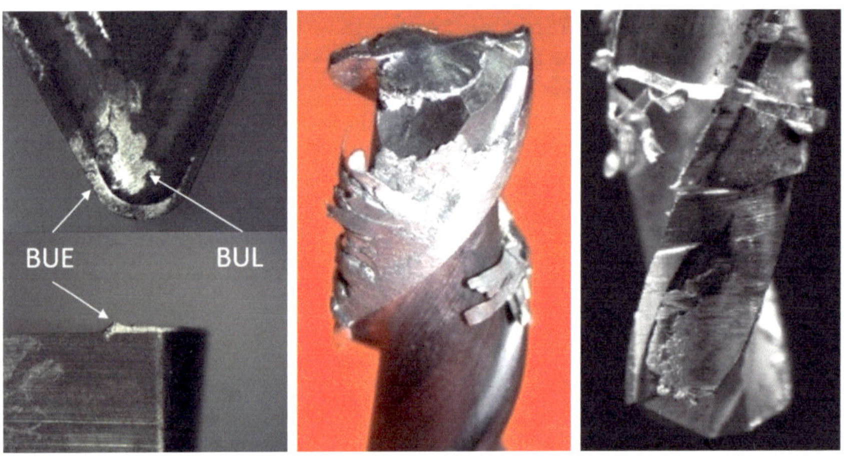

Figure 12.
Secondary adhesion in turning insert, milling and drilling cutting tools.

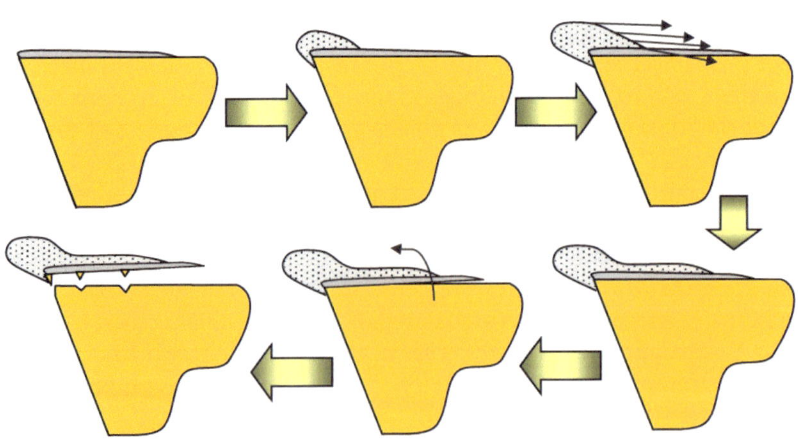

Figure 13.
Scheme of the secondary adhesion mechanism.

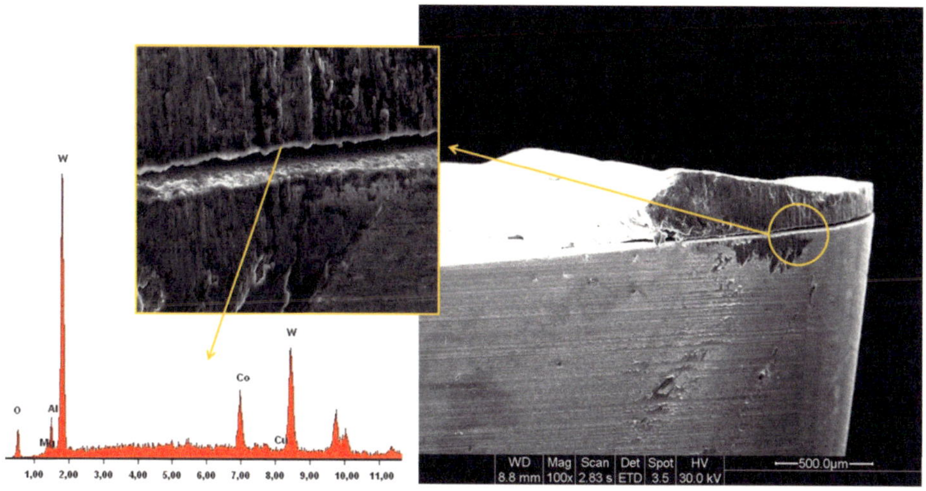

Figure 14.
Detachment of adhered material in the dry turning of AA2024.

If the temperature achieved is low the adhesion is not very significant, no matter if the chips are long or short. Otherwise, when a critical temperature is reached, other types of wear mechanisms such as diffusion may appear, increasing the synergy effect previously described [68, 69].

6. General remarks

The use of aluminum alloys, mostly forged, remains essential to build aircrafts. Al-Cu and Al-Zn are the most used alloys due to their excellent physicochemical-cost ratio properties. They come as raw material as sheets, blocks or cylinders that have to be drilled, milled or turned, in order to give them a final geometry.

Drilling, milling and turning are complex machining processes whose fundaments are based on the theory of metal cutting. The drilling process is fundamental in the manufacture of aircraft for the assembly of the structures using rivets. For specific applications with high quality requirements, OD and VAS techniques are used. Milling produces light components with a high accurate dimension, being mainly applied to monolithic parts that present deflection, overcuts, residual stresses and part deformation issues if the parameters are not properly selected. Turning generates surfaces of revolution, used to manufacture non-critical elements as shafts, fasteners and spacers.

The parameters that governs the machining processes, mainly cutting speed and feed rate, are highly related to the quality features usually required in aeronautics, surface quality, burr formation, macro geometrical deviations, form errors, etc. Generally, feed rate increase cutting forces and roughness while the effect of the cutting speed is related to thermal phenomenon and its influence depends on the machining regime. Feed rate selection usually comes with an agreement between different quality criteria and the process efficiency and high cutting speeds are the best choice whenever they are possible. Finally, both affects the tool wear produce by the secondary adhesion mechanism creating a BUL/BUE, which affects the macro and micro geometrical deviation. However, these effects can be diminished in different ways, using advanced tool coatings or projecting harmful cutting fluids in the cutting zone. In more advanced systems, machine intelligence is commonly used to look for adaptive control responses that auto-regulate the cutting parameters after the measure of the system state.

Conflict of interest

The authors declare no conflict of interest.

Author details

Jorge Salguero*, Irene Del Sol, Alvaro Gomez-Parra and Moises Batista
Department of Mechanical Engineering and Industrial Design, University of Cadiz, Spain

*Address all correspondence to: jorge.salguero@uca.es

References

[1] Kalpakjian S, Schmid S. Manufacturing Engineering and Technology. 7th ed. New York: Pearson Prentice Hall; 2014. p. 1167

[2] Schmitz C. Handbook of Aluminium Recycling. 1st ed. Vulkan-Verlag: Essen; 2006. p. 1886

[3] Constellium. About Aluminium [Internet]. 2020. Available from: https://www.constellium.com/about-constellium/about-aluminium [Accessed: 21 May 2020]

[4] Askeland DR, Fulay PP, Wright WJ. The Science and Engineering of Materials. 6th ed. Stamford: Cengage Learning; 2010. p. 923

[5] ASM International. ASM Handbook: Volume 2, Properties and Selection, Nonferrous Alloys and Special-Purpose Materials. 1st ed. ASM International: Ohio; 1990. p. 1328

[6] Dashwood RJ, Grimes R. Structural materials: Aluminum and its alloys – Properties. In: Blockley R, Shyy W, editors. Encyclopedia of Aerospace Engineering. 1st ed. Chichester: Wiley; 2010. pp. 1-12. DOI: 10.1002/9780470686652.eae195

[7] Hinrichsen J, Bautista C. The challenge of reducing both airframe weight and manufacturing cost. Air & Space Europe. 2001;**3**(3-4):119-124. DOI: 10.1016/S1290-0958(01)90072-3

[8] Santos MC, Machado AR, Sales WF, Barrozo MAS, Ezugwu EO. Machining of aluminum alloys: A review. The International Journal of Advanced Manufacturing Technology. 2016;**86**:3067-3080. DOI: 10.1007/s00170-016-8431-9

[9] Cantor B, Assender H, Grant P. Aerospace Materials. 1st ed. Philadelphia: CRC Press; 2001. p. 312

[10] Taylor FW. On the Art of Cutting Metals. 1st ed. New York: American Society of Mechanical Engineers; 1907. p. 248

[11] Barz A, Buer T, Haasis HD. A study on the effects of additive manufacturing on the structure of supply networks. In: Proceedings of the 7th IFAC Conference on Management and Control of Production and Logistics (MCPL); 22-24 February 2016. Bremen. Germany: Elsevier; 2016. pp. 72-77

[12] Sharman ARC, Amarasinghe A, Ridgway K. Tool life and surface integrity aspects when drilling and hole making in Inconel 718. Journal of Materials Processing Technology. 2008;**200**(1-3):424-432. DOI: 10.1016/j.jmatprotec.2007.08.080

[13] Ucun I. 3D finite element modelling of drilling process of Al7075-T6 alloy and experimental validation. Journal of Mechanical Science and Technology. 2016;**30**(4):1843-1850. DOI: 10.1007/s12206-016-0341-0

[14] Nouari M, List G, Girot F, Gehin D. Effect of machining parameters and coating on wear mechanisms in dry drilling of aluminium alloys. International Journal of Machine Tools and Manufacture. 2005;**45**(12-13):1436-1442. DOI: 10.1016/j.ijmachtools.2005.01.026

[15] Bono M, Ni J. The location of the maximum temperature on the cutting edges of a drill. International Journal of Machine Tools and Manufacture. 2006;**46**(7-8):901-907. DOI: 10.1016/j.ijmachtools.2005.04.020

[16] Ralph WC, Johnson WS, Toivonen P, Makeev A, Newman JC. Effect of various aircraft production drilling procedures on hole quality. International Journal of Fatigue. 2006;**28**(8):943-950. DOI: 10.1016/j.ijfatigue.2005.09.009

[17] Yaşar N. Thrust force modelling and surface roughness optimization in drilling of AA-7075: FEM and GRA. Journal of Mechanical Science and Technology. 2019;**33**:4771-4781. DOI: 10.1007/s12206-019-0918-5

[18] Ralph WC, Johnson WS, Makeev A, Newman JC. Fatigue performance of production-quality aircraft fastener holes. International Journal of Fatigue. 2007;**29**(7):1319-1327. DOI: 10.1016/j.ijfatigue.2006.10.016

[19] Giasin K, Hodzic A, Phadnis V, Ayvar-Soberanis S. Assessment of cutting forces and hole quality in drilling Al2024 aluminium alloy: Experimental and finite element study. The International Journal of Advanced Manufacturing Technology. 2016;**87**:2041-2061. DOI: 10.1007/s00170-016-8563-y

[20] Nouari M, List G, Girot F, Coupard D. Experimental analysis and optimisation of tool wear in dry machining of aluminium alloys. Wear. 2003;**255**(7-12):1359-1368. DOI: 10.1016/S0043-1648(03)00105-4

[21] Sheth S, George PM. Experimental investigation, prediction and optimization of cylindricity and perpendicularity during drilling of WCB material using grey relational analysis. Precision Engineering. 2016;**45**:33-43. DOI: 10.1016/j.precisioneng.2016.01.002

[22] Bu Y, Liao WH, Tian W, Shen JX, Hu J. An analytical model for exit burrs in drilling of aluminum materials. The International Journal of Advanced Manufacturing Technology. 2016;**85**:2783-2796. DOI: 10.1007/s00170-015-8125-8

[23] Whinnem E, Lipczynski G, Eriksson I. Development of orbital drilling for the Boeing 787. International Journal of Aerospace. 2009;**1**(1):811-816. DOI: 10.4271/2008-01-2317

[24] Qiang F, Zemin P, Shaohua F, Xiangnan X, Yinglin K. A novel helical milling end-effector and its application. IEEE/ASME Transactions on Mechatronics. 2015;**20**(6):3112-3122. DOI: 10.1109/TMECH.2015.2409986

[25] Sadek A, Meshreki M, Attia MH. Characterization and optimization of orbital drilling of woven carbon fiber reinforced epoxy laminates. CIRP Annals. 2012;**61**(1):123-126. DOI: 10.1016/j.cirp.2012.03.089

[26] Amini S, Paktinat H, Barani A, Tehran AF. Vibration drilling of Al2024-T6. Materials and Manufacturing Processes. 2013;**28**(4):476-480. DOI: 10.1080/10426914.2012.736659

[27] Barani A, Amini S, Paktinat H, Tehrani AF. Built-up edge investigation in vibration drilling of Al2024-T6. Ultrasonics. 2014;**54**(5):1300-1310. DOI: 10.1016/j.ultras.2014.01.003

[28] Herranz S, Campa FJ, López de Lacalle LN, Rivero A, Lamikiz A, Ukar E, et al. The milling of airframe components with low rigidity: A general approach to avoid static and dynamic problems. Proceedings of the Institution of Mechanical Engineers, Part B. 2005;**219**:789-801. DOI: 10.1243/095440505x32742

[29] Del Sol I, Rivero A, Gamez AJ. Effects of machining parameters on the quality in machining of aluminium alloys thin plates. Metals. 2019;**9**(9):1-11. DOI: 10.3390/met9090927

[30] Campbell FC. Manufacturing Technology for Aerospace Structural Materials. Amsterdam: Elsevier Science; 2006. p. 616. DOI: 10.1016/B978-1-85617-495-4.X5000-8

[31] Ratchev S, Liu S, Huang W, Becker AA. A flexible force model for end milling of low-rigidity parts.

Journal of Materials Processing Technology. 2004;**153-154**:134-138. DOI: 10.1016/j.jmatprotec.2004.04.300

[32] Zhang J, Lin B, Fei J, Huang T, Xiao J, Zhang X, et al. Modelling and experimental validation for surface error caused by axial cutting force in end-milling process. The International Journal of Advanced Manufacturing Technology. 2018;**99**:327-335. DOI: 10.1007/s00170-018-2468-x

[33] Elbestawi MA, Sagherian R. Dynamic modelling for the prediction of surface errors in the milling of thin-walled sections. Journal of Materials Processing Technology. 1991;**25**:215-228. DOI: 10.1016/0924-0136(91)90090-2

[34] Wan M, Zhang WH. Calculations of chip thickness and cutting forces in flexible end milling. The International Journal of Advanced Manufacturing Technology. 2006;**29**:637-647. DOI: 10.1007/s00170-005-2572-6

[35] Yan Q, Luo M, Tang K. Multi-axis variable depth-of-cut machining of thin-walled workpieces based on the workpiece deflection constraint. Computer-Aided Design. 2018;**100**: 14-29. DOI: 10.1016/j.cad.2018.02.007

[36] Aijun T, Zhanqiang L. Deformations of thin-walled plate due to static end milling force. Journal of Materials Processing Technology. 2008;**206**:345-351. DOI: 10.1016/j.jmatprotec.2007.12.089

[37] Del Sol I, Rivero A, Lopez de Lacalle LN, Gamez AJ. Thin-wall machining of light alloys: A review of models and industrial approaches. Materials. 2019;**12**(12):1-28. DOI: 10.3390/ma12122012

[38] Yang Y, Xu D, Liu Q. Milling vibration attenuation by eddy current damping. International Journal of Advanced Manufacturing Technology. 2015;**81**(1):445-454. DOI: 10.1007/s00170-015-7239-3

[39] Meshreki M, Attia H, Kövecses J. Development of a new model for the varying dynamics of flexible pocket-structures during machining. Journal of Manufacturing Science and Engineering. 2011;**133**(4):041002. DOI: 10.1115/1.4004322

[40] Jiang X, Lu W, Zhang Z. An approach for improving the machining efficiency and quality of aerospace curved thin-walled parts during five-axis NC machining. The International Journal of Advanced Manufacturing Technology. 2018;**97**:2477-2488. DOI: 10.1007/s00170-018-2129-0

[41] Wu Q, Li DP, Zhang YD. Detecting milling deformation in 7075 aluminum alloy aeronautical monolithic components using the quasi-symmetric machining method. Metals. 2016;**6**(4): 1-14. DOI: 10.3390/met6040080

[42] Ning H, Zhigang W, Chengyu J, Bing Z. Finite element method analysis and control stratagem for machining deformation of thin-walled components. Journal of Materials Processing Technology. 2003;**139**:332-336. DOI: 10.1016/S0924-0136(03)00550-8

[43] Zuperl U, Cus F. Optimization of cutting conditions during cutting by using neural networks. Computer-Integrated Manufacturing. 2003;**19**(1-2):189-199. DOI: 10.1016/S0736-5845(02)00079-0

[44] Cus F, Milfelner M, Bali J. Determination of cutting forces in ball-end nulling with neural networks. In: Proceedings of the 11th International Scientific Conference Achievements in Mechanical & Materials Engineering (AMME). Poland: Polish Academy of Sciences; 2002. pp. 59-62

[45] Interempresas Metalmecanica. Mecanizado de alta velocidad [Internet]. Available from: http://www.interempresas.net/MetalMecanica/Articulos/26127-Area-tematica-Mecanizado-de-alta-velocidad.html [Accessed: 18 May 2020]

[46] Muhammad BB, Wan M, Liu Y, Yuan H. Active damping of milling vibration using operational amplifier circuit. Chinese Journal of Mechanical Engineering. 2018;**1**(31-90):1-8. DOI: 10.1186/s10033-018-0291-9

[47] Wang G, Li W, Tong G, Pang C. Improving the machining accuracy of thin-walled parts by online measuring and allowance compensation. The International Journal of Advanced Manufacturing Technology. 2017;**92**:2755-2763. DOI: 10.1007/s00170-017-0358-2

[48] Rubio A, Rivero A, Del Sol I, Ukar E, Lamikiz A. Capacitation of flexibles fixtures for its use in high quality machining processes: An application case of the industry 4.0 paradigm. DYNA. 2018;**93**:608-612. DOI: 10.6036/8824

[49] Sebastian MA, Sanchez-Sola JM, Carrilero MS, Gonzalez JM, Alvarez M, Marcos M. Parametric model for predicting surface finish of machined UNS A92024 alloy bars. Journal for Manufacturing Science and Production. 2002;**4**:181-188. DOI: 10.1515/IJMSP.2002.4.4.181

[50] Belgasim O, El-Axir MH. Modeling of residual stresses induced in machining aluminum magnesium alloy (Al-3Mg). In: Proceedings of the World Congress on Engineering (WCE 2010); June 30–July 2 2010. London: International Association of Engineers (IAEng); 2010. pp. 1-6

[51] Gomez-Parra A, Alvarez-Alcon M, Salguero J, Batista M, Marcos M. Analysis of the evolution of the built-up edge and built-up layer formation mechanisms in the dry turning of aeronautical aluminium alloys. Wear. 2013;**302**:1209-1218

[52] Trujillo FJ, Sevilla L, Salguero J, Batista M, Marcos M. Parametric potential model for determining the microgeometrical deviations of horizontally dry-turned UNS A97075 (Al-Zn) alloy. Advanced Science Letters. 2013;**19**(3):731-735. DOI: 10.1166/asl.2013.4818

[53] Martin S, Trujillo FJ, Bermudo C, Sevilla L. Fatigue behavior parametric analysis of dry machined UNS A97075 aluminum alloy. Meta. 2020;**10**(631):1-22. DOI: 10.3390/met10050631

[54] Marcos M, Sebastian MA, Contreras JP, Sanchez-Carrilero M, Sanchez M, Sanchez-Sola JM. Study of roundness on cylindrical bars turned of aluminium–copper alloys UNS A92024. Journal of Materials Processing Technology. 2005;**162-163**:644-648. DOI: 10.1016/j.jmatprotec.2005.02.061

[55] Salguero J, Sanchez-Sola JM, Gómez A, Rubio E, Batista E, Sebastian MA, et al. An exponential parametric model for evaluating parallelism deviation in the dry turning of aerospace alloys. In: Proceedings of the 14 International Conference on Advances in Materials & Processing Technologies (AMPT 2011); 13-16 July 2011. Turkey. Istambul: AMTP; 2011. pp. 1-6

[56] Rubio EM, Camacho AM, Sanchez-Sola JM, Marcos M. Surface roughness of AA7050 alloy turned bars: Analysis of the influence of the length of machining. Journal of Materials Processing Technology. 2005;**162-163**:682-689. DOI: 10.1016/j.jmatprotec.2005.02.096

[57] Salguero J, Batista M, Carrilero MS, Alvarez M, Marcos M. Sustainable

manufacturing in aerospace industry. Analysis of the viability of intermediate stages elimination in sheet processing. Advanced Materials Research. 2010;**107**:9-14. DOI: 10.4028/www.scientific.net/AMR.107.9

[58] Marksberry PW, Jawahir IS. A comprehensive tool-wear/tool-life performance model in the evaluation of NDM (near dry machining) for sustainable manufacturing. International Journal of Machine Tools and Manufacture. 2008;**48**:878-886. DOI: 10.1016/j.ijmachtools.2007.11.006

[59] Carrilero MS, Bienvenido R, Sanchez JM, Alvarez M, Gonzalez A, Marcos M. A SEM and EDS insight into the BUL and BUE differences in the turning processes of AA2024 Al–Cu alloy. International Journal of Machine Tools and Manufacture. 2002;**42**:215-220. DOI: 10.1016/S0890-6955(01)00112-2

[60] Trent EM, Wright PK. Metal Cutting. 4th ed. Woburn: Elsevier; 2000. p. 464. DOI: 10.1016/B978-0-7506-7069-2.X5000-1

[61] ASM Metals Handbook Volume 16: Machining. Ohio: ASM International; 1990. p. 944

[62] Jeelani S, Musial S. Effect of cutting speed and tool rake angle on the fatigue life of 2024-T351 aluminium alloy. International Journal of Fatigue. 1984;**6-3**:169-172. DOI: 10.1016/0142-1123(84)90034-3

[63] Batista M, Salguero J, Gomez-Parra A, Carrilero MS, Alvarez M, Marcos M. Identification, analysis and evolution of the mechanisms of wear for secondary adhesion for dry turning processes of Al-Cu alloys. Advanced Materials Research. 2010;**107**(1):141-146. DOI: 10.4028/www.scientific.net/AMR.107.141

[64] Stephenson DA, Agapiou JS. Metal Cutting Theory and Practice. 3rd ed. Boca Raton (USA): CRC Press; 2006

[65] Klocke F, Eisenblätter G, Krieg T. Machining: Wear of tools. In: Jürgen Buschow KH, Cahn RW, Flemings MC, Ilschner B, Kramer EJ, Mahajan S, Veyssière P, editors. Encyclopedia of Materials: Science and Technology. 2nd ed. Amsterdam: Elsevier; 2001. pp. 4708-7711. DOI: 10.1016/B0-08-043152-6/00821-4

[66] Batista M, Del Sol I, Fernandez-Vidal SR, Salguero J. Experimental parametric model for adhesion wear measurements in the dry turning of an AA2024 alloy. Materials. 2018;**11**(9):1-16. DOI: 10.3390/ma11091598

[67] Batista M, Del Sol I, Gomez-Parra A, Ramirez-Peña M, Salguero J. Study of the tool wear process in the dry turning of Al–Cu alloy. Metals. 2019;**9**(10):1-12. DOI: 10.3390/met9101094

[68] Girot F, Calamaz M, List G, Coupard D, Gehin D, Gutierrez-Orrantia ME. Towards dry machining of aluminum alloys and titanium alloys. In: 3rd Manufacturing Engineering Society International Conference (MESIC 2009); 17-19 June 2009. Alcoy. Madrid: SIF; 2009. pp. 1-15

[69] Santos MC, Machado AR, Barrozo MAS. Temperature in machining of aluminum alloys. In: Stanimirovic I, Stanimirovic Z, editors. Temperature sensing. London: Intech open; 2018. pp. 71-89. DOI: 10.5772/intechopen.75943

Chapter 7

Analysis of Surface Roughness of EN AW 2024 and EN AW 2030 Alloys after Micromachining

Francisco Mata and Issam Hanafi

Abstract

Micromachining is the most suitable technology for the production of very small components (micro-components) in the industry. It is a high-precision manufacturing process with applications in various industrial sectors, including machine building. This chapter presents the experimental study of the roughness (Ra and Rt) of aluminum alloys using a specific micro-turning process. The roughness measurements carried out show how it is possible to achieve very good surface qualities up to 0.05 mm diameter. For lower diameters, the surface quality worsens and the shape defects increase (conicity) due to the very low rigidity of the workpiece, which makes it very sensitive when passing through the forming process. The fundamental objective of this research is to analyze the surface quality of the finishes obtained in these micromachining processes and to evaluate their suitability to the specifications required by the mechanical industry (roughness, presence of burrs, shape and geometry, etc.). Predictive roughness models are proposed, with a good degree of approximation, to help characterize micromachining processes.

Keywords: micromachines, micro-turning, miniaturization, models

1. Introduction

The traditional machine concept, consisting of large mechanical elements with high energy consumption, low precision, and little degree of automation, has given way to much more complex realities in the integration of technologies (mechanical, pneumatic, electronic, etc.), in the reduction of sizes and consumption, and of course, in the possibility of performing high-precision operations with exhaustive controls. In this context, it is possible to speak of micromachines, which are small machines to develop operations of precision and are integrated by mechanisms and components of very small size. Consider, for example, the machinery of a watch or the injection systems used in the automotive industry. It is therefore necessary to manufacture components of very small dimensions, that is, micro-components for applications in the mechanical, biomedical, automotive, or mechatronic industries, among others.

The development of new materials with high performance, their study at micro- and nanotechnological scale, and the continuous attempt to achieve as compact designs as possible support the need to advance manufacturing processes on this scale [1]. The miniaturization of components has become a specialty within the design and micro-manufacturing, a discipline also differentiated, involving machines, tools, and measuring equipment that must necessarily be adapted to the particularities of this type of conformation. This implies a growing demand for the integration needs of microscale components manufactured using different materials, including metals and their alloys [2–5].

Micro-fabrication can be carried out using adapted traditional shaping techniques (microinjection), by chip removal in machine tools (using tools of appropriate geometry), or using other advanced conformation techniques (laser, ultrasound, photochemical forming, thermal diffusion, electric discharge, etc.) [5]. Machining meso (<10 mm) and micro (<1 mm) are the most suitable technologies for the production of very small components in the industry [1–7], especially when high surface finishing and dimensional tolerances are required. These are high-precision manufacturing processes with applications in leading sectors such as the following [7, 8]:

- Medical industry: Production of small components that can be used as parts of bone and dental prostheses (implants and micro-implants) or as essential devices for performing surgical operations.

- Electronics industry: Instrumentation and control equipment (sensors, actuators, etc.), telephony components, etc.

- Aerospace industry: Flow control elements, control devices, etc.

- Automotive industry: Small actuators, fuel injection nozzles and injectors, micro-valves and other small components (screws, pins, bearing needles, micro-axes, micro-wheels, etc.), and microsensors for pressure signal collection, temperature, proximity, angular speed, or acceleration. Indeed, the evolution of automobiles in recent decades toward the integration of mechanical and electronic technologies (mechatronics) demands components of very small dimensions for stability control systems, electronic injection, air bag, etc.

- Watchmaking: Micro-mechanisms (wheels, etc.).

- Micromachines: Integrated by microsensors, micro-motors, micro-gears, micro-actuators, etc.

In elements and microelements that have to integrate mechanisms, where there must be contact and relative movement between them, the surface finish becomes critical in addition to the precise dimensional adjustments. The small size of some of these elements and their peculiar shape, together with the specification of a good surface finish that prevents premature wear, requires opting for micro-conforming processes such as turning or milling, essentially. It should be added that the machining of micro-workpieces does not necessarily involve the use of small-volume machine tools ("micro-lathes" or "micro-milling machines"); micromachining operations can be developed in the conventional CNC machines and therefore it is important to carry out experimental studies leading to the characterization of these processes (establishment of limits to ensure the integrity and quality of machined workpieces, definition of appropriate functional cutting parameters, study of the geometry of cutting tools, determination of achievable surface quality levels, etc.)

The study of surfaces is a technique for characterizing materials, which is very useful in practice. Surface roughness is a parameter that has a great influence on the behavior and functionality of mechanical components and on production costs [9, 10], constituting an important quality control variable. Roughness is critical in mechanical contacts in addition to other fields such as fluid circulation or biomedical applications. The surface roughness obtained during the machining process is affected by the cutting parameters, tool wear, and material hardness. To achieve the desired roughness, it is necessary to know the mechanisms of the cutting and detachment of the material and the kinetics of machining processes, which affect the performance of the cutting tools [11].

In most applications of the machined micro-workpieces, high quality is required on shaped surfaces, including dimensional accuracy and surface integrity. For this reason, it is necessary to carry out various investigations with the aim of optimizing the cutting parameters in order to obtain a certain roughness [9, 12] and tolerable formal characteristics (cylindricity).

Conventional micromachining is a flexible approach that can use any material that can be machined [13]. However, it has some restrictions on the development of components [14]. It is therefore necessary to investigate, to develop models that allow to optimize the processes, and to improve the quality of the products and to lower the production costs.

In order to reduce the number of tests required for a complete characterization of micromachining processes, statistical nonlinear regression techniques, numerical strategies based on neural networks or other similar techniques can be used [15–19], which will allow us to establish prediction models and help us to better understand cutting mechanisms. Based on these models, the experimental program can be completed in order to know the particularities of the micromachining processes and to define the corresponding physical cutting models.

In recent times, due to environmental requirements, these processes are carried out in the absence of cutting fluids [20]. However, the total suppression of fluids results in very aggressive process conditions [21–23]. Each alloy has its own characteristics of machinability, which will mark the operational limits of the process. Although conventional machining processes for metals and their alloys are therefore well known, it is not possible to make a direct extrapolation and anticipate what the behavior will be in the event of chip removal operations on very small workpieces. In these cases, the geometry of the tool and the stability of the workpiece (lean:stiff ratio) can significantly influence the results and differentiate the behavior pattern from the conventional machining of standard size workpieces.

The main objective of this research is to analyze the surface quality of the finishes obtained in these micromachining processes and to evaluate their suitability to the specifications required in micro-components of different devices and machines (roughness, presence of burrs, shape and geometry, etc.). Nonlinear models are proposed in this study that can help in the characterization of micromachining processes, depending on the diameter required for a given application as well as the necessary surface quality.

2. Materials

Aluminum (Al) has a combination of properties that make it very useful in mechanical engineering, such as its low density (2700 kg/m3) and its high resistance to corrosion. Its mechanical strength (up to 690 Mpa) can be significantly increased by suitable alloys. Aluminum alloys are a viable alternative in improving flexibility and competitiveness. The current trend is its gradual incorporation into the definition of industrial cycles incorporating high-speed machines, advanced CNC, and specific aluminum alloys. The intrinsic characteristics of aluminum alloys favor high-speed machining with feed and cut speeds much higher than those achieved with ferrous alloys.

	Cu	Mn	Zn	Mg	Fe	Cr	Si	Ti
EN AW 2024	4.15	0.65	0.5	0.69	0.7	0.1	0.45	0.2
EN AW 2030	4.5	1	0.5	1.3	0.7	0.1	0.8	0.2

Table 1.
Chemical composition of tested aluminum alloys.

	EN AW 2024	EN AW 2030
Designation (ISO)	AlCu4Mg1	AlCu4PbMg
Density	2.77	2.82
Modulus of elasticity in N/mm^2	72.4×10^3	73.6×10^3
Mean coefficient of expansion in m/m.°C	22.9×10^{-6}	23×10^{-6}
Thermal conductivity in $W.m/m^2.°C$	120	190

Table 2.
Mechanical properties of tested aluminum alloys.

As far as possible, metal carbide tools should be used for turning, as they offer higher productivity and a longer service life.

This type of alloy has applications in dental prostheses, micro-valves, actuators and other instrumentation components, injectors of different motors, precision mechanisms, and in general, components of micromachines.

Commercial EN AW 2024 and EN AW 2030 aluminum alloys are used for the experiments. Chemical composition and some mechanical properties of tested aluminum alloys are given in **Tables 1** and **2**.

3. Experimental procedure

The micro-turning process has been carried out on an Eclipse CNC Lathe, with a power of 1.5 Kw and 4000 rpm maximum rotation speed of the head (**Figure 1**).

To measure roughness (mean roughness: Ra, and maximum roughness: Rt), given the very small diameter of the machined workpieces, a Talysurf CLI roughness meter (**Figure 2**) has been used for topographic exploration, using an inductive contact sensor or noncontact laser meter. A conventional optical microscope (**Figure 2**) has been used for the observation of conicity and cylindricity.

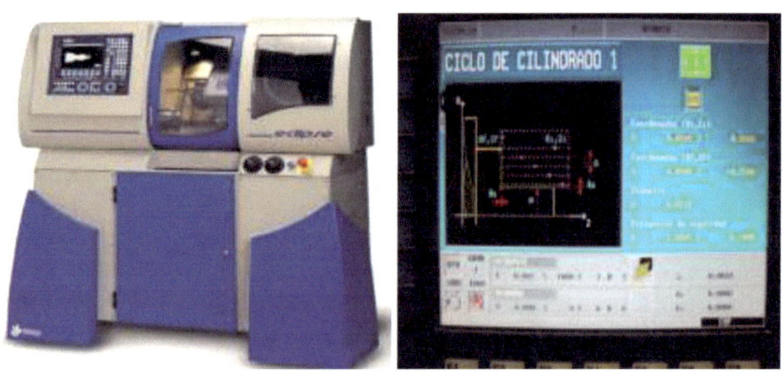

Figure 1.
Eclipse lathe used in micro-turning tests. Left: Equipment. Right: Example of turning program followed.

For cutting tools, SDCR2020K-07 and finishing insert have been used, with the characteristics reflected in **Figure 3**.

The cutting parameters used were as follows: cutting speed: 500 rpm, feed rate: 0.002 mm/rev, and depth of cut: 0.001 mm. The diameter of the test pieces ranged from 0.5 to 0.025 mm, with lengths ranging from 10 to 5 mm, in order to keep a minimum value of the L/D ratio (length/diameter). **Figure 4** shows some details of turning tests and shows the relative size of machined workpieces.

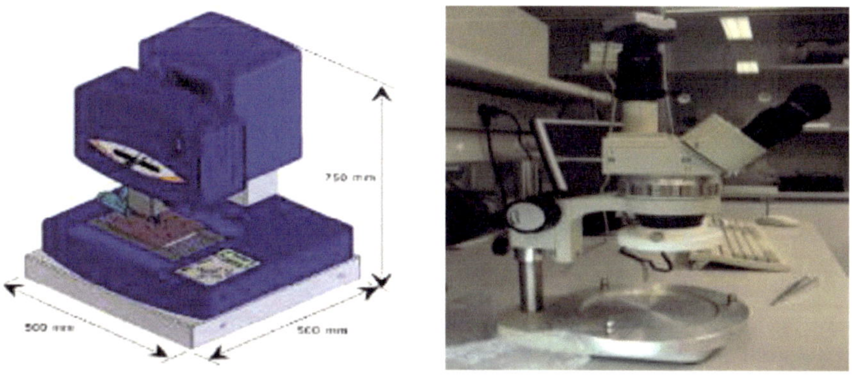

Figure 2.
Measuring equipment. Left: Talysurf CLI roughness meter. Right: Optical microscope.

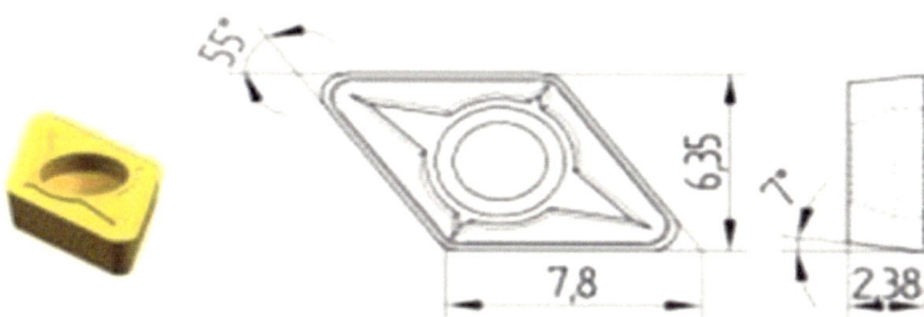

Figure 3.
Cutting tool geometry.

Figure 4.
Details of the experiment. Display of the machined test pieces (0.2 mm in diameter, as shown in the image).

In certain applications, such as those where precise adjustments have to be made in the component assembly process, it is essential to be able to obtain accurate diameters in accordance with the technical specifications laid down in the relevant project. For this purpose, it is necessary to use techniques to characterize the quality of the parts, including the study of roughness, the determination of the degree, or percentage of conicity-cylindricity and dimensional precision.

4. Mathematical models for prediction

Finally, as regards the treatment of the results, it is possible to establish mathematical models of prediction, which can be very useful to characterize the processes, while serving as a practical guide for the development of the same, setting certain cutting conditions, materials, tools, etc. When measuring the accuracy of the estimate and the predictability, account shall be taken of the following:

The sum of squares of errors (SSE), which is the sum of squares of the deviations of the residue values from their sample mean.

Multiple coefficient determination, R^2, and R^2 adjusted, are some common measures in regression analysis, denoting the percentage of variance justified by independent variables. The adjusted R^2 takes into account the size of the data set, and its value is slightly lower than its corresponding R^2.

The validation process was performed by comparing the observed values and the estimated values using the different methods through the mean quadratic error (RMSE).

5. Results and discussion

5.1 Analysis of experimental data

After the micro-turning tests were carried out, the roughness was measured and the cylindrical properties of the machined parts were studied. **Table 3** shows the experimental results of the roughness parameters, Ra and Rt, for the two materials, depending on the diameter of the workpiece.

As can be seen, and especially significantly in the mean roughness (Ra), the roughness is maintained at low and almost constant values as we reduce the diameter of the workpiece, up to 0.2 mm. For lower values, there is an increase in roughness, especially for values of 0.05 mm and below. This increase is due to the low rigidity of the workpieces and the sensitivity to vibrations.

Figure 5 shows an example of roughness profile obtained.

Diameter (mm)	EN AW 2024		EN AW 2030	
	Ra (μm)	Rt (μm)	Ra (μm)	Rt (μm)
0.4	1.4514	26.145	1.4031	21.327
0.2	1.3011	22.308	1.3170	20.196
0.1	1.3821	25.236	1.3392	23.376
0.05	1.4928	22.032	1.4581	22.814
0.025	5.5316	46.173	3.958	38.471

Table 3.
Experimental roughness values.

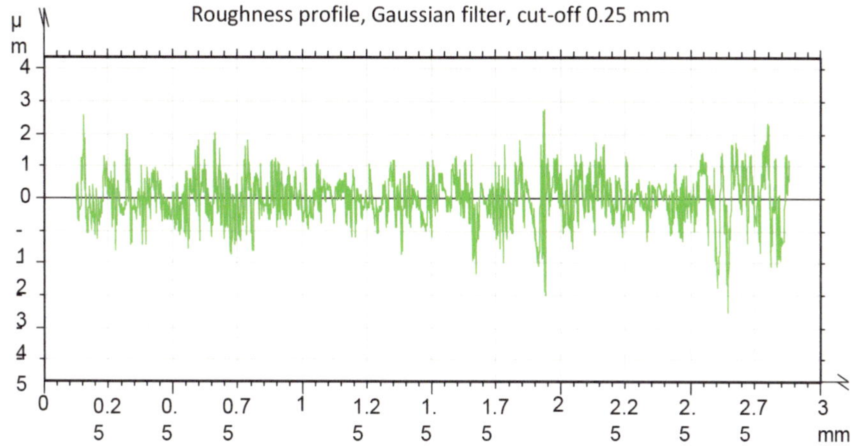

Figure 5.
Example of the obtained roughness profile.

The appearance of the surface and the machining marks can be seen in **Figure 6**.

As regards the preservation of the formal characteristics of the workpieces, the conicity (%) has been calculated with the help of increased images. The conicity, as a function of length and diameter, is calculated using Eq. (1). The results are presented in **Table 4**.

$$c[\%] = 100\frac{D}{L} \tag{1}$$

It is observed that the conicity values are generally low, although they may not become manageable in certain applications. The conicity or "cylindricity defect" is increased by reducing the diameter of the workpiece, mainly due to the increased sensitivity of the workpiece as it passes through the machining process. It can also be seen how the conicity values are always lower in the case of titanium alloy, which may be due to its greater rigidity and its greater ease for micromachining (an interaction of the cutting tool with the micro-workpiece is expected to be somewhat more "fluid," which should probably translate into lower values of the cutting forces).

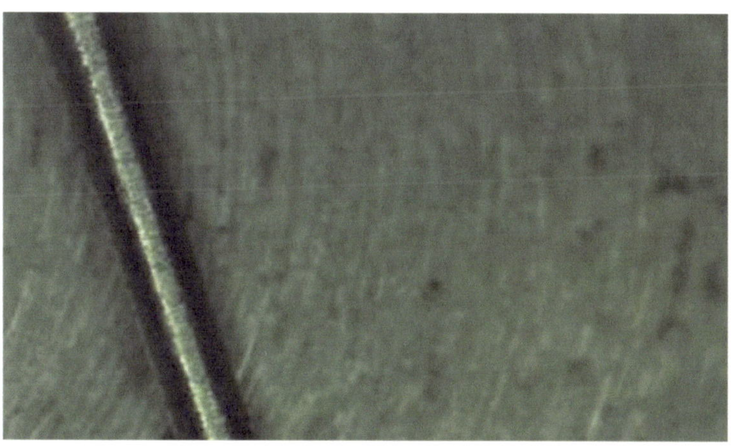

Figure 6.
Workpiece (450 magnifications), diameter: 0.2 mm.

Diameter (mm)	EN AW 2024	EN AW 2030
	Conicity (%)	Conicity (%)
0.4	6.25	5.85
0.2	6.66	6.21
0.1	6.94	6.32
0.05	7.26	6.83
0.025	8.42	7.91

Table 4.
Evolution of conicity according to the diameter of the workpiece.

Figure 7 shows the detail of the graphical treatment performed to measure conicity. It is important to use an appropriate "length-to-diameter" ratio to reduce conicity, vibration and improve the surface quality of machined workpieces.

5.2 Prediction models

The mathematical models developed for the prediction of the roughness parameters, Ra and Rt, are presented below.

$$Ra = 2{,}188 \times 10^{-8} \times d^{-5.167} + 1{,}377 \tag{2}$$

$$Rt = 63{,}98 \times 10^{10} \times \exp(-868{,}9 \times d) + 22{,}4 \times \exp(0{,}3462 \times d) \tag{3}$$

$$Ra = 1{,}35 + \exp\left(4{,}16 + (-128{,}47) \times d\right) \tag{4}$$

$$Rt = 21{,}62 + \exp\left(5{,}45 + (-105{,}037) \times d\right) \tag{5}$$

Table 5 reflects the measures of the goodness of the estimate and the predictability, using the parameters SSE, RMSE, R^2, and R^2 adjusted.

As can be seen, the values indicate that the estimates made are generally very good. **Figures 8** and **9** show the experimental roughness values of micromachined

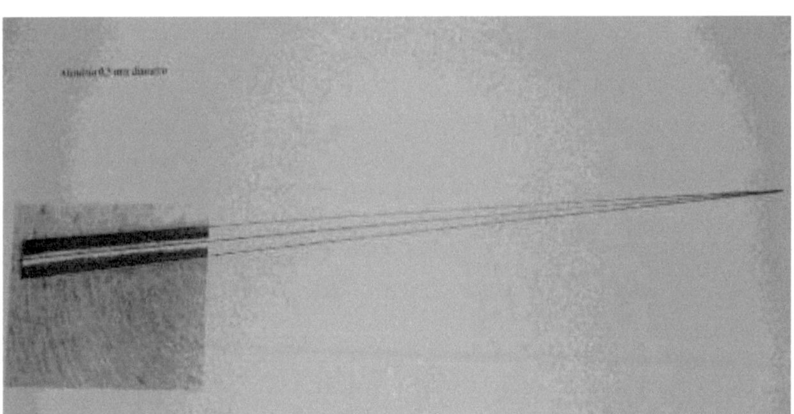

Figure 7.
Obtaining conicity.

Models	SSE	R^2	R^2 adjusted	RMSE
(2)	0.011	0.99	0.99	0.075
(3)	7.824	0.98	0.92	2.797
(4)	0.0040	0.99	0.99	0.0283
(5)	5.1754	0.98	0.98	1.0174

Table 5.
Validation of proposed nonlinear models.

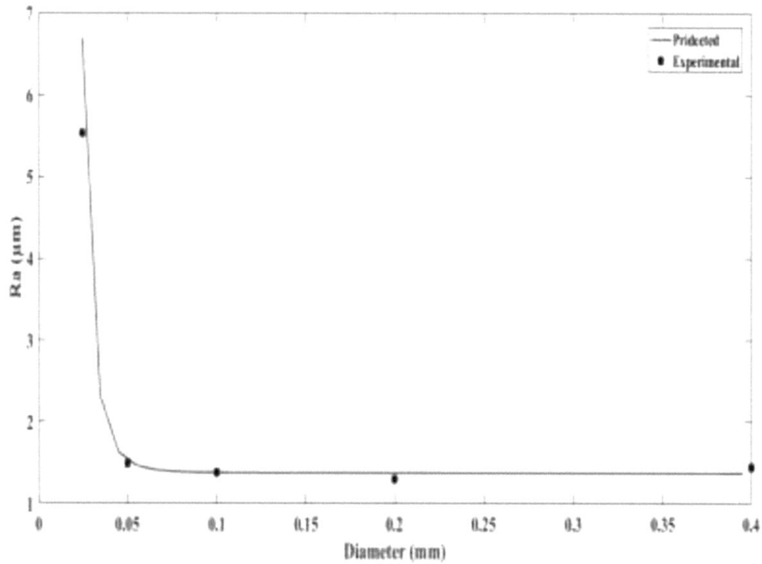

a)

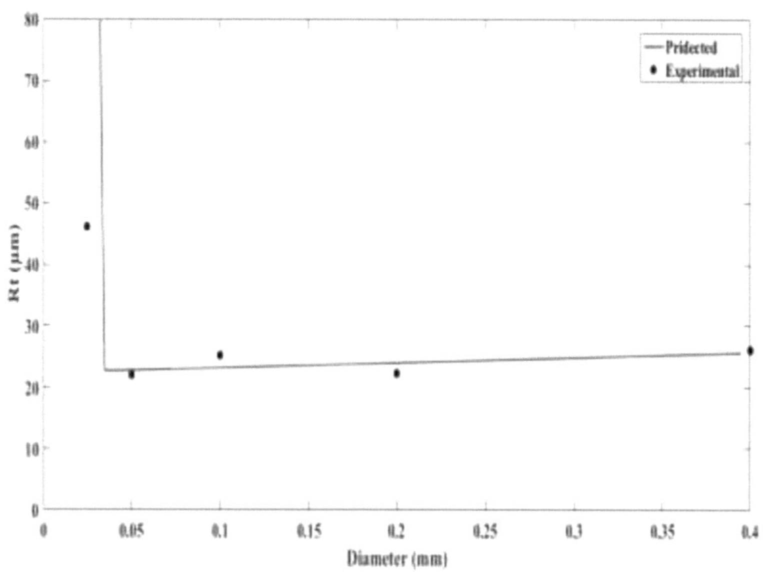

b)

Figure 8.
Experimental and predicted roughness values as a function of the diameter of the workpiece. (a) Ra and (b) Rt for EN AW 2024.

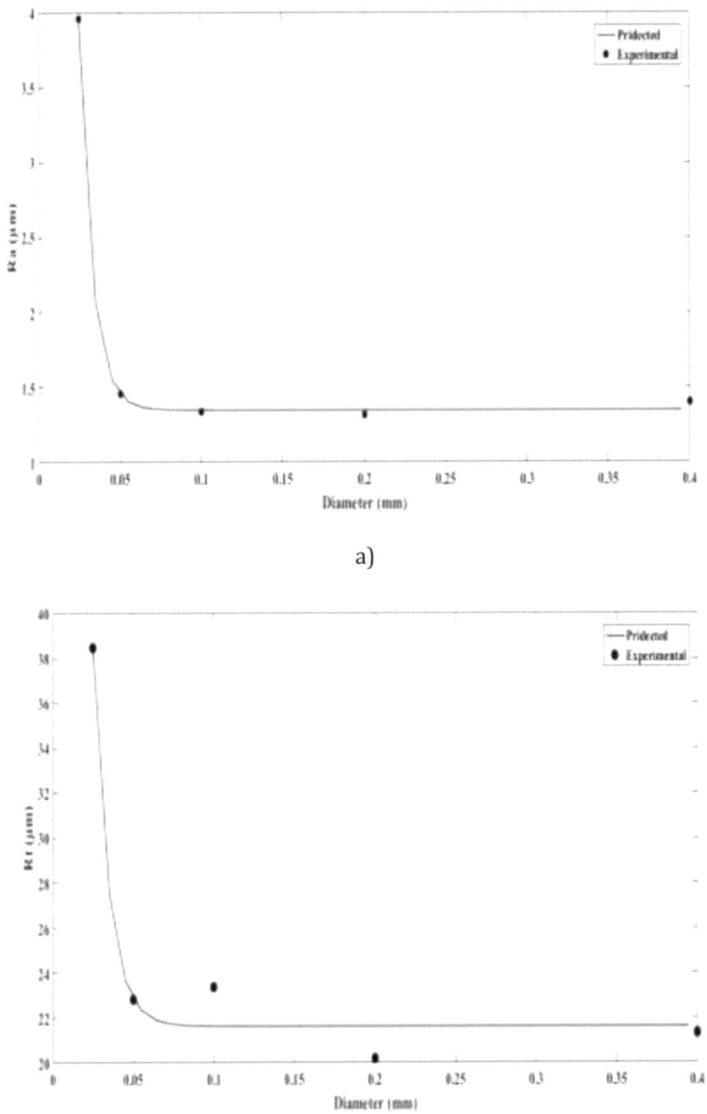

a)

b)

Figure 9.
Experimental and predicted roughness values as a function of the diameter of the workpiece. (a) Ra and (b) Rt for EN AW 2030.

surfaces (Ra and Rt) for EN AW 2024 and EN AW 2030 as one with the representation of prediction models versus the diameter of the workpiece. In all cases, it is observed how the proposed models adjust the range of the experimental values with great approximation.

6. Conclusions

The roughness of machined surfaces is the best indicator of product quality and provides relevant information on their potential for application in different sectors. This work is part of a research on the micro-turning of these materials with the aim of evaluating up to what diameters it is possible to work with conventional finishing tools. Certainly, better behavior is to be expected when specific tools are used for micro-turning. However, low surface roughness values are generally obtained, enabling the specifications of a significant number of practical applications where dimensional accuracy is critical. Among these applications are components and micro-components used in the construction of micromachines (micro-axes, bearing needles, etc.) and, in general, small instrumentation parts, injection systems, etc.

The results obtained allow us to conclude that it is possible to conform by chip removal very small revolution pieces (0.05 mm in diameter) with these alloys, guaranteeing very good surface qualities according to the typical specifications of these applications. It is important to note that the deformation progress used is very low, which has undoubtedly contributed to low roughness values. The use of low values of feed rate and low depths of cut allows cutting forces of very small values in order to guarantee the integrity of the machined workpieces (there is obvious risk of plastic or even fracture if not). On the other hand, the proposed models show very good adjustments as corroborated by the indicators of goodness.

Author details

Francisco Mata[1*] and Issam Hanafi[2]

1 University of Castilla-La Mancha, Almadén, Spain

2 National School of Applied Sciences Al-Hoceima (ENSAH), Al-Hoceima, Morocco

*Address all correspondence to: francisco.mcabrera@uclm.es

References

[1] Chern GL, Wu E, Cheng J, Jao J. Study on burr formation in micro-machining using micro-tools fabricated by micro-EDM. Precision Engineering. 2007;**32**:122-129

[2] Aronson RB. The new world of micromanufacturing. Manufacturing Engineering. 2003:81-92

[3] Azizur M, Rahman M, Senthil Kumar A, Lim HS. CNC microturning: An application to miniaturization. International Journal of Machine Tools and Manufacture. 2005;**45**:631-639

[4] Hocheng H, Chang JH, Jadhav U. Micromachining of various metals by using *Acidithiobacillus ferroxidans* 13820 culture supernatant experiments. Journal of Cleaner Production. 2012;**20**:180-185

[5] Groover MP. Fundamentals of Modern Manufacturing Materials. Process and Systems. New Jersey, United States of America: Prentice Hall International Editions; 1996

[6] Wuele H, Hüntrup H, Tritscher H. Micro-cutting of steel to meet new requirements of miniaturization. Annals CIRP. 2001;**50**:61-64

[7] Chae J, Park SS, Freiheit T. Investigations of micro-cutting operations. International Journal of Machine Tools and Manufacture. 2006;**46**:313-332

[8] Masuzawa T. State of the art of micromachining. Annals CIRP. 2000;**49**:473-488

[9] Abouelatta OB, Mádl J. Surface roughness prediction based on cutting parameters and tool vibrations in turning operations. Journal of Materials Processing Technology. 2001;**118**:269-277

[10] Petropoulos G, Davim JP, Mata F, Pandazaras C. New considerations of evaluating the anisotropy of machined surfaces. Journal of the Balkan Tribological Association. 2006;**12**:1-6

[11] Sreejith PS, Krishnamurthy R, Malhota SK, Narayanasamy K. Evaluation of PCD tool performance during machining of carbon/phenolic ablative composites. Journal of Materials Processing Technology. 2000;**104**:53-58

[12] Eriksen E. Influence from production parameters on the surface roughness of a machined short fiber reinforced thermoplastic. International Journal of Machine Tools and Manufacture. 1999;**39**:1661-1618

[13] Liow J. Mechanical micromachining: A sustainable micro-device manufacturing approach. Journal of Cleaner Production. 2009;**17**:662-667

[14] Christenson T, Guckelh R, Markuks W. Micromachining and Microfabrication Process Technology. In: Proceedings of Spie, Austin, Texas, United States of America; 1995. p. 134

[15] Ozel T. Predictive modeling of surface roughness and tool wear in hard turning using regression and neural networks. International Journal of Machine Tools and Manufacture. 2005;**45**:467-479

[16] Sahin Y, Motorcu AR. Surface roughness model for machining mild steel with carbide tool. Materials and Design. 2005;**26**:321-326

[17] Kohli A, Dixit US. A neural-network based methodology for the prediction of surface roughness in a turning process. International Journal of Advanced Manufacturing Technology. 2005;**25**:118-129

[18] Jiao Y, Lei ST, Pei ZJ. Fuzzy adaptive networks in machining process modeling: Surface roughness prediction for turning operations. International Journal of Machine Tools and Manufacture. 2004;**44**:1643-1651

[19] Feng CX, Wang X. Development of empirical models for surface roughness prediction in finish turning. International Journal of Advanced Manufacturing Technology. 2002;**20**:348-356

[20] Sulliman SMA, Abukari MI, Mirghani EF. Microbial contamination of cutting fluids and associated hazards. Tribology International. 1997;**30**:753-757

[21] Kelly JF, Cotterell MG. Minimal lubrication machining of aluminium alloys. Journal of Materials Processing Technology. 2002;**120**:327-334

[22] Nouari M, List G, Girot F, Coupard D. Experimental analysis and optimisation of tool wear in dry machining of aluminium alloys. Wear. 2003;**255**:1359-1368

[23] Carrilero MS, Bienvenido R, Sánchez-Sola JM, Alvarez M, Gonzalez A, Marcos M. A SEM and EDS insight into the BUL and BUE differences in the turning processes of AA2024 Al-Cu alloy. International Journal of Machine Tools and Manufacture. 2002;**42**:215-220

Chapter 8

Resistance Welding of Aluminium Alloys with an Electromechanical Electrode Force System

Zygmunt Mikno

Abstract

The idea presented in this chapter is an innovative welding machine electrode force system. The operation, advantages of the new solution and the optimisation of the welding process were illustrated by the welding of aluminium bars (5182) (ø 4 mm). The solution involves controlling the force and/or displacement of welding machine electrodes. The modulation of electrode force significantly improves welding, particularly as regards aluminium alloys (requiring a very short welding process). The tests involved the numerical analysis of two electrode force systems, i.e. a conventional Pneumatic Force System (PFS) and an Electromechanical (Servomechanical Force) System (EFS). The numerical tests were performed using SORPAS software. FEM calculation results were verified experimentally. The technological welding tests were conducted using inverter welding machines (1 kHz) equipped with various electrode force systems. The research included metallographic and strength (peeling) tests and measurements of characteristic parameters. The welding process optimisation based on the EFS and the hybrid algorithm of force control resulted in i) more favourable space distribution of welding power, ii) energy concentration in the central weld zone, iii) favourable melting of the material within the entire weld transcrystallisation zone, iv) obtainment of the full weld nugget and v) longer weld nugget diameter.

Keywords: resistance welding of aluminium, electromechanical force system, cross-wire welding, projection welding, electrode force, FEM

1. Introduction

Force constitutes one of the most important parameters in the resistance welding process. The remaining parameters include current and current flow time. During cross-wire projection welding (particularly of aluminium alloys) involving the use of a conventional application, i.e. the pneumatic force system (PFS), it is very difficult, nearly impossible, to make a weld containing the full weld nugget. Aluminium, when subjected to welding, gets plasticised very quickly, which is responsible for the formation of the excessively large area of contact between welded elements and, consequently, results in a rapid decrease in current density. These are not favourable conditions for the melting of materials. In addition, the PFS is characterised by high inertia and the impossibility of performing fast changes in force during current flow. For this reason, the value of preset force is usually constant and unfavourably too high. If the aforesaid force is excessively high, the high deformation of welded elements (bars) may occur as a result. The overly low force may lead to the formation of projection joint imperfections (such as expulsion caused by high temperature in contact areas) [1]. In the PFS, force applied during the welding process results from specific force preset by a pneumatic cylinder. The displacement of electrodes results from the action of this force and the changeable mechanical resistance of materials subjected to welding. A significant disadvantage of the above-presented method of control is the fact that neither force nor displacement (during the flow of current) is actually controlled.

An alternative solution requires another method making it possible to carry out faster changes in force during the welding of materials [1–3]. In publication [1], the authors emphasise the growing popularity of the electromechanical (servomechanical) force system (EFS) and an advantage consisting in an increase in an electrode displacement rate during welding. In publication [2] the authors inform about the possible extension of the window of technological parameters, improving the weldability of materials. In work [3, 4] the authors mention the possible modulation of force and its fast changes, particularly at the final stage of the welding process. The authors stress an increase in electrode service life in spot resistance welding and the application of servomotors in the riveting technology [5]. In publications [3, 6] authors state that the EFS has eliminated the dynamic impact of electrodes against a welded material (during the exertion of initial force), which was characteristic of pneumatic actuators. The EFS has enabled a gentle "touch" of an electrode against a material being welded. In work [3] the authors enumerate other advantages of the EFS including (i) superior (faster) operation of a welding gun (servo) in space, (ii) greater repeatability of force, (iii) reduced noise, (iv) shorter welding time and (v) shorter movement during the closing and opening of the electrodes, extending the service life of related mechanisms.

The tests discussed in the article aimed at replacing the PFS with the EFS. It was also important to appropriately control the servomotor in order to perform the controlled movement/shift of electrodes, particularly during the flow of current. The control process has changed considerably, i.e. the displacement of electrodes is a preset parameter and resultant force depends on the displacement of electrodes and the resistance of the deformation of a contact area being heated. Available reference publications do not contain information about such a method of electrode movement control as that presented in this study.

The authors [7, 8] describe a new control system and the results of its operation, particularly noticeable in projection welding. In [7, 9] the authors refer to a new control system applied when welding sheets with an embossed projection. Another use of the new solution, i.e. cross-wire welding, and the welding of nuts are presented in publications [10–12] respectively.

In publications [7–12] the authors present a completely different solution, i.e. the slowing down of the displacement of an electrode during the projection welding of sheets with an embossed projection. This approach is new and characterised by advantages which are definitely worth mentioning. The above-named idea can be used in relation to aluminium alloys as these materials require a very short welding time (50 ms). It is possible to decrease the penetration of bars and to generate more energy in the optimum place, i.e. in the contact area between the bars. The new idea of electrode displacement control significantly alters the previous approach to the course of the resistance welding process (projection cross-wire welding) and considerably influences the development of the entire research area (pressure welding).

2. Characteristics of welding machine electrode force systems (EFS and PFS)

The essence of the EFS (in comparison with that of the PFS) involves a significantly higher rate of changes, i.e. changes in the force and/or displacement of electrodes. During the resistance welding process, the aforesaid approach is of significant importance because of the fact that the time of welding current flow in typical applications is very short and amounts to 0.2 s (200 ms). The optimisation (improvement) of the welding technology requires the modulation (change) of electrode force during the above-named time. Regrettably, as regards the conventional electrode force system (PFS), common in industrial applications, such modulation is impossible because of the significant inertia (delay) of this solution. **Figure 1** presents exemplary courses of electrode force and displacement in relation to the PFS (dashed line) and EFS (full line). The aforesaid courses refer to two operating modes, i.e. the approach mode and the force mode. The time necessary to obtain previously adjusted electrode force, i.e. electrode force stabilisation time (EFST), by the PFS exceeds 200 ms. The EFST parameter related to the EFS is significantly shorter and restricted within the range of 50–80 ms, depending on the configuration of the EFS (servomotor power, gear etc.). An important characteristic of this solution is the possibility of modulating the course of electrode force during the flow of current, which is nearly impossible as regards the PFS.

The EFS can be controlled in two different manners, i.e. using an algorithm enabling the control of force and an algorithm enabling the control of the (electrode) displacement rate. The first of the algorithms is already used in industrial practice. The time of delay in the stabilisation of preset electrode force is restricted within the range of 50–80 ms. In such an operating mode, it is possible to modulate force and obtain two or three different values (in CFT amounting to 200 ms).

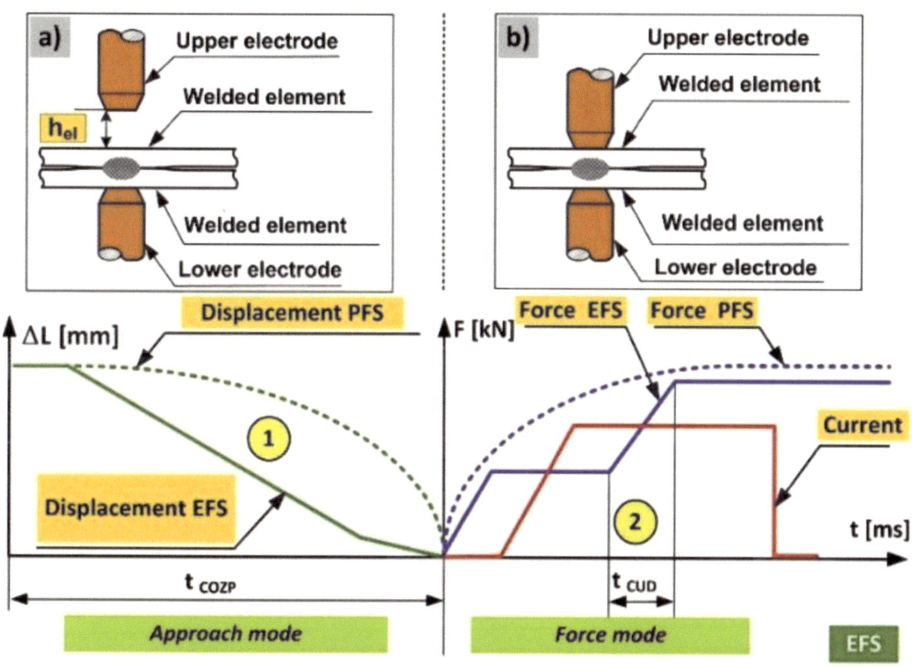

Figure 1.
Comparison of the EFS and PFS based on exemplary courses of electrode force and displacement in relation to operating modes: (a) approach mode and (b) force mode [13].

The aforesaid time (50–80 ms) depends on the configuration of the EFS (motor, gear) and on the preset value of force. In turn, the second algorithm (developed by the author) has been used to weld demanding (in the aforesaid respect) materials, i.e. aluminium alloys. Until today, the author has not come across any information concerning the method of control presented in this chapter.

The unique characteristic of the EFS and of the solution is a special algorithm, where the control of the displacement of electrodes results in the exertion of electrode force. In the above-named algorithm of control, delays between preset and actual values are counted in milliseconds, making it possible to develop a very fast algorithm enabling the exertion of variable (electrode) force [9]. The above-presented manner of controlling the force of electrodes through the control of their displacement alters previous views on methods enabling the control of force (movement of electrodes) in the resistance welding process.

3. Methodology of numerical and experimental tests

The crosswise projection welding of aluminium bars (Al 5182) performed using the PFS was subjected to numerical analysis verified experimentally and aimed to subsequently optimise the welding process performed using the EFS system. The assumed acceptance criteria included (i) obtainment of the full weld nugget having a diameter of not less than 1.5 mm, (ii) lack of deformation and the penetration of the bars less than 20% of the diameters of elements subjected to welding (Δl_{PP} = 1.6 mm), (iii) lack of overheating in the area of contact between the electrode and the material being welded ($T_{e\text{-}m\ max} \leq 500°C$), (iv) lack of expulsion and (v) maximum current flow time $t_{PP\ max}$ = 63 ms. An additionally expected result was the reduction of (welding) current flow time.

The material of bars subjected to welding and adopted in FEM-based calculations was aluminium alloy grade 5182 with solidus (temperature) being 577°C and liquidus amounting to 638°C [14]. The chemical composition of the aluminium alloy grade Al 5182 used in the bars is presented in **Table 1**.

3.1 FEM calculations

The numerical calculations were performed using the SORPAS® 3D software program [15]. The calculations were carried out for ¼ of the model and its mirror reflection in relation to the plane determined by x-z-axes and y-z-axes (**Figure 2**). The mesh in the area of contact between the elements (bars) subjected to welding was concentrated in order to provide the appropriate accuracy of calculations. The lack of proper mesh density resulted in the lack of contact between the elements subjected to welding and, consequently, incorrect calculations.

3.2 Numerical (FEM) model

The numerical model of the crosswise welding of bars is presented in **Figure 2**. The calculations were performed using the 3D model [15].

Al (max)	Mn (max)	Mg (max)
95.2	0.35	4.5

Table 1.
Chemical composition of the materials subjected to welding, i.e. bars made of aluminium alloy grade Al 5182.

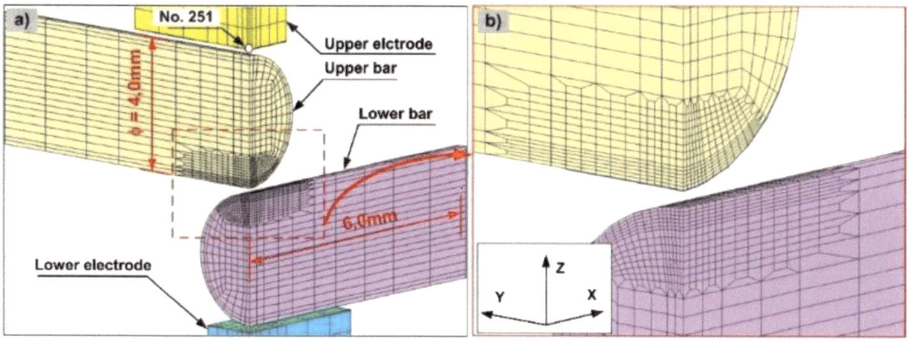

Figure 2.
Model (3D) of the crosswise welding of aluminium bars (Al 5182).

The numerical calculations included the analysis of (i) waveforms of dynamic resistance and momentary power, (ii) energy supplied to the weld, (iii) diameter and volume of the molten material of the weld nugget, (iv) displacement of electrodes (penetration of bars), (v) expulsion (if any) and (vi) temperature in the electrode bar contact area (Te-m, point 251—**Figure 2a**). The primary objective included the determination of the most favourable space distribution of welding power enabling the melting of the material in the central zone of the joint (to obtain the full weld nugget). As in all other cases of projection welding, the aspect of particular importance was the beginning of the welding process, i.e. the beginning of welding current flow.

3.3 Process parameters

The assumptions adopted in the numerical model included (i) copper electrodes (A2/2) and (ii) elements subjected to welding, i.e. aluminium (grade Al 5182) bars having a diameter of 4 mm and a length of 12 mm (**Figure 2a**). The 3D model was composed of approximately 9000 elements. To ensure the required accuracy of calculations, it was necessary to concentrate the mesh in the area of contact between the bars (**Figure 2b**).

Data related to the electrodes and materials subjected to welding and used in the FEM calculations were obtained from the SORPAS software program database (**Table 2**) [14]:

- Aluminium bars Al 5182—material database designation SORPAS AA5182(O): Al95, Mn0.25, Mg4.5, solidus (577°C), liquidus (638°C) (**Table 3**)

- Electrodes of class A2/2 CuCrZr (**Table 4**)

Based on the present recommendations and guidelines concerning the crosswise projection welding of bars, the following ranges of parameters were adopted for:

- PFS [16–18]: (i) welding current I = 8.0–12.0 kA, (ii) welding current flow time t_{PP} = 3 ms (upslope) + 60 ms (primary welding time) and (iii) electrode force F = 0.5–1.5 kN

- EFS: (i) welding current I = 8.0 kA, (ii) welding current flow time t_{PP} = 3 ms (upslope) + 35 ms and (iii) control of electrode displacement during the flow of welding current [11].

Time step increment	Squeeze	Upslope	Weld	Hold	Unit
Pneumatic force system	500	10–70	200	500	ms
Electromechanical force system	200	10	250–300	500	ms
Time step	0.1	0.1	0.1	1	ms
Convergence control					
		Convergence accuracy			
Electrical model		1.00E−5			
Thermal model		1.00E−5			
Mechanical model		1.00E−5			
Welding parameters					
Welding current		DC			
Heat loss to the environment					
Air temperature		20		°C	
Heat transfer rate		300		W/m^2 K	
Electrode dimensions					
Length × width		10.0 × 8.0		mm	
Electrode height		5		mm	
Contact between welded elements		Sliding			

Table 2.
Parameters of the SORPAS software program used in numerical (FEM) calculations.

The numerical calculations were performed in relation to a DC inverter welding machine (1 kHz). The remaining welding cycle parameters are presented in **Table 2**. **Table 5** presents the preset parameters of the welding cycle and the parameters characteristic of variants selected for FEM calculations.

The PFS variants are designated as P1 ÷ P9 (P, pneumatic system), whereas the EFS variants are designated as E1 ÷ E3 (E, electromechanical system). The analysis of the welding process performed in relation to the PFS aimed to investigate and depict the variability of process parameters and determine the most favourable welding conditions (MFWC). The results of the analysis revealed the lack of the monotonicity of the weld nugget growth (**Figure 4a**) visible in relation to a force of 0.75 kN. For this reason it was necessary to perform additional calculations in this area, i.e. for a value of 0.7 kN and that of 0.8 kN. In total, the analysis of the process was focused on 35 points (I = 8/9/10/11/12 kA, F = 1.5/1.25/1.0/0.8/0.75/0.7/0.5 kN).

The numerical optimisation concerning the process involving the use of the EFS was performed for lower values of current than those analysed in relation to the PFS (8.0 ÷ 10.0 kA). The numerical calculations were continued until the occurrence of one (of six) previously adopted boundary conditions.

3.4 Experimental tests

The experimental tests were performed using inverter welding stations (DC 1 kHz) shown in **Figure 3a** (PFS) and **Figure 3b** (EFS). The welding parameters were recorded using a LogWeld 4 measurement device.

The results obtained in the numerical calculations were verified experimentally. The experimental tests involved nine variants (P1–P9) from **Table 5** (PFS). All of the variants (**Figure 6**) were subjected to destructive tests (peeling), confirming the

Temperature (°C)	Thermal conductivity (W/m·K)	Temperature (°C)	Heat capacity (J/kg·K)	Temperature (°C)	Resistivity (mΩ·m)	Temperature (°C)	Mass density (kg/m^3)	Temperature (°C)	Thermal expansion coefficient (10^{-6}/°C)	Temperature (°C)	Young's modulus of elasticity (kN/mm^2)
25	123.0	23	794	25	0.056	25	2660	25	23.9	25	70.0
100	134.0	50	825	100	0.068			250	25.0		
200	147.6			200	0.079						
400	160.9			400	0.103						
500	164.3			500	0.115						
600	163.3			600	0.131						

Table 3.
Material parameters of welded materials (bars Al 95 Mn0.25, Mg4.5) [14].

Temperature (°C)	Thermal conductivity (W/m·K)	Temperature (°C)	Heat capacity (J/kg·K)	Temperature (°C)	Resistivity (mΩ·m)	Temperature (°C)	Mass density (kg/m^3)	Temperature (°C)	Thermal expansion coefficient (10^{-6}/°C)	Temperature (°C)	Young's modulus of elasticity (kN/mm^2)
20	326.6	20	372	20	0.022	20	8890	25	16.5	25	117.0
100	342.1	127	402	100	0.027	1080	8320				
300	338.0	327	422	200	0.038						
500	340.3	527	438	300	0.042						
700	332.0	727	456	400	0.049						
900	321.8	927	485	500	0.057						
				600	0.065						
				700	0.073						
				800	0.082						
				900	0.091						
				1000	0.102						
				1100	0.220						

Table 4.
Material parameters of electrodes ISO 5182 A2–2 electrode CuCrZr [14].

No.	Variant	Current	Welding time	Force	PenetrationΔ l	Weld diameter	Weld volume	Energy	Remarks
		kA	ms	kN	mm		mm³	kJ	
1		2	3	4	5	6	7	8	9
Pneumatic system (PFS)									
1	P1	8.0	63	1.5	1.47	0.1	0.0	0.17	Overly small weld nugget diameter
2	P2	10.0	46		2.38	0.1	0.0	0.17	Excessive penetration of bars
3	P3	12.0	29		1.85	0.0	0.0	0.15	
4	P4	8.0	63	1.0	1.00	0.2	0.0	0.20	Overly small weld nugget diameter
5	P5	10.0	59		2.13	1.5	0.9	0.23	Most favourable welding conditions in spite of significant penetration of bars
6	P6	12.0	46		2.54	0.2	0.0	0.30	Overly small weld nugget diameter
7	P7	8.0	63	0.5	0.57	0.3	0.1	0.25	Overly small weld nugget diameter
8	P8	10.0	8		0.19	1.5	0.7	0.07	Unfavourably short welding time, high dynamics of the force system required
9	P9	12.0	5		0.15	0.8	0.1	0.05	Overly small weld nugget diameter
Electromechanical (servo) system (EFS)									
10	E1	8.0	38	Servo force	0.6	1.95	4.5	0.16	OK, full weld nugget, nugget diameter > 1.6 mm, penetration of bars <1.6 mm
11	E2	9.0	25		1.2	2.00	5.3	0.10	
12	E3	10.0	20		1.5	2.20	5.6	0.13	

Red colour, unacceptable param.; green colour, acceptable param.; orange colour, the most favourable welding conditions.

Table 5.
Preset welding cycle parameters and parameters characteristic of selected variants in the FEM calculations related to the PFS and EFS (FEM).

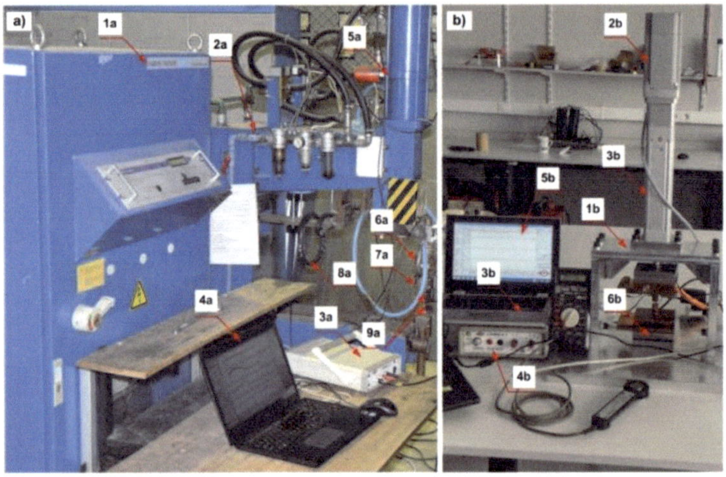

Figure 3.
Welding machine stations: (a) SPD: (1a) inverter welding power source Harms & Wende (25 kA, 1 kHz), (2a) welding machine housing ASPA (5.5 kN), (3a, 4a) measurement device LogWeld 4, (5a) pneumatic actuator, (6a) head for measurements of electrode force, (7a) laser sensor for displacement measurements, (8a) welding current measurement sensor and (9a) leads for measurements of welding voltage. (b) SED: (1b) electromechanical welding machine F = 2 kN, (2b) servomotor, (3b) linear gear, (4b, 5b) measurement device LogWeld 4 and (6b) electrode force measurement module.

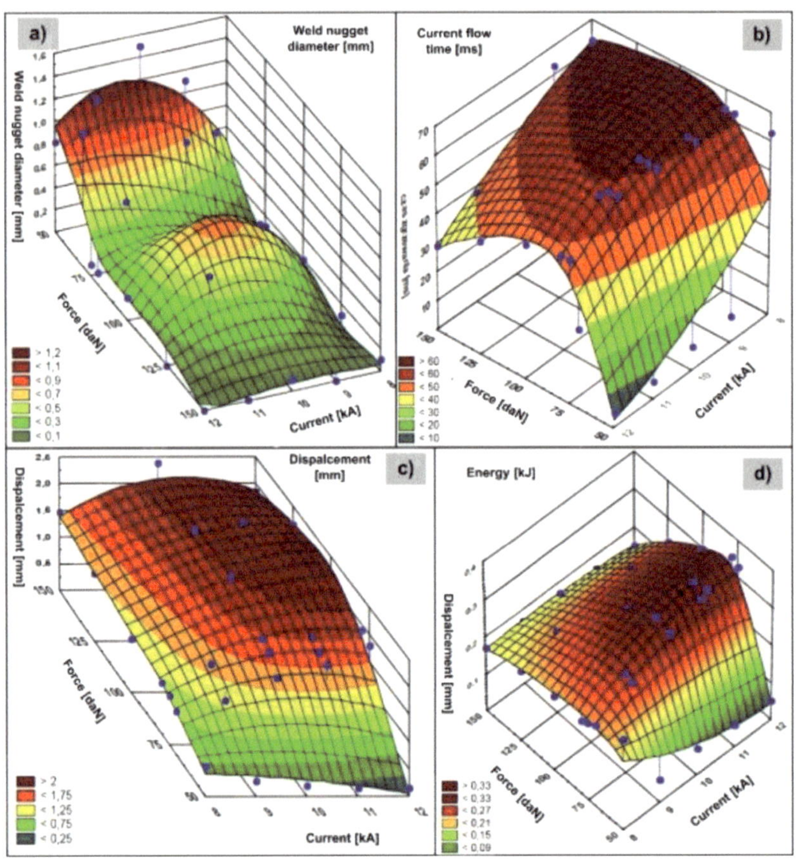

Figure 4.
Variability of characteristic parameters in relation to the PFS (Al 5182, ϕ = 4 mm, MES) [11]: (a) weld nugget diameter, (b) current flow time, (c) displacement (of electrodes - penetration of bars), (d) energy (of welding).

formation of the ring weld. However, none of the PFS variants satisfied the previously assumed criteria. Nonetheless, in spite of the exceeding of the previously assumed value of bar penetration and the obtainment of the ring weld, variant P5 was adopted as the reference variant for further optimisation-related activities. The reason for such a choice resulted from the fact that the aforesaid variant (P5) enabled the obtainment of the longest weld nugget diameter. Parameters similar to those used with reference to variant P5 were used in additional technological welding tests (**Table 6**, PE1–PE3). The results related to the preset parameters of the technological cycle and characteristic parameters of selected welding tests involving the use of the PFS and variants PE1–PE3 are presented in **Table 6**.

The welding cycle parameters used in relation to variants PE1–PE3 included electrode force F = 1.0 kN, welding current I = 9.0–10 kA and welding time t_{PP} = 43–63 ms. The metallographic tests involving the above-named variants confirmed the results obtained in the numerical calculations, e.g. the ring-like shape of the weld nugget.

Key: I_{rms}, root-mean-square current; PE, pneumatic experiment.

4. Process optimisation

The optimisation of the crosswise projection welding of bars was performed using the EFS. The primary criterion of the optimisation process involved the obtainment of the full weld nugget having a diameter of not less than 1.5 mm. The optimisation process assumed the use of the EFS and, in addition, the adjustment of the lowest possible value of welding current.

The optimisation process also aimed to adjust appropriate and lower electrode force than that applied initially in the process performed using the PFS and to control the displacement of the electrodes so that it could be possible to obtain the most favourable space distribution of welding power, i.e. ensuring the emission of appropriately more heat (energy) in the central part of the contact area between elements being welded (in order to melt the material of these elements) [10, 11].

The preset welding cycle parameters (grey) and the parameters characteristic of the technological welding tests performed using the EFS are presented in **Table 7**. The technological welding tests were performed using a current of approximately 8.0 kA, i.e. the lowest value analysed in relation to the welding process performed using the PFS. The aforesaid value of current applied in the PFS, within the entire range of analysed values of electrode force (0.5 kN ÷ 1.5 kN), was insufficient to melt the material of the elements subjected to welding. The welding process was optimised using the EFS and a welding current of 8.0 kA and that of 8.5 kA as well as the appropriate profile of electrode force (variants EE1 and EE2, **Table 7**).

5. Results

5.1 FEM calculation results

The PFS-related numerical calculation results are presented in **Figure 4** and **Table 8**. The results are presented in spatial diagrams developed using the Statistica software program [19]. **Figure 4** presents (in the form of a spatial diagram) the formation of the weld nugget diameter (**Figure 4a**), welding time (**Figure 4b**), bar

No.	Variant no.	Preset parameters							Recorded parameters		
		Electrode force	Upslope		Main welding time		Total current (I_{rms})	Welding energy	Bar penetration (displacement of electrode)	Weld diameter	Number of tests
			Current	Time	Current	Time					
		kN	kA	ms	kA	ms	kA	kJ	mm	mm	pcs
		A	B1	B2	C1	C2	D	E	F	G	H
1	PE1	1.0	10.0	3	10.0	40–60	10.0	0.23	1.50	1.5	20
2	PE2	1.0	9.5	3	9.5	50–70	9.5	0.21	1.38	1.3	20

Table 6.
Preset and characteristic parameters of the PFS (experiment) [11].

No.	Variant no.	Preset parameters								Recorded parameters							
		Force			Upslope		Main welding time		Total current (I_{rms})	Electrode displacement and time				Welding energy	Bar penetration (displacement of electrode)	Weld diameter	Number of tests
		Initial	Min.	Max.	Cur	Time	Cur.	Time		$t_1/\Delta l_1$	$t_2/\Delta l_2$	$t_3/\Delta l_3$	$t_4/\Delta l_4$				
		kN			kA	ms	kA	ms	kA	ms/mm				kJ	mm	mm	pcs
		A1	A2	A3	B1	B2	C1	C2	D	E0	E1	E2	E3	F	G	H	I
1	EE1	1.0	0.4	1.0	8.0	3	8.0	45	8.0	10 0.08	30 0.25	10 0.05	30 0.25	0.16	0.70	1.87	20
2	EE2	1.0	0.4	1.0	8.5	3	8.5	40	8.5	10 0.08	25 0.25	7 0.05	30 0.25	0.20	0.75	1.92	20

Key: I_{rms}, root-mean-square current; EE, electromechanical experiment.

Table 7.
Preset and characteristic parameters of the EFS (experiment) [11].

penetration depth (electrode displacement) (**Figure 4c**) and welding energy (**Figure 4d**). The correlations are presented in relation to various values of welding current and electrode force.

Numerical values related to the graphic representation of the results presented in **Figure 4** are presented in **Table 8(a-d)**, containing, in addition, information about the following:

- Unsatisfied criterion (**Table 8e**), i.e.:
- $t_{PP\ max}$—maximum time of welding current flow ($t_{PP\ max}$ = 63 ms)
- ΔL—displacement of electrodes (penetration of bars, ΔL_{max} = 1.6 mm, 20% of the diameters of the bars)
- W—expulsion
- Weld itself (**Table 8f**), i.e.:
- D—weld nugget diameter: below 0.7 mm; (*S*, *small nugget diameter*)
- R—ring-shaped weld nugget: 0.7 mm < D ≤ 1.5 mm; (*R*, *ring weld nugget*)
- F—full weld nugget: D > 1.5 mm; (*F*, *full weld nugget*)
- Weld nugget volume (**Table 8g**)

    a. current flow time

    b. displacement (of electrodes - penetration of bars)

    c. welding energy.

The results presented in **Table 8** supplement the information concerning the (course of) variability of the characteristic parameters from **Figure 4**.

Table 8 also contains the numerical calculation results obtained for the EFS (green). In relation to all of the previously assumed parameters, the conditions concerning the optimised method of control were satisfied.

The comparison of the FEM calculation results (in the form of the distribution of temperature) related to the two (i.e. PFS and EFS) electrode force systems, different values of welding current (8.0 and 10.0 kA) and various ranges of temperature (20–638°C and 577–638°C) is presented in **Figure 5**.

5.2 Experimental test results

The PFS-related technological welding tests involving the aluminium bars were performed in relation to all of the nine variants P1–P9 from **Table 5**. The results after the peeling tests are presented in **Figure 6**.

Parameters similar to those used with reference to variant P5, i.e. in relation to which the longest weld nugget diameter was obtained, were used in additional technological welding tests performed in relation to a wider welding current range of 9.0–10.0 kA. Results (in the form of metallographic structures) related to the above-presented parameters are presented in **Figure 7**. The preset

Current [kA]	**(a) Weld nugget diameter [mm]**							
	0.5 mm < d < 1.5 mm				**d < 0.5 mm**			**Servo**
	Force [kN]							
	1.50	**1.25**	**1.00**	**0.80**	**0.75**	**0.70**	**0.50**	
8	0.1	0.1	0.2	0.0	0.1	0.0	0.3	1.95
9	0.0	0.0	0.6	0.0	0.0	0.0	1.1	2.00
10	0.1	0.1	1.5	1.1	0.1	0.0	1.5	2.20
11	0.0	0.7	0.3	0.0	0.2	0.5	1.1	
12	0.0	0.0	0.2	0.1	0.1	1.2	0.8	
Current [kA]	**(b) Current flow time [ms]**							
	t_weld >63 ms				**t_weld <63 ms**			**Servo**
	Force [kN]							
	1.50	1.25	1.00	0.80	0.75	0.70	0.50	
8	63	63	63	63	63	63	63	38
9	63	63	63	63	63	63	9	25
10	46	56	59	63	63	63	8	20
11	38	48	53	60	63	63	6	
12	29	39	46	49	49	27	5	
Current [kA]	**(c) Final electrode displacement [mm]**							
	D < 1.6 mm				**D > 1.6 mm**			**Servo**
	Force [kN]							
	1.50	1.25	1.00	0.80	0.75	0.70	0.50	
8	1.47	1.25	1.00	0.95	0.9	0.80	0.57	0.60
9	1.60	1.62	1.50	1.30	1.10	1.00	0.28	1.20
10	2.38	2.07	2.13	1.75	1.70	1.40	0.19	1.50
11	2.00	2.20	1.90	1.80	1.70	1.65	0.16	
12	1.85	2.15	2.54	1.90	0.80	0.70	0.15	
Current [kA]	**(d) Energy [kJ]**							
	E < 0.16		**0.16 < E < 0.32**			**E ≥ 0.32**		**Servo**
	Force [kN]							
	1.50	1.25	1.00	0.80	0.75	0.70	0.50	
8	0.17	0.19	0.20	0.20	0.21	0.22	0.25	0.16
9	0.17	0.22	0.25	0.27	0.27	0.26	0.06	0.10
10	0.17	0.23	0.29	0.33	0.29	0.32	0.07	0.13
11	0.16	0.24	0.30	0.35	0.32	0.36	0.06	
12	0.15	0.24	0.30	0.34	0.33	0.22	0.05	
Current [kA]	**(e) Achieved criterion**							
	D > 1.6 mm		**t, welding time (63 ms)**			**E, expulsion**		**Servo**
	Force [kN]							
	1.50	1.25	1.00	0.80	0.75	0.70	0.50	
8	t	t	t	t	t	t	t	OK

Current [kA]	(e) Achieved criterion							
	D > 1.6 mm		t, welding time (63 ms)			E, expulsion		Servo
	Force [kN]							
9	t	t	t	t	t	t	E	OK
10	D	D	D	t	t	t	E	OK
11	D	D	D	P	t	t	E	
12	D	D	D	D	D	E	E	
Current [kA]	**(f) Type of nugget**							
	F, full nugget		**R, ring nugget**			**L, low nugget (f < 0.7 mm)**		**Servo**
	Force [kN]							
	1.50	1.25	1.00	0.80	0.75	0.70	0.50	
8	L	L	L	L	L	L	L	F
9	L	L	L	L	L	L	R	F
10	L	L	R	R	L	L	R	F
11	L	L	L	L	L	L	R	
12	L	L	L	L	L	R	R	
Current [kA]	**(g) Weld nugget volume [mm^3]**							
	0.5 mm < V (vol.) < 1.5 mm^3				**V (vol.) < 0.5 mm^3**			**Servo**
	Force [kN]							
	1.50	1.25	1.00	0.80	0.75	0.70	0.50	
8	0.0	0.0	0.0	0.0	0.0	0.0	0.1	4.50
9	0.0	0.0	0.0	0.0	0.0	0.0	0.2	5.25
10	0.0	0.0	0.9	0.3	0.0	0.0	0.7	5.60
11	0.0	0.2	0.0	0.0	0.0	0.2	0.5	
12	0.0	0.0	0.0	0.0	0.0	1.3	0.1	

Table 8.
FEM calculation results concerning the crosswise projection welding of bars in relation to the PFS and EFS (AL 5182) [11].

welding cycle parameters in relation to variants PE1 and PE3 are presented in **Table 6**.

In terms of the EFS, the technological welding tests were performed in relation to a current of 8.0 kA and that of 8.5 kA (**Table 7**, variants EE1 and EE2). The comparative results in the form of the metallographic structures of the joints are presented in **Figure 8a1–a2** and **b1–b2** (in relation to the PFS and EFS, respectively).

6. Discussion

6.1 FEM calculations

The PFS-related conclusions based on the analysis of the results presented in **Figure 4** and **Table 8** are the following:

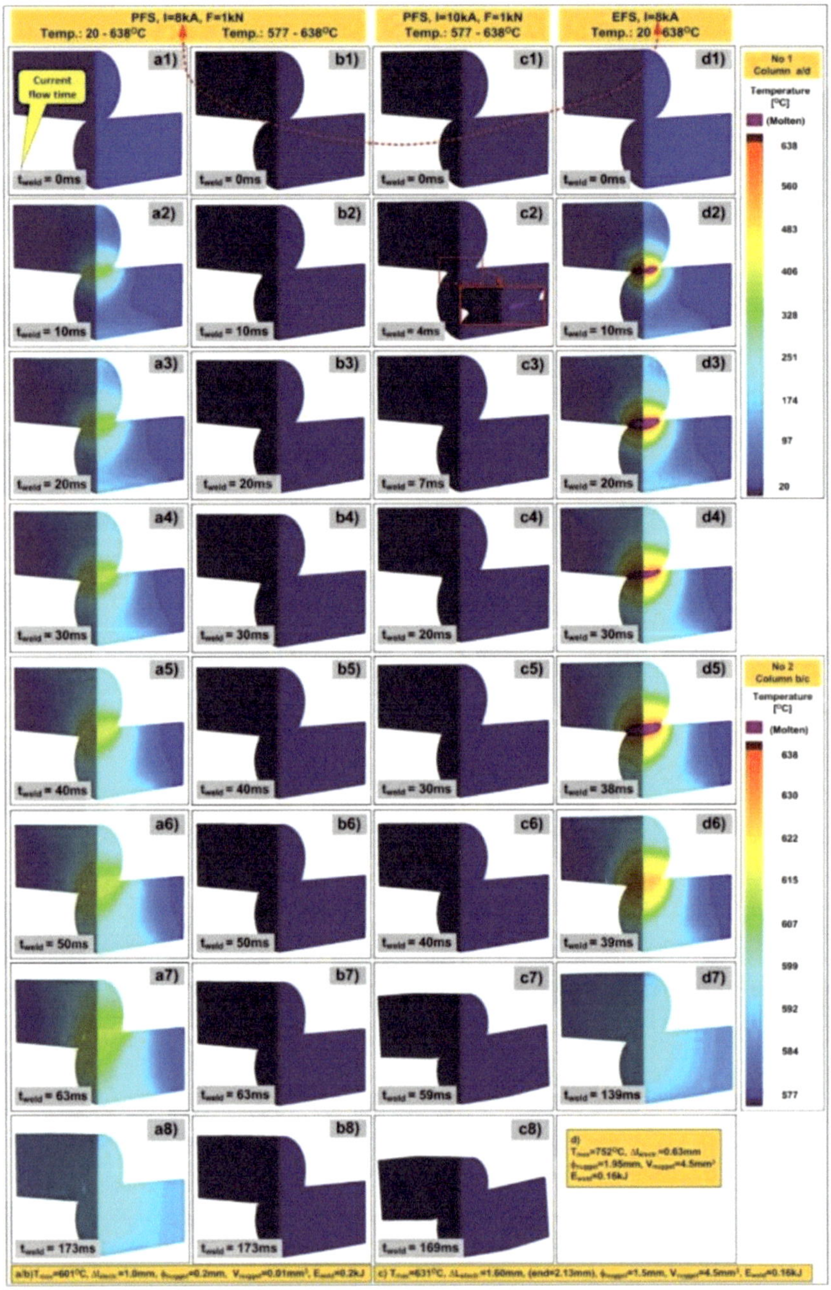

Figure 5.
Distribution of temperature in the welding area (FEM) in relation to: (a/b) PFS (I = 8 kA, F = 1.0 kN), (c) PFS (I = 10 kA, F = 1.0 kN) and (d) EFS (I = 8.0 kA, force exerted by the servomotor).

- Maximum obtainable weld nugget diameter amounted to 1.5 mm (**Table 8a**, parameter field 1).

- Ring-shaped weld was formed (**Table 8f**, parameter field 2) within the entire range of the variability of welding current parameters and that of electrode force, also as regards the longest obtained weld nugget diameter (i.e. restricted within the range of 1.0–1.5 mm).

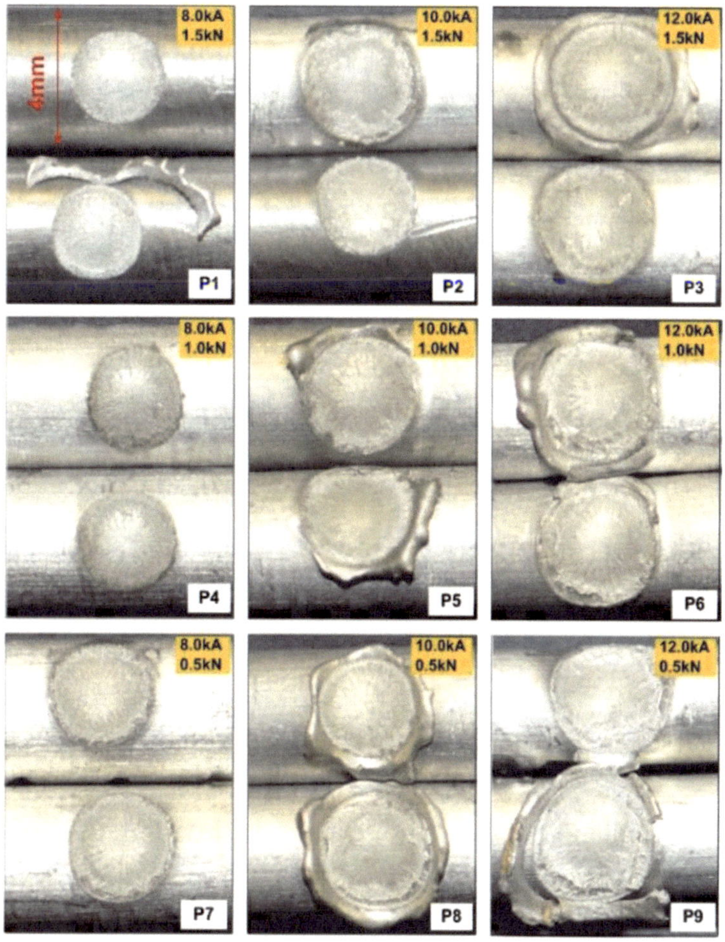

Figure 6.
Specimens after the peeling tests (PFS, variants P1–P9).

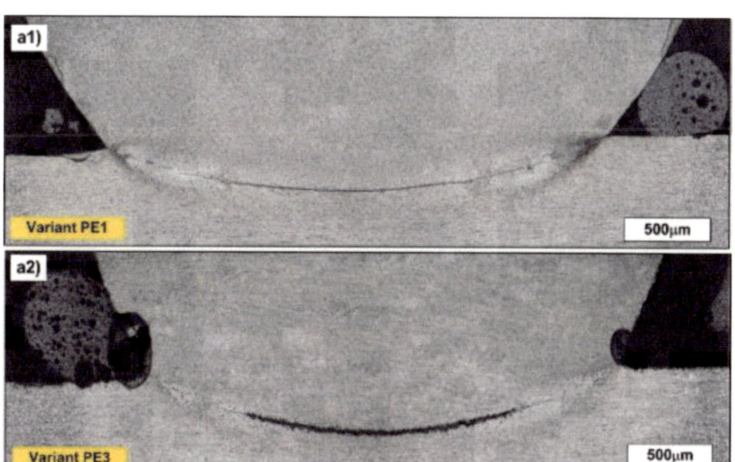

Figure 7.
Results of the metallographic tests for the PFS (F = 1 kN, I = 9.0/10.0 kA).

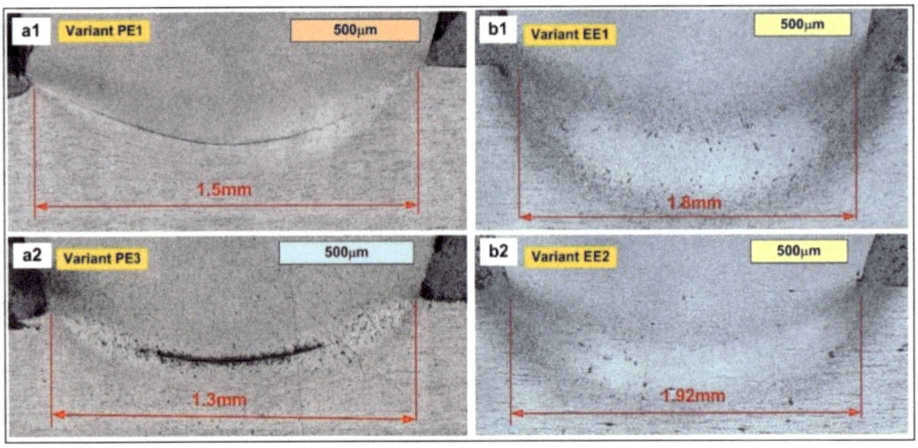

Figure 8.
Metallographic test results in relation to [11]: (a) PFS, (a1) variant PE1 and (a2) variant PE3; (b) EFS, (b1) variant EE1 and (b2) variant EE2.

- Criterion concerned with the exceeding of the maximum welding current flow time ($t_{PP\ max}$ = 63 ms) was observed in relation to the lower value of welding current (**Table 8e**, parameter fields 3a and 3b, respectively).

- The highest volume of the molten material was observed in relation to the highest values of welding current and the lowest values of electrode force (**Table 8g**, parameter field 4). The above-named parameters were also connected with relatively low welding energy (**Table 8d**, parameter field 5). However, welding time was relatively short and amounted to a few milliseconds. In the aforesaid case, even the slight exceeding of the welding time resulted in expulsion (**Table 8e**, parameter field 7).

- Excessive penetration of the bars (above the acceptable value) was related to the obtainment of the higher value of welding current and the higher value of electrode force (**Table 8c**, parameter field 6a (final penetration), and **Table 8e**, parameter field 6b (penetration value $\Delta l_{PP} > 1.6$ mm)).

- Risk of expulsion was accompanied by the lowest value of electrode force and the higher value of welding current (**Table 8e**, parameter field 7).

The crucial aspect which remained was the failure to satisfy the principal criterion, i.e. the obtainment of the full weld nugget having a diameter of 1.5 mm.

The analysis of the FEM-based calculation results, presented in **Figure 5**, is as follows:

- In relation to the PFS, a welding current of 8.0 kA and a force of 1.0 kN, **Figure 5** presents the distribution of temperature within the entire range of temperature subjected to analysis, i.e. from ambient temperature to the melting point (liquidus) (**Figure 5a**). In such an approach, within the range of temperature, the melting of the material did not take place, and the weld nugget diameter calculated by the SOPRPAS software program amounted to a mere 0.2 mm. **Figure 5b** presents the distribution of temperature within the range of *solidus* (577°C) to *liquidus* (638°C). In the above-presented approach, within the entire range of welding time, it was impossible to obtain the melting of the material. As a result, the solid-state joint was formed within the entire area of contact.

- In relation to a higher current of 10.0 kA (F = 1.0 kN) and the PFS, energy supplied to the weld was higher. However, the plasticisation of the welding area combined with the exertion of constant and excessively high (electrode) force led to the situation where the material was melted and pushed outside. This, in turn, resulted in the increasingly large area of contact between the elements subjected to welding (bars), leading to the abrupt decrease in current density and, consequently, the immediate cooling of the weld material. Although it was possible to observe the melting of the material, the process was extremely short (2 ms) **Figure 5c2**. Precisely after two milliseconds, the temperature in the entire welding area decreased below the melting point (**Figure 5c3**).

- In relation to the lowest analysed current value amounting to 8.0 kA (**Figure 5d**) and the EFS, because of a different manner of electrode force control, it was easily possible to observe the welding of the material (subjected to welding) and the formation of the full weld nugget from the very beginning of the flow of welding current. In terms of the case under analysis, i.e. the welding of aluminium bars (Al 5182), the shutdown of current resulted in the immediate (within 1 ms) lowering of temperature below the melting point (**Figure 5d5–d6**). The foregoing indicated the intense discharge of heat from the welding area and, consequently, demanded process control-related parameters, i.e. welding current and electrode force. It should be noted that in relation to a current of 8.0 kA and the PFS, the joint formed within the entire area of contact (between the elements being welded) was in the solid state and no visible melting of the material had taken place (**Figure 5a**).

6.2 Experimental tests

In relation to the EFS and variants EE1 and EE2 from **Table 7**, it was possible to obtain the melting of the material within the entire area of the weld. Importantly, the melting of the material took place in the central (most favourable) part of the welded joint. The obtained weld nugget diameter exceeded the previously assumed value amounting to 1.5 mm (**Figure 8b1–b2**).

6.3 Comparison of results

The comparative metallographic test results concerning the PFS and EFS are presented in **Figure 8**. In relation to the PFS (**Figure 8a1–a2**), it was possible to observe the formation of the ring-shaped weld nearly within the entire range of technological cycle parameters (**Figures 6** and 7). In terms of the EFS, the material subjected to welding was melted in the central part of the joint, and the weld nugget "grew" from inside towards outside.

7. Optimisation of the projection welding process illustrated with an example of the crosswise projection welding of bars

Based on the FEM calculation and experimental test results, the optimisation of the crosswise projection welding of (aluminium) bars could be characterised as presented below. The process of optimisation was performed on the basis of characteristic courses/waveforms of related parameters (electrode force, momentary power, electrode displacement and the weld nugget diameter) in relation to the two (i.e. pneumatic and electromechanical) electrode force systems (**Figure 9**).

To present the issue in a more convenient manner, the comparison was based on the same value of welding current, i.e. 8.0 kA. It should be emphasised that in relation to the PFS, the aforesaid value was insufficient to obtain a proper joint. The melting of the material was nearly invisible (**Figure 5b**). In turn, as regards the EFS, it was possible to obtain the full weld nugget having the previously assumed diameter exceeding a minimum of 1.5 mm (**Figure 5d**).

Curves 1 and 3 in **Figure 9** refer to the PFS, whereas curves nos. 2 and 4 are related to the EFS. Curves 3 and 4 present the welding current waveform in relation to the PFS and EFS, respectively.

There was a strict correlation between the characteristic process parameters, where the change of one of them immediately led to changes in the remaining parameters. To explain the existing correlations, it was necessary to divide the analysis of the process into stages.

The PFS-related process could be described as follows. After adjusting the preset constant electrode force (**Figure 9a**, curve 1) as well as the specific value of welding current and the time of current flow (**Figure 9a**, curve 3), during the first stage subjected to analysis (K1), specific welding energy was generated (**Figure 9b**, curve 1). The waveform of the welding power (stage K2) had a direct effect (ultimately) on the specific displacement of the electrodes (**Figure 9d**, curve 1). At the subsequent stage (K3), the effect of the above-named factors led to the obtainment of the weld nugget characterised by a specific shape and the diameter of a mere 0.2 mm (**Table 5**, variant P4; **Figure 9c**, curve 1).

As regards the use of the PFS, the value of welding current amounting to 8.0 kA was overly low, only enabling the plasticisation of the material and resulting in an excessive increase in the area of contact between the elements subjected to welding.

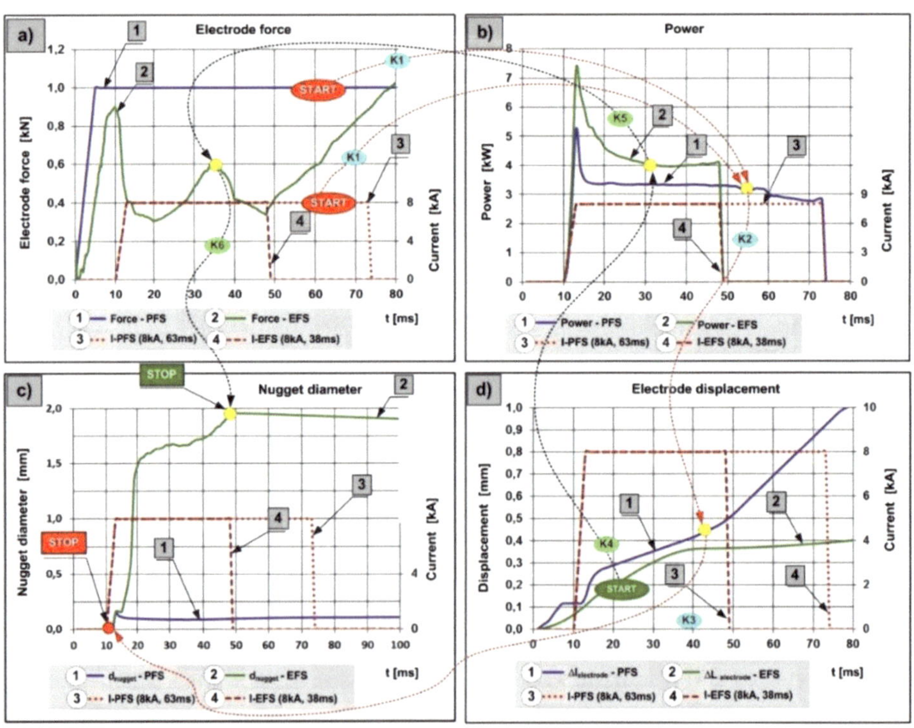

Figure 9.
FEM calculation results: (a) electrode force, (b) momentary power, (c) weld nugget diameter, (d) displacement of electrodes (bar penetration depth): —Curves 1 and 3, PFS (variant P1, I = 8.0 kA, F = 1.0 kN); —Curves 2 and 4, EFS (variant E1, I = 8.0 kA, force exerted by the servomotor).

The foregoing led to a decrease in current density and, consequently, precluded the melting of the material subjected to welding. The material in the contact area was only heated and plasticised, whereas the maximum welding time amounting to 63 ms was exceeded.

The starting point for the optimisation of the crosswise projection welding of bars involved a *proper* change in the course of the displacement of electrodes (bar penetration depth) (**Figure 9d**, curve 2) resulting from the use of the EFS and the application of an appropriate algorithm enabling the control of the electrodes [9]. The essence of the new method of control, i.e. a change in the course of the displacement of electrodes, consisted in the direct control of the aforesaid parameters, particularly during the flow of welding current. The new method of control involved the exertion of lower electrode force at the beginning of current flow and a decrease (slowing down) in the rate of displacement (of electrodes) aimed to obtain, at the subsequent stage (K4), the more favourable distribution of power density as well as to generate higher welding power (**Figure 9b**, curve 2) in comparison with those accompanying the use of the PFS [10]. The slowed down displacement of the electrodes combined with constant welding current led to an increase in resistance in the contact area between the elements (materials) subjected to welding and, consequently, an increase in welding power.

At the subsequent stage (K5), the above-presented method of control translated into the more favourable course of electrode force (**Figure 9a**, curve 2). The obtained values of electrode force were lower than those accompanying the use of the PFS. It should be noted that electrode force directly affected the value of resistance in the contact area (particularly in the *welded bar-welded bar* configuration), which, in turn, led to the proper space distribution of welding power and energy. As a result, the area of contact between the elements subjected to welding was smaller, the resistance in the aforesaid contact area was higher, and the distribution of temperature in the welding area was more favourable. All of the above-presented factors made it possible (at the final stage (K6)) to obtain the full wed nugget having the nominal diameter exceeding 1.5 mm (**Figure 9c**, curve 2). The aforesaid favourable outcome resulted from the more favourable distribution of temperature in the welding area, ultimately leading to the melting of the material and the formation of the weld nugget having the appropriately longer diameter.

The summary of the above-presented analysis concerning a welding current of 8.0 kA identified as overly low to obtain a proper joint using the PFS should contain a statement saying that the use of the EFS and the application of the appropriate control of electrode force and/or displacement (after satisfying the remaining requirements (quality-related criteria)) made it possible to significantly improve the welding process and obtain the full weld nugget having the diameter of a previously assumed length (> 1.5 mm).

8. Summary

The adjustment of the most favourable parameters in the crosswise projection welding of bars performed using the PFS is extremely difficult, if not impossible, particularly as regards soft materials such as aluminium alloys. Electrode force is unfavourably excessively high in relation to necessarily short welding time (bars Al 5182 – 40–60 ms) and high welding current. Such conditions are mutually exclusive and constitute a significant obstacle when adjusting welding parameters. The primary limitation is the dynamics of the electrode force system, i.e. the impossibility of quickly controlling electrode force in short time, particularly during the flow of current.

A characteristic of the PFS is the fact that the preset parameter is (electrode) force and the resultant parameter is the displacement (of electrodes), not controlled in any way.

The improvement of the welding process (extension of the window of parameters) requires the use of the EFS. In the operating mode involving the displacement of electrodes, it is possible to set a more convenient trajectory of electrode movement, enabling the obtainment of the more favourable distribution of current density and the more favourable space distribution of welding power. This, in turns, translates into the generation of higher energy in the central area of the joint and, as result, the generation of higher temperature in the aforesaid area and, consequently, the obtainment of the full weld nugget having larger dimensions that those obtainable using the PFS.

The use of the EFS makes it possible to control the displacement of the electrodes during the flow of current, reach the final, previously assumed, position of the electrodes and exert lower final pressure (force) by the electrodes.

The FEM calculation results indicate the possibility of successful welding using even lower welding current than that used in the experimental verification (8.0 kA).

Acknowledgements

This work was supported by the Polish National Centre for Research and Development (NCBR) under project no. TANGO1/267374/NCBR/2015 and co-funded by Łukasiewicz – Instytut Spawalnictwa, Poland.

Acronyms

PFS	pneumatic (electrode) force system
EFS	electromechanical (electrode) force system
t_{PP}/CFT	(welding) current flow time
upslope	time of current upslope from the initial to the final value of the cycle range
MFWC	most favourable welding conditions
EFST	electrode force stabilisation time
$\Delta L_{PP\ max} - \Delta L_{PB\ max}$	maximum displacement of electrodes (penetration of bars)
$T_{e\text{-}m\ max}$	maximum temperature in the *electrode – welded material* contact area

Glossary

Weld nugget is a part of the spot, projection or seam weld molten during the welding process [20].
Expulsion signifies the expulsion (during welding) of the molten metal from the area of contact between elements subjected to welding or from the area of contact between the electrode and a given element subjected to welding [21].
Welding area in spot or projection welding (e.g. of sheets with an embossed projection) is the area including the weld nugget, heat affected zone (HAZ) as well as the area of contact between the electrode and the material subjected to welding along with adjacent areas.

Author details

Zygmunt Mikno
Research Network ŁUKASIEWICZ, Institute of Welding, Gliwice, Poland

*Address all correspondence to: zygmunt.mikno@is.gliwice.pl

References

[1] Zhang H, Senkara J. Resistance Welding Fundamentals and Applications. USA: CRC Press, Taylor & Francis Group; 2011

[2] Zhang X, Chen G, Zhang Y, Lai H. Improvement of resistance spot weldability for dual-phase (DP600) steels using servo gun. Journal of Materials Processing Technology. 2009; **209**(5):2671-2675

[3] Slavick SA. Using servo guns for automated resistance welding. Welding Journal. 1999;**78**(7):29-32

[4] Tang H, Hou W, Hu S. Forging force in resistance spot welding. Proceedings of the Institution of Mechanical Engineers, Part B: Journal of Engineering Manufacture. 2002;**216**(7):957-968. DOI: 10.1243/09544050260174166

[5] Gould JE. Joining aluminum sheet in the automotive industry—A 30 year history. Welding Journal (Welding Research). 2012;**91**:23-34

[6] Zhang XQ, Chen GL, Zhang YS. On-line evaluation of electrode wear by servo gun. International Journal of Advanced Manufacturing Technology. 2008;**36**:681-688

[7] Mikno Z. Projection welding with pneumatic and servomechanical electrode operating force systems. Welding Journal (Welding Research). 2016;**95**:286-299

[8] Mikno Z, Stepien M, Grzesik B. Optimization of resistance welding by using an electric servo actuator. Welding in the World. 2017. DOI: 10.1007/s40194-017-0437-x

[9] Mikno Z, Bartnik Z, Ambroziak A, Pietras A. Method for projection resistance welding of steel plates with embossed projections. In: Patent P. 220870. Polish Patent Office; 2012

[10] Mikno Z, Grzesik G, Stepien M. A manner of resistance projection welding in configuration of the cross, mainly for aluminum wires. In: Patent P. 228089. Polish Patent Office; 2015

[11] Mikno Z. Cross-wire projection welding of aluminium alloys in relation to pneumatic and electromechanical electrode force systems. International Journal of Advanced Manufacturing Technology. 2019;**102**:4167-4178. DOI: 10.1007/s00170-019-03443-5

[12] Mikno Z. Projection welding of nuts involving the use of electromechanical and pneumatic electrode force. International Journal of Advanced Manufacturing Technology. 2018:1-21. DOI: 10.1007/s00170-018-2525-5

[13] Mikno Z, Gradkowski M. System pneumatycznego docisku w zgrzewaniu rezystancyjnym. Biuletyn Instytutu Spawalnictwa. 2019;**5**(2019):13-18

[14] The database of the material and electrode parameters: model 3D Version 4.0x64 of the Swantec Inc., SORPAS Software. Available from: http://swantec.com/

[15] User Manual SORPAS® 3D Version 4.0 x64, SWANTEC Software and Eng. ApS; 2015

[16] Projection welding, Part 2 Chapter 2. Welding Processes. AWS Welding Handbook, Vol. 3. 9th ed.

[17] Gould JE. Projection welding. In: ASM Handbook. Vol. 6: Welding, Brazing, and Soldering. USA: ASM International; 1993. pp. 230-237

[18] Papkala H. Resistance Welding of Metals. Krosno: Publishing House KaBe; 2003

[19] Statistica 12. Available from: www.statsoft.pl

[20] Szczeciński Z et al. wytyczne nr W-90/IS-45, SPAWALNICZY SŁOWNIK TERMINOLOGICZNY. 1990

[21] Wyroby płaskie walcowane na gorąco ze stali o podwyższonej granicy plastyczności do obróbki plastycznej na zimno, Część 2: Warunki techniczne dostawy wyrobów walcowanych termomechanicznie. PN-EN 10149-2: 2014-02

Section 3

Aluminium Composites and Alloys Application

Chapter 9

Applications of Aluminum Alloys in Rail Transportation

Xiaoguang Sun, Xiaohui Han, Chaofang Dong and Xiaogang Li

Abstract

This chapter focus on the latest applications of aluminum alloys in rail transportation field. The typical high-strength aluminum alloys used on high speed train is introduced. The unique properties of aluminum alloys are analyzed. The detailed application is illustrated including car-body, gear box and axle box tie rod. The main challenges encountered in the application are also mentioned. The key manufacturing techniques, such as casting, forming, welding, are analyzed. Finally, the future improvement directions for better application is summarized. It is expected to set up a bridge for materials providers, equipment manufacturers and end-users, thereby promoting the advance of manufacturing technology and application of aluminum alloys in wider fields.

Keywords: aluminum alloy, application, rail transportation

1. Introduction

With the rapid development of human civilization, the consequent air pollution and greenhouse gas (GHG) emissions have threaten human being for years. Energy conservation and emission reduction is an increasing priority in the development of transportation industry.

A number of approaches can be used to improve energy efficiency and reduce CO_2 emission, such as reducing aerodynamic resistance, transmission loss, tire rolling resistance, and weight [1]. Among these options, lightweight structure is currently considered as one of the most efficient solutions. In recent years, lightweighting has become a major research theme in the transport industry around the world.

Material substitution appears a promising option for lightweighting. Traditionally, steel, such as low-carbon steel and stainless steel, is used as main material to build major structure of transportation equipment. Titanium, aluminum and magnesium are promising lightweight metallic materials as alternatives to steel and cast iron. Among these materials, aluminum shows a balanced performance, such as light weight, good corrosion resistance, good formability, high specific strength and relatively low cost. Density of aluminum alloy is only one third of steel. Considering structural optimization due to material replacement, the overall weight of rail car-body is decreased by 50% when aluminum is used. This degree of reduction deserves an effort in engineering application.

Although aluminum accounts for ~8 wt% (by mass) of the earth's crust, the high affinity of aluminum for oxygen, as well as the stability of aluminum oxides and silicates hindered its separation for a long time. For this reason, aluminum became economic for engineering applications only at the end of the nineteenth century [2].

Aluminum alloys was originally used in aviation industry which is an important part of the transportation industry. It was firstly used on Junker F13 fuselage with the invention of 2017-T4 alloy in 1920. It began to be used in the manufacture of train with increase of the train speed until 1980s. So far, it has been widely used in passenger cars with speeds above 200km/h, such as German ICE series high-speed EMU car-body, French ALSTOM double-decker TGV high-speed EMU car-body, Italy Pendolino(ETR) series pendulum high-speed EMU car-body, Japanese Shinkansen, and Chinese CRH high-speed EMU car-body and so on. Especially with the huge expansion of high-speed rail lines in China in the last 20 years, the development and application of aluminum alloy prospered.

A series of new manufacturing processes, such as friction stir welding and laser welding, were developed and applied. These technical progress promoted application of aluminum alloy not only in high speed trains but also urban rail vehicles, which continuously contributes to energy conservation and climate change. This chapter will introduce the application status and key technologies of aluminum alloy for manufacturing high speed train. It is expected that the technical analysis and summary will inspire and encourage material scientist and engineers to vigorously push technological innovation for comfortable travel experience in an environmental-friendly mode.

2. Characteristics of aluminum alloy

The special operating environment of trains and unique properties of aluminum alloy promotes the their combination, which favored the popularization of high speed train and in-depth application of aluminum alloy in transportation industry. For engineering application on rail transit vehicles, aluminum has the following advantages:

(1) The net weight of the vehicle can be greatly reduced. The use of aluminum alloy can greatly reduce the net weight of rail passenger car while it meets the safety requirements in the aspect of strength and rigidity. Generally speaking, car-body made of aluminum alloy is 30%~50% lighter than that of steel. For high-speed and double-deck trains, the most effective way to make vehicles light is to increase the proportion of aluminum used in vehicles as much as possible.

(2) Aluminum alloy has excellent fire resistance. Although the melting point of aluminum (660 °C) is much lower than that of steel (1530 °C), the fire-resistant of the car body is not only related to the melting point of the material, but also to the thermal conductivity of the material. Compared with steel, aluminum alloy has excellent thermal conductivity and better heat dissipation.

(3) Aluminum alloy has good corrosion resistance. The surface of aluminum alloy is easy to form a layer of dense oxide film, which has a good anti-oxidation ability in the atmosphere. Therefore, car-body made of aluminum alloy has better corrosion resistance than that of steel, especially in the components that are not easy to be coated, such as the box structure and some of its internal beams and columns, aluminum counterpart show obvious advantages. At the same time, aluminum alloy surface can be colored, painted, sprayed, through chemical methods to greatly improve the corrosion resistance of the components, together with improved decorative effect.

(4) Aluminum alloy is easy to process, manufacture and maintain. With the development and application of large hollow and complex section aluminum

profiles, aluminum welding technology is constantly improving, and vehicle manufacturing technology is becoming more and more mature. Aluminum alloy parts is easy to be replaced, and suitable for all kinds of surface treatment. The workload need for manufacturing rail car-body is also greatly reduced than the steel one.

(5) The price of aluminum alloy is moderate. The high price of aluminum material increases the manufacturing cost of vehicles, but aluminum alloy also makes vehicles light, which leads to the increase of transport capacity, reduction of energy consumption and reduction of maintenance costs. Taking the comparison of maintenance hours of vehicles leaving the factory at 10 A as an example, steel car is 100%, aluminum car is 52%.The recovery value of scrapped vehicles is 100% for steel vehicles and 480% for aluminum vehicles. From the perspective of comprehensive economic benefits, the use of aluminum vehicles is economical and reasonable. Therefore, the final cost of using aluminum alloy is moderate.

3. Typical aluminum alloys and their properties

Aluminum alloys commonly used in high-speed trains include 5000 series, 6000 series and 7000 series. 5000 series alloys show maximum strength and high corrosion resistance among the typical non-heat treatable alloy, which adapt to welded structure. 6000 series alloys show moderate strength and good corrosion resistance, together with perfect extrusion formability making complex and thin-wall hollow section possible. 7000 series alloys show excellent strength among age-hardening aluminum alloy, which provide wide space for weight reduction. Typical aluminum alloys for high speed train car-body are shown in **Table 1**.

Series	Designation	State	Mechanical properties			Application area	Note
			Tensile strength/ MPa	Yield strength/ MPa	Elongation/ %		
5000	5083	O	≥275 ~ 350	125 ~ 200	≥16	Front skin	Plate
6000	6005A	T6	≥270	≥225	≥8	Underframe, sidewall, roof, end wall, beam	Plate, 3 ~ 5 mm
	6005A	T6	≥255	≥215	≥6	Underframe, sidewall, roof, end wall, beam	Section, 3 ~ 5 mm
	6082	T6	≥310	≥260	≥10	Underframe, end beam	Plate, 3 ~ 6 mm
	6106		≥250	≥200	≥8	End wall	Section
7000	7B05	T4	≥315	≥195	≥11	Underframe, end beam	P, 2.9 ~ 6.5
	7B05	T5	≥325	≥245	≥10	Underframe, end beam, traction beam	Section

Table 1.
Typical aluminum alloys for high speed train car-body [3, 4].

4. Detailed application

In railway vehicles, aluminum alloy is primarily used in manufacturing car-body and its ancillary structure. The car-body is a prolonged hexahedral structure. Aluminum is used to manufacture all parts of the structure, including roof, underframe, end wall and side wall. However, different aluminum designations are selected for different parts of the car-body based on their properties shown in section 3. Both extruded section and plate are used. Extruded profiles account for about 70% of the total weight of the aluminum alloy car-body, while the plates account for about 27%, and the castings and forgings account for about 3%. On the other hand, there is an emerging trend that aluminum alloy come into application to other structures such as gear-box and axle box than car-body. The detailed introduction is as follows.

4.1 Car-body

The development of aluminum alloy materials and large extruded profiles paves the way for the modernization and lightening of railway vehicles, In recent years, with the popularity of lightweight design concept for railway vehicles, as well as the requirement of simplified construction and maintenance, large integral thin plate and hollow complex thin wall profiles has been developed successfully. In Japan, 6N01(6005 alloy) alloy with better extrusion, welding and corrosion properties has been developed to produce porous complex thin-wall hollow profiles, widely replacing 7N01 and 7003 profiles as the floor, side plate and roof structure of the car-body. In Western Europe, aluminum alloy body is mainly made of 6005A extruded profiles, the main reason of which is that the extrusion performance of 6005A is better, the production process is more simplified, and the stress corrosion problem of 7000 series alloy can be avoided. The application of typical aluminum alloys on 300 series Shinkansen high speed train is shown in **Figure 1**. A complete car-body is shown in **Figure 2**. As is shown, the car-body is mainly composed of extruded section.

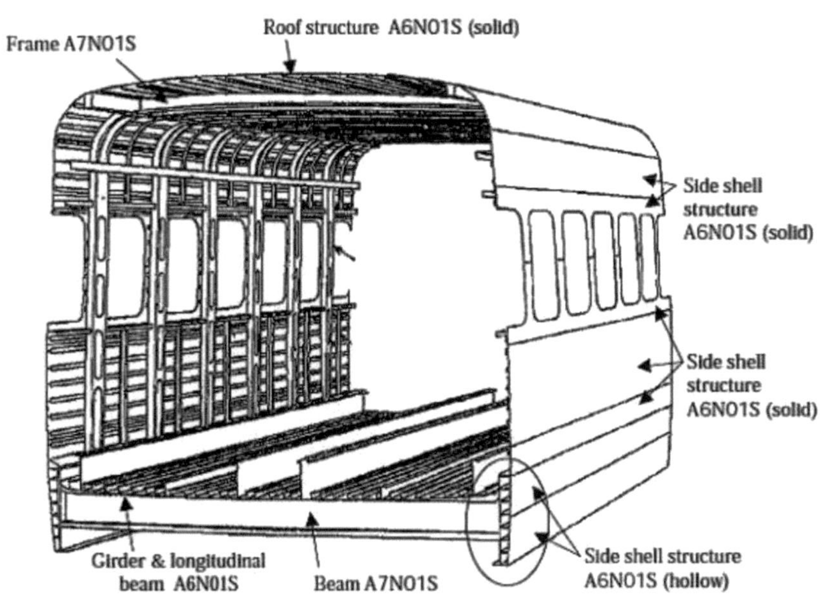

Figure 1.
Application of aluminum alloys on car-body of Shinkansen 300 series high speed train.

Figure 2.
Typical car-body of high speed train made of aluminum alloy.

Figure 3.
Typical sidewall of high speed train composed of extruded sections.

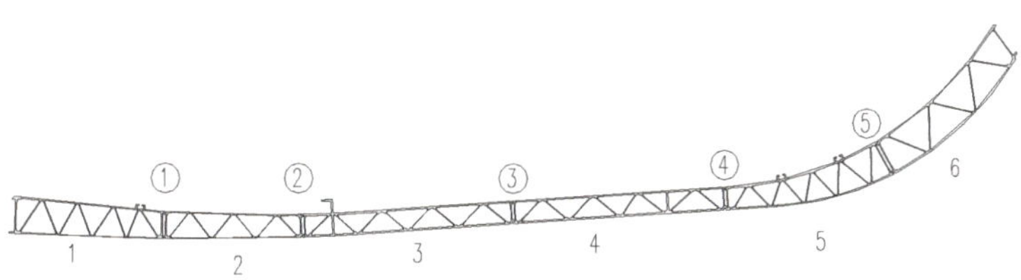

Figure 4.
Part of a section structure of high speed train car-body.

The car-body can be easily welded automatically with through - length welds, as shown in **Figure 3**. The aluminum section profiles can be designed according to the section structure of the car-body, as shown in **Figure 4**. Typical extruded section profiles for high speed train are shown in **Figure 5**.

However, when it comes to the head car, the structure is quite different. In order to achieve optimal aerodynamic performance running at high speed, a streamlined

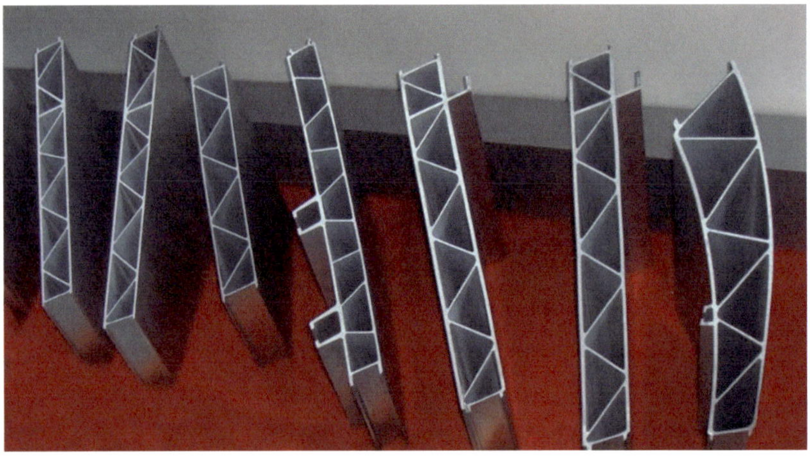

Figure 5.
Typical extruded section profiles for high speed train.

Figure 6.
Streamlined head car of CRH 380A high speed train.

design was given to the head car, as shown in **Figure 6**. This unique shape make the it impossible to manufacture with relatively regular and straight sections. Therefore, beam and slab structure became the optimum option for head car. As shown in **Figure 7**, a framework is designed based on the requirement for stiffness and strength to support the front skin against plastic deformation. It is welded with hundreds of beam made of aluminum plates prior to skin fixation. Afterwards, the skin is divided into small pieces based on the principle of good workability. Each piece is deformed to specific shape based on the design profile. Then the piece is fixed on the framework one after another, shown in **Figure 8**.

4.2 Gear box

For further reduction of the weight of the train, it is obviously not enough by reducing the weight of the car-body because the car-body accounts for only about 20% of the total mass of the train. Key components of bogie including traction motor, wheel-sets, frame and braking system attracted attention of proponents of lightweighting.

Figure 7.
Internal structure of the head car of high speed train.

Figure 8.
Head car of high speed train.

The lightweighting of gear box can help to reduce unsprung mass and wear or damage to rail. In this part, the application of aluminum alloy on gear box is introduced.

Gear box of high speed train is manufactured by casting aluminum rather than wrought alloy due to complex and unequal thickness. Low pressure casting is widely used in non-ferrous alloy casting because of its high feeding pressure and temperature gradient and stable filling, which can effectively improve the density of castings and product yield.

AlSi7MgA and AlCu4Ti are commonly used as casting materials for gear box due to good flowability, low thermal expansivity and shrinking percentage. Typical aluminum gear boxes on CRH series high speed train are shown in **Figure 9**.

4.3 Axle box

Axle box is one of the important bearing parts of trian bogie and transfer joint of motion. The left axle box part is installed on the axle journal through a rolling bearing, and the right swivel arm is connected with the positioning swivel seat on the

Figure 9.
Typical aluminum gear boxes on CRH series high speed train.

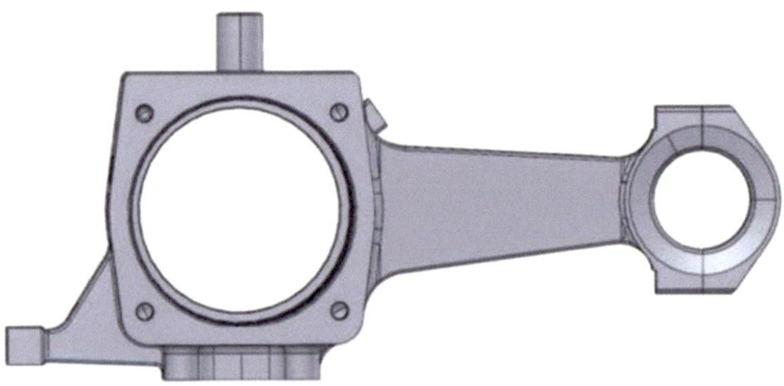

Figure 10.
3D model of an aluminum alloy axle box.

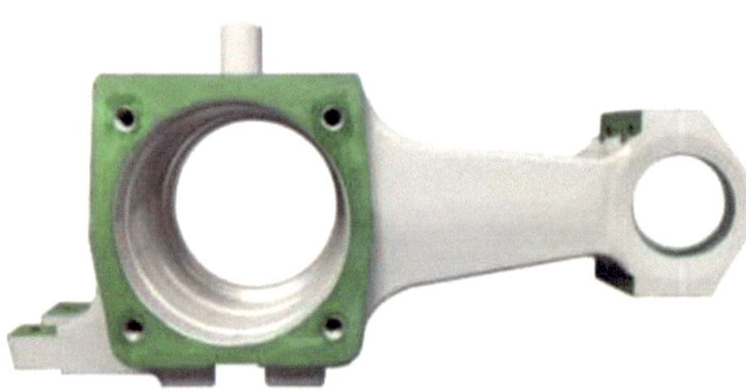

Figure 11.
Finished aluminum alloy axle box of high-speed train.

frame through an elastic node. When the train is running, axle box bears the action of vertical force, longitudinal force and transverse force.Therefore, the bearing condition of the axle box body is complex, and its structure and performance stability are very important for the safe operation of the train. 7050 aluminum alloy forgings shows high strength and toughness, which can significantly reduce unsprung weight. The weight of forged aluminum product decrease 62.5% as compared to the traditional carbon steel one. Therefore, forging aluminum alloy axle box is widely used on high-speed train. 3D model of an aluminum alloy axle box is shown in **Figure 10**. A typical finished aluminum alloy axle box of highspeed train is shown in **Figure 11**.

5. Key manufacturing techniques

The engineering application of aluminum alloy in the field of rail transportation encounter a series of challenges, which promote the development of a series of key manufacturing techniques including casting, forming welding and anti-corrosion processes.

5.1 Casting

Casting is an important process for the manufacture of complex structures such as gear box. The main challenges of gearbox casting process are as follows:

(1) Ensure there is no cold partition, porosity and other defects in the thin-walled area, so as to meet the requirements of casting appearance quality.

(2) Ensure there is no excessive shrinkage porosity, shrinkage cavity and other defects on the box surface, flange, hanging and other key parts, so as to meet the internal quality requirements of castings radiographic inspection.

(3) Ensure a high requirements for machining surface pinhole and non-machining surface, especially for aluminum alloy castings that are easy to produce pinhole defects.

In the production of aluminum alloy gearbox for high-speed EMU, the common casting defects include porosity, pinhole and shrinkage cavity.

In order to eliminate those pores, two measures need to be taken on the premise of controlling the air production content of molding sand. Firstly, the venting near the inner runner and riser should be improved by opening more air hole and hollowing out the loam core of the outer molding. Secondly, thickness of the coating should be guaranteed to decrease surface void on the sand (core) by using coating with high thermal conductivity such as zircon powder. The filling pressure and holding pressure are increased appropriately, so as to increase the resistance of gas entering the metal liquid.

The key of eliminating pinhole mainly relies on the control of hydrogen content in liquid aluminum. The refining process can reduce the oxidation inclusion and hydrogen content in liquid aluminum, and thus effectively reduce pinhole forming tendency.

Regarding to shrinkage cavity at the top of gear box, it is proved effective by simulating the solidification process with MAGMAsoft. Chilling block and riser locating can be optimized to ensure the feeding channel of the top riser unblocked under reasonable temperature distribution.

5.2 Forming process

There are four different technologies available to manufacture the front skin panels of the head car. The most commonly used one is the hammer press where a hammer machine is used conveniently to produce the target shapes. However, the dimensional tolerance of the produced product heavily depends on the worker's experience. After installing the panels on the structural frame, any further modifications of the geometrical features can only be completed by using the hand tools such as hammer. Such a manual process renders the high repeatability of manufactured components almost impossible. The second technique is the expanding-stretching process. It is applicable for the panels with curved profile but only to a certain extent. Additionally, the rotating press machine, which mainly aims for manufacturing a panel with small and uniform curvature, is used, while a process called mould press that uses mould to produce the target shape is employed for the panel with complicated and small curvature.

As shown in **Figure 12**, the front skin of a typical CRH 380A high speed train is divided into around 70 small pieces which are joined together through a total of 170 meters long welding line. As the length of each panel is limited to only 1 meter, the manufacturing process becomes time consuming and low efficient for producing a considerable amount of small components. The product quality of the front panel is also compromised due to the increased residual stress resulted from the uneconomical and complicated assembling process.

In order to ensure the assembly precision of each piece, a commercial finite element analysis simulating the skin drawing and springback process based on flexible multi-point die is necessary. In order to improve the computational efficiency and obtain satisfactory

Figure 12.
Front panel of CRH 380A high speed train.

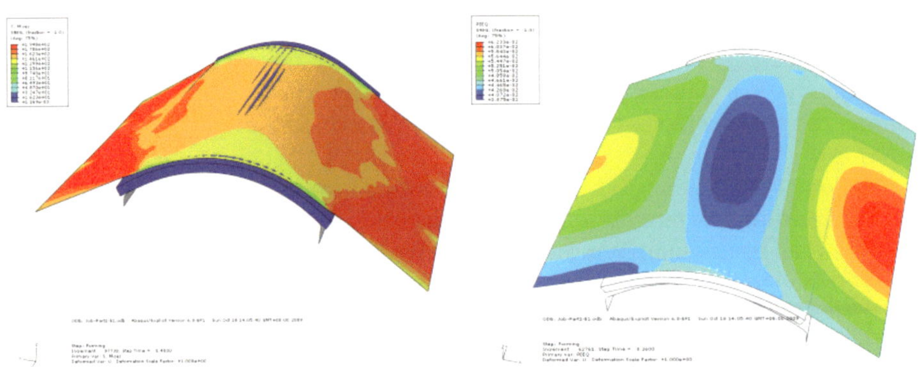

Figure 13.
Distribution of MISES stress and equivalent plastic strain for forming of a single piece.

Figure 14.
Skin drawing based on flexible multi-point die.

computational accuracy, the dynamic explicit algorithm is used to simulate the drawing process, and the static implicit algorithm is used to calculate the springback. **Figure 13** shows the Distribution of MISES stress and equivalent plastic strain for forming of a single piece. The forming process is developed based on the simulation results which can save experimental time and improve adaptability to different products, shown in **Figure 14**.

5.3 Welding

Due to special thermal physical properties and welding characteristics of aluminum alloy, such as low melting point, thermal conductivity, high electrical conductivity and the expansion and shrinkage, high content of alloy elements as compared to carbon steel and stainless steel [5, 6], it is much more easily to produce pores, crack, lack of penetration, incomplete fusion, large welding deformation, bite edge and slag when aluminum alloys are welded. The assembly precision, quality and performance can be severely affected by welding process. Therefore, the welding process is quite crucial in the manufacture of high speed train.

The following aspects should be considered as the basis of selecting welding method:

(1) Aluminum alloy is coated with a dense oxide film which can easily adsorb moisture and bring hydrogen to molten pool. The aluminum can also become oxidate slag existing in the weld which affecting the performance. Therefore, it is very important to remove the oxide on the base metal and groove surface of aluminum alloy before welding.

(2) The thermal conductivity of aluminum alloy is five times that of low carbon steel. High power or energy concentrated welding heat source should be a preferential option. And preheating is necessary for thick plate welding.

(3) The thermal expansion coefficient and cooling shrinkage rate of aluminum alloy are two times that of steel. Therefore, aluminum alloy melting welding deformation is serious. Deformation control measures such as reverse deformation and reinforcement constraint should be considered.

(4) Aluminum alloy material has different kinds of alloying elements, and the loss of alloying elements during fusion welding is easy to lead to the decrease of joint strength and corrosion resistance, and the weld metal and heat affected zone are easy to produce intergranular cracks. Cracking susceptibility should be considered.

In this section, three typical welding processes which are widely used in manufacturing process of high speed train are introduced.

5.3.1 Arc welding MIG

Pulse MIG welding is the most widely used and developed method of aluminum alloy welding which is characterized by large thermal power, high linear energy, good energy concentration and good protection effect. The above features are suitable for welding aluminum alloy based on its unique thermal physical properties. Pulsed MIG welding can be used to control wire melting and droplet transition, improve arc stability and achieve droplet jet transition with small average current, thus suitable for all-position welding.

Considering the cathodic atomization effect of MIG welding on removing aluminum oxide film, DCEP (Direct Current Electrode Positive) is commonly used for pulse MIG welding. Semi-automatic pulse MIG welding are fit for irregular short welds while regular long straight welds are usually automatically MIG welded with laser tracking, shown in **Figure 15**.

Typical defect includes poor formability, burning through, excessive penetration, cracking, pore and slag, as shown in **Figure 16**.

5.3.2 FSW welding

In addition to the traditional MIG welding, friction stir welding(FSW) has been widely used. This process and MIG welding process are not compatible on the design groove. Groove designed according to MIG welding process cannot be used for friction stir welding, so the promotion of this technology is subject to certain restrictions. However, FSW shows unique advantages as follows:

(1) Low manufacture cost. No consumable welding materials, such as electrode, wire, flux and protective gas, are required during the welding process. Welding stirring head is the only consumption. It is not necessary to remove the oxide film before welding, which reduce cleaning time and improve production efficiency.

(2) Good welding quality. The temperature of friction stir welding is relatively low, so the microstructure change of the heat affected zone is negligible, and the residual stress is low leading to a low deformation. The joint efficiency is high as compared to MIG weld.

(3) Environment-friendly. The welding process is safe. There is no pollution, no smoke, no radiation, etc.

(4) Less energy consumption. Because friction stir welding solely depends on the welding head rotation and movement, it saves more energy than fusion welding or even conventional friction welding.

(5) High welding efficiency. It can complete the welding of long weld, large section and different position at one time.

The above advantages promote its application in rail transportation field, shown in **Figure 17**. It can be seen that FSW showed flat weld than MIG which requires little post-processing (**Figure 18**).

Figure 15.
Automatically MIG welding process.

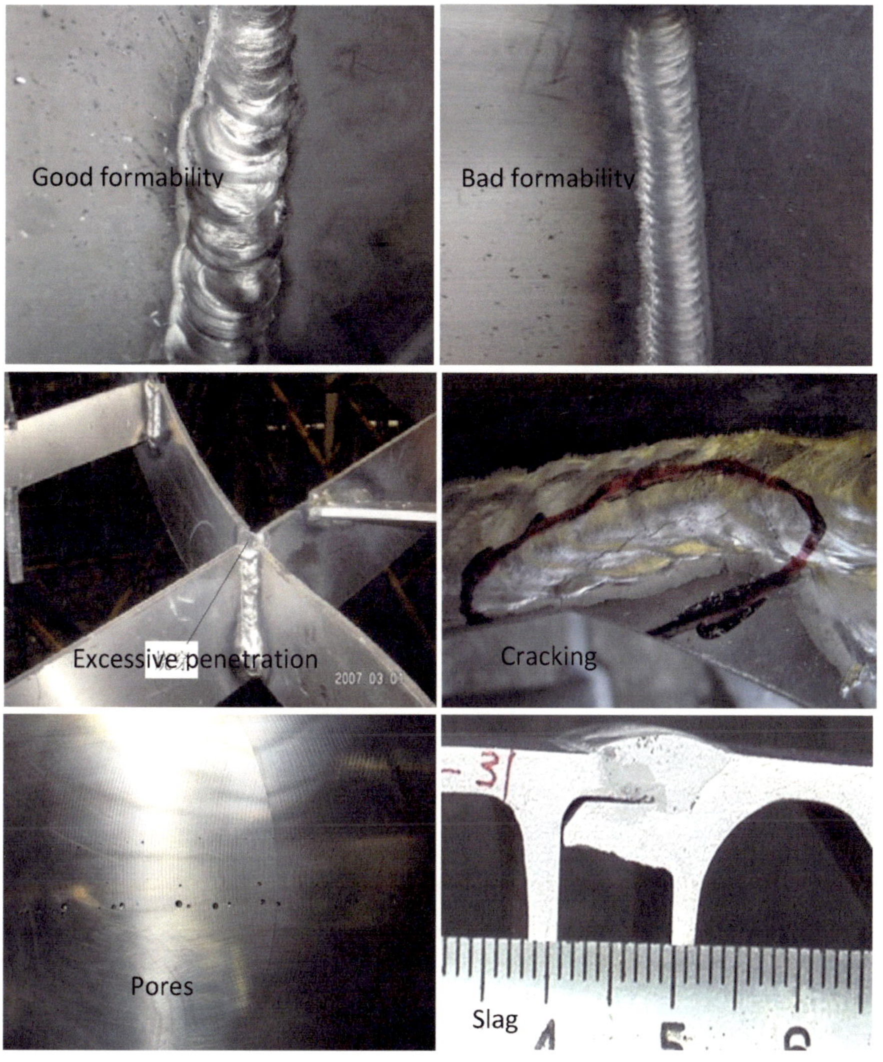

Figure 16.
Typical defect in aluminum alloy weld.

Figure 17.
FSW process of a sleeper beam.

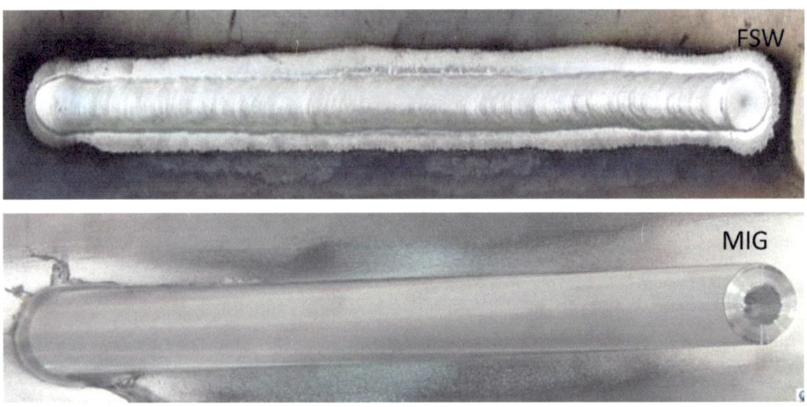

Figure 18.
A comparison between MIG and FSW welds.

The defects in friction stir welding joints mainly include holes, unwelded joints, flaps and grooves. Defects are mainly caused by the fact that in the welding process, different parts of the weld metal have undergone different thermomechanical processes, and thus bring overheating or insufficient flow of plastic materials. The top of the weld is subjected to the strong friction and stirring effect of the stirring needle and the shaft shoulder at the same time, even if the welding speed is very high or the stirring head speed is not high enough, it can still ensure a certain heat input and form a defect free connection; In the middle of the weld, the heat input is less than the top, but the heat output is also less than the top and bottom, so the total heat absorption is greater than the top and bottom, and the material softening degree is the highest. The heat input at the bottom of the welding seam is the least and the output is the largest. so the welding defects will appear at the bottom of the welding seam when the process parameters are not properly selected or the size of the welding tool is not appropriate.

5.3.3 Laser-MIG welding

Laser-MIG welding is a new welding technology, which has a wide development prospect. The laser -MIG hybrid welding technology combines the laser welding technology and MIG welding technology organically, which overcomes each other's shortcomings, and thus favor to obtain high quality welding joint.

Laser -MIG welding uses both laser beam and arc, which has the characteristics of high welding speed, stable welding process, high thermal efficiency and allowing greater welding assembly clearance. The laser -MIG welding pool is smaller than that of MIG welding. As compared to MIG welding, laser-MIG welding shows lower heat input, smaller heat affected zone and smaller work deformation.

Based on the characteristics of concentrated heat source, strong penetration and arc wire filling welding, a new design of joint and groove of laser-MIG hybrid welding was carried out through experimental optimization and verification. Compared with the traditional MIG welding, the upper groove angle is smaller, the depth is smaller for laser-MIG hybrid welding.

The wide application range and high efficiency of laser-MIG welding enhance its competitiveness in terms of investment cost, reduced production time, reduced production cost and improved productivity. Currently It is in the stage of small-scale application in the manufacture of high speed train and relevant component, shown in **Figure 19**.

Figure 20 shows a comparison between MIG and laser-MIG hybrid welds. The laser-MIG hybrid welds is flat which reflect a good formability.

Figure 19.
Lase-MIG weld machine for roof of aluminum car-body of high speed train.

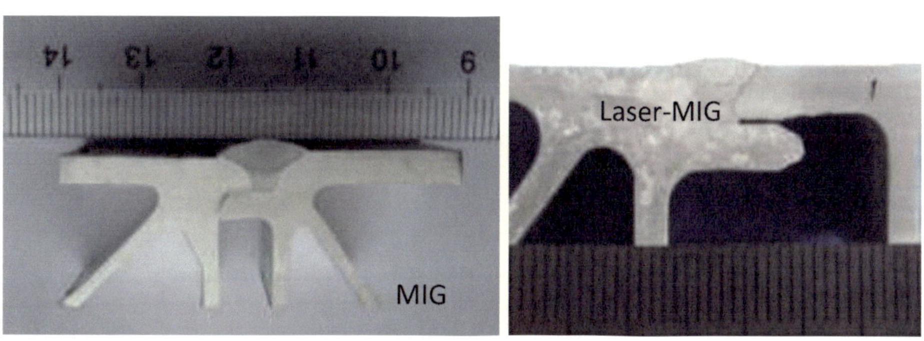

Figure 20.
A comparison between MIG and laser-MIG hybrid welds.

5.4 Anti-corrosion process

The increasing operation speed of train make it experience the baptism of different environments within hours. The environmental change is quite complex when the train goes from inland to coastal cities, from high to low latitudes, from low to high altitude. The weather may change dramatically from sunny to rainy. The temperature may change from subzero to 40°C, The humidity may change from very dry to very damp (~100%). The air may change from fresh and clean to polluted. It may contain dust, oxysulfide, oxynitride, or chloridion. These ingredients would lead to corrosion to aluminum alloy which is detrimental to the safety and long-term reliability of the train, especially when it's running at speed higher than 200 km/h.

As is known, aluminum shows good corrosion performance since it can form passive film in atmosphere. However, the corrosion resistance is also threatened by alloying elements and aggressive environmental factors. Pitting, galvanic and stress corrosion are common types of corrosion for engineering structure made of aluminum used in atmospheric environment.

For rail vehicle, an organic coating system is used to protect aluminum against corrosion. In order to deal with different environments, the coating system for outer surface is different from internal surface. The outer coating system is used to fight against harsh natural environment while the internal coating tackle the condensing water and leaking water from washing room. Therefore, it has higher requirement for the outer coating system. It needs to be evaluated by a series of accelerated corrosion test including salt spray test, damp heat test and high-low temperature test. The outer coating system consists of sand blasting pretreatment, epoxy primer, polyurethane putty, polyurethane interlayer, polyurethane finishing coat and varnish. The internal coating system consists of cleaning, etch coating, rust inhibiting primer and polyurethane top-coat. In case aluminum component joints with other alloys, a surface pre-treatment accompanied by rust inhibiting primer is necessary to ensure physical isolation from each other and against galvanic corrosion (**Figure 21**).

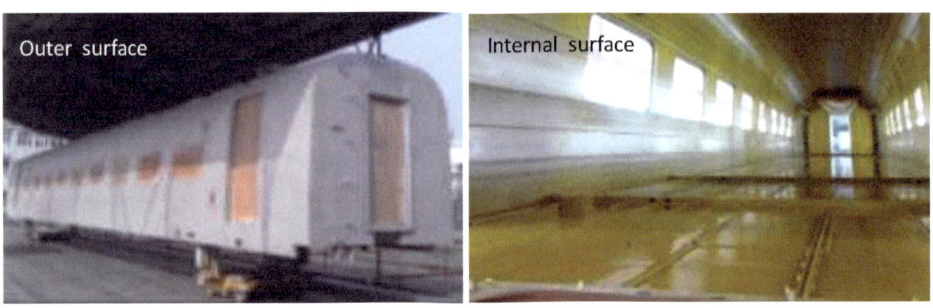

Figure 21.
Coated outer surface and internal surface of aluminum car-body.

6. Conclusion

In this chapter, the latest applications of aluminum alloys in rail transportation field is introduced. The typical high-strength aluminum alloys used on high speed train was illustrated combined with the unique characteristics of aluminum alloys. The detailed application on key part of rail vehicle including car-body, gear box and axle box tie rod, were introduced. The main challenges and engineering experience were also mentioned. The key manufacturing techniques, such as casting, forming, welding, and anti-corrosion were analyzed. Hopefully, the chapter can promote the development and application of advanced materials, especially aluminum alloy, and continuously contribute to sustainable development of human civilization through technological innovation.

Acknowledgements

Authors would like to thank Mr. LIU Xuezhi, Mr. ZHANG Shilei and Mr. LI Shuaizhen from CRRC Qingdao Sifang Co., Ltd for providing figures of the chapter.

Conflict of interest

There is no conflict of interest regarding to the content of this chapter.

Author details

Xiaoguang Sun[1,2]*, Xiaohui Han[1], Chaofang Dong[2] and Xiaogang Li[2]

1 CRRC Qingdao Sifang Co., Ltd, Qingdao, China

2 Corrosion and Protection Center, University of Science and Technology Beijing, Beijing, China

*Address all correspondence to: sunx_sf@126.com; sunxiaoguang@ustb.edu.cn

References

[1] Kawajiri, K.; Kobayashi, M.; Sakamoto, K., Lightweight materials equal lightweight greenhouse gas emissions?: A historical analysis of greenhouse gases of vehicle material substitution. Journal of Cleaner Production. 2020;253:119805. DOI: 10.1016/j.jclepro.2019.119805

[2] Ashkenazi, D., How aluminum changed the world: A metallurgical revolution through technological and cultural perspectives. Technological Forecasting and Social Change, 2019; 143: 101-113. DOI: https://doi.org/10.1016/j.techfore.2019.03.011

[3] GB/T 32182-2015 Aluminium and aluminium alloy plates and sheets for railway application.

[4] GB/T 26494-2011 Aluminium alloys extruded profiles used for structural material of railway vehicle carbodies.

[5] Mallick, P. K., Designing lightweight vehicle body. In Materials, Design and Manufacturing for Lightweight Vehicles, 2021; pp 405-432

[6] Mallick, P. K., Joining for lightweight vehicles. In Materials, Design and Manufacturing for Lightweight Vehicles, 2021; pp 321-371